# REINFORCED CONCRETE TECHNOLOGY

# REINFORCED CONCRETE TECHNOLOGY

Samuel E. French, Ph.D., P.E.

# REINFORCED CONCRETE TECHNOLOGY

Samuel E. French, Ph.D., P.E.
*Structural Engineer*

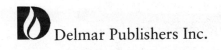 Delmar Publishers Inc.

I(T)P™

## NOTICE TO THE READER

Cover photo courtesy of Portland Cement Association
Cover design by Cheri Plasse

**Delmar Staff**
New Product Acquisitions: Mark W. Huth
Developmental Editor: Cathy Carter
Project Editor: Carol Micheli
Production Supervisor: Wendy Troeger
Art Director: Russell Schneck
Art Supervisor: Judi Orozco
Design Supervisor: Susan C. Mathews

For information, address Delmar Publishers Inc.
3 Columbia Circle, Box 15-015
Albany, New York 12212-5015

Copyright © 1994
The trademark ITP is used under license.
by Delmar Publishers Inc.

printed in the United States of America
published simultaneously in Canada
by Nelson Canada,
a division of The Thomson Corporation

1 2 3 4 5 6 7 8 9 10 XXX 00 99 98  97  96  95  94

**Library of Congress Cataloging-in-Publication Data**
French, Samuel E., 1930–
    Reinforced concrete technology / Samuel E. French.
        p.      cm.
        Includes bibliographical references and index.
        ISBN 0-8273-5495-9
        1. Reinforced concrete construction.   I. Title.
TA683.F827   1994
624.1'8341—dc20
                                        92-26646
                                        CIP

# CONTENTS

# PREFACE

Over a period of some twenty years in engineering practice, the author has repeatedly heard the complaint "Why can't concrete be designed by M-c-over-I like everything else?" Very early, the author came to respect that complaint and over the years has developed a rational, simple and accurate procedure for designing concrete " . . . by M-c-over-I like everything else." This textbook utilizes that simplified approach to the flexural design of reinforced concrete.

The basis of the many complaints about design methods stems primarily from the highly specialized formulas and procedures that are commonly used for the design of reinforced concrete. The use of such specialized procedures is unique to concrete; such procedures are not used for any of the other common construction materials. Not surprisingly, these specialized procedures have their greatest value to those who are regularly engaged in the design of large, complex concrete structures. Such designers will of course maintain complete familiarity with these specialized formulas and procedures simply through their frequent use of them.

There are many other designers of concrete, however, who are not involved in the design of large, complex structures every day. For these designers, it is simply not practical to maintain familiarity with the intricate and specialized procedures used by design specialists. Those who design smaller, more routine structures, for example, need nothing more than the classical but easily remembered M-c-over-I procedures. Also, there are those who work primarily in steel or timber construction and who encounter concrete only intermittently; such designers will find that the classical M-c-over-I procedures presented in this text are quite similar to their usual design practices and are consequently much easier to remember and apply than the more unique methods.

The classical methods presented in this textbook are already in broad use in modern practice, but generally at more advanced levels. It is the section constants from the classical analysis, for example, that are currently used in more advanced studies to find the elastic response of concrete structures to earthquake loads. Too, the classical design methods are currently being used to design both prestressed concrete sections and composite concrete sections (also to be undertaken in more advanced studies). Bringing the classical design methods forward into elementary levels of design is thus not a drastic departure from present practices; it does, however, permit a distinct simplification over current practices and it improves the continuity with future studies.

The classical methods presented in this textbook are intended for use at introductory levels of design in civil engineering, civil engineering technology, and construction technology. Calculus is not used. A student familiar with algebra, statics, and strength of materials will be quite comfortable with the theoretical developments presented in this textbook.

It is emphasized that the approach used here is valid at any level of engineering or technology. This particular textbook, however, is directed specifically toward undergraduate students who are taking their first course in concrete design. Accordingly, the topics included in this textbook are limited to those topics that are commonly presented in a one-semester course in elementary concrete design.

The text is specifically directed toward the design of routine concrete structures less than 70 feet in height, having a type of structural system (diaphragm and shearwall) that excludes bending in the columns. They are the type of structural system that are intended by the American Concrete Institute to be designed using shear and moment coefficients, sidesway prevented. Such structures constitute probably 85% of all concrete construction in the world.

The concentration on this type of structure provides several benefits to the student just beginning the study of concrete. First, the structure and the structural system are small enough to be comprehensible; the concentration of attention can therefore be focused on concrete rather than on structural systems. Second, the absence of large column loads and moments considerably simplifies the study of columns; the concentration of attention can be focused on understanding routine column behavior rather than being diffused into a confusion of specialized cases of bending. And third, the type of structures that will be encountered by the engineering and construction technologist, by an overwhelming margin, will be these simple, routine ACI structures; the design of large, heavily loaded, high-rise concrete rigid frames subject to wind and earthquake are relegated to specialists in structural design.

It is intended that this textbook will be used without additional references or design manuals. It is recommended, however, that anyone engaged in concrete design on a regular basis keep on hand a personal marked-up copy of the design code, ACI 318-89, *Building Code Requirements for Reinforced Concrete*. When the meaning of a particular provision in the design code becomes clear only after repeated study, a marginal note will save time several weeks or months hence when the provision is encountered again.

Design tables are included in the back of the book, along with detailed derivations of the equations that were used to develop the tables. Those who wish to enlarge the tables can do so simply by evaluating the given equations to any desired extent on a personal computer. The equations are general and may be applied either in Imperial units or in SI units.

The Imperial (British) system of weights and measures is used exclusively in this text. It should be noted, however, that the beam and column design constants given in the tables of the appendix are dimensionless and are therefore valid for both the Imperial system and the SI system. But it should also be noted that the standard concrete strengths commonly used in the Imperial system (2500, 3000, 4000, 5000, 6000 $lb/in^2$) do not correspond to the standard concrete strengths commonly used in the SI system (15, 20, 25, 30, 35, 40 $N/mm^2$). Nor do the yield strengths of steel commonly used in the Imperial system (40,000, 50,000, 60,000 $lb/in^2$) correspond to the yield strengths of steel commonly used in the SI system (300 and 400 $N/mm^2$). As a consequence, two independent sets of design tables would be required if the SI system were to be added to this text. For an elementary text such as this, it was elected to use only the Imperial system and thereby avoid any complications in having two sets of dimensions and two sets of design tables.

As with his earlier books, the author is indebted to his wife, Sherry, who typed the manuscript of this text; her quiet and continuing support in these speculative ventures has helped immeasurably to make the ventures pleasant ones.

Samuel E. French, Ph.D., P.E.
Martin, Tennessee

# 1

# CONCRETE AND ITS REINFORCEMENT

To be a good structural material, a material should be homogeneous, isotropic, and elastic. Portland cement concrete is none of these. It is, nonetheless, a very popular construction material.

Portland cement concrete is manufactured using a coarse aggregate of stone, a fine aggregate of sand, and a cementing paste of portland cement and water. The end result is a man-made stone that can be shaped while in its plastic state and allowed to harden into its final configuration. Compared to other construction materials, such as steel, aluminum, or timber, portland cement concrete is weak and heavy, but its characteristics of durability, adaptability, and availability have made it a popular material of construction.

Portland cement concrete is quite weak in tension. In most structural applications, its tensile strength is assumed to be zero and steel is provided at all points where the structure will experience tension. The result is concrete reinforced for tension or, more simply, reinforced concrete.

## Coarse Aggregate

The coarse aggregate used in making concrete may be crushed stone or a natural gravel. The size of an individual particle may range from the size of a pea to that of a boulder, depending on whether the concrete is being cast into a thin decorative panel or a large dam. Since the remaining constituents are proportioned to fill the void spaces between the particles of coarse aggregate, the coarse aggregate will obviously form the bulk of the volume of the concrete. The quality of the finished concrete is heavily dependent on the quality of the coarse aggregate. A hard, durable concrete can only be made from a hard, durable coarse aggregate.

## Fine Aggregate

The fine aggregate used in making concrete is usually a sharp silica sand. Calcareous sands (derived from limestone) may be somewhat weaker and, although frequently used, can impart a limit on the ultimate strength of the concrete. The mixture of sand, cement, and water is called the *paste* or the *matrix* of the mix.

## Portland Cement

Portland cement is a hydraulic cement developed in 1824 by an English bricklayer, Joseph Aspdin. The name "portland" was applied because the color of the finished concrete resembled a building stone quarried on the Isle of Portland, off the coast of Dorset, England. The cement is manufactured from a mixture of about 4 parts limestone and 1 part clay, heated almost to the melting point of the mixture (about 2700°F), and then cooled and ground to a fine powder. A small amount of gypsum is added to control set.

Portland cement works by hydration; that is, it forms a chemical bond with the water in the mix and, in doing so, forms a bond to other cement particles, to the aggregate, and to any reinforcement that it contacts. It does not need air to harden or "set"; it will harden as well under water as when exposed to air.

Five standard types of portland cement are manufactured:

Type I:     regular portland cement
Type II:    increased resistance to sulfate attack
Type III:   high early strength
Type IV:    lowered heat of hydration
Type V:     high resistance to sulfate attack

By itself, type I portland cement concrete is reasonably resistant to most forms of chemical attack that might occur naturally, but high concentrations of waterborne sulfates can have very deleterious effects on it. Commonly, such concentrations can occur in sewage, in groundwater carrying dissolved gypsum, and in some seawater exposures (from sulfur-producing marine organisms). For castings in a high-sulfate environment, special cements are manufactured—types II and V—which have an increased resistance to sulfate attack. Type III cement achieves its specified strength in 7 days rather than in 28 days. It does so at a penalty in increased heat of hydration as well as in higher cost.

Heat is generated by all five types of cement during hydration. For the more common sizes of structural members this heat is readily dissipated and presents no problems. For extremely large castings such as heavy equipment foundations, the trapped heat can pose serious problems and must be provided for. For such large castings a special cement is manufactured—type IV—which has a lower heat of hydration than other types and helps to alleviate the problem.

## Water

As a general rule, any water suitable for drinking is suitable for making concrete. It should be remembered, however, that the hydration of portland cement is a very complex chemical reaction and that undesirable results can occur from even small amounts of certain compounds in solution. Nonetheless, if drinking water of standard quality is used as the mixing water, the chances of introducing a deleterious compound through the water is minimal.

## Water/Cement Ratio

The strength of portland cement concrete depends on the number of pounds of water used per pound of cement. This ratio of water to cement, shown as the water/cement (W/C) ratio in Figure 1-1, is the single most important parameter used to control the strength of concrete. As indicated in the figure, there is a minimum W/C ratio required for hydration of all the cement molecules. Any water in the mix in excess of that amount will reduce the final strength of the concrete. It is assumed, of course, that the aggregate is at least as strong as the sand-cement paste.

In introductory comments such as these, it is difficult to overstate the importance of the W/C ratio. An excess of water reduces not only strength but hardness, durability, resistance to chemical attack, resistance to freeze-

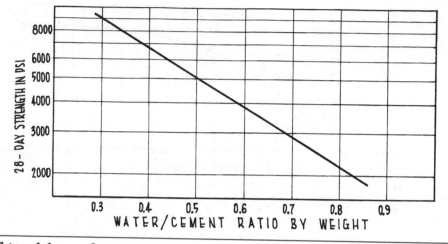

**Figure 1-1** Concrete strength.

thaw, and all other desirable properties of the concrete. Thus the construction worker who adds water to the mix to make it more workable is, by this single act, significantly reducing all the desirable characteristics of the finished concrete.

At the higher W/C ratios, the "yield" or volume of usable concrete per pound of cement is higher, affording a more economical mix. Unfortunately, as water content increases, the concrete has a higher tendency to shrink as it hardens. This tendency to shrink, combined with the reduced strength, durability, and hardness, imposes a practical upper limit on the W/C ratio of about 0.80.

At the lower W/C ratios, the shrinkage of the concrete can be minimized, but the mix is expensive and workability during placement is reduced. Thus the lower limit of the W/C ratio is limited by practical considerations, usually to about 0.40. Almost all reinforced concrete used in today's practice will have W/C ratios between these general limits of 0.40 to 0.80, although some prestressed concrete units may have lower values. In this book, consideration is limited to concretes having W/C ratios within the foregoing limits.

## Strength Gain with Time

As with other structural materials, the strength of a concrete and the configuration of its stress-strain curve form the basis of all structural calcula-

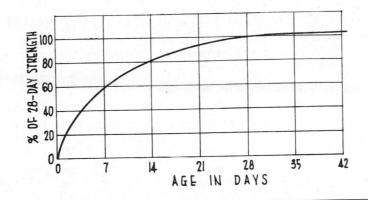

**Figure 1-2**    Strength gain.

tions. The strength of a concrete is gauged at 28 days by its compressive stress at failure. For measuring the stress at failure, a concrete cylinder 12 in. long and 6 in. in diameter is cast from a sample of the concrete. The cylinder is allowed to cure under controlled conditions for 28 days and is then placed in a compression test machine and loaded to failure. The compressive stress computed for the highest load is the "ultimate" stress, or 28-day strength, of that concrete.

A typical strength-time curve is shown in Figure 1-2. It should be noted that the gain in strength after 28 days is not acknowledged, but that concrete continues to gain strength for months or even years after manufacture. It should also be noted that the concrete attains about 60% of its design strength at 7 days, a useful bit of information to know when precast elements are to be lifted or forms are to be removed.

There is no immediate test that can be performed at the time the concrete is being cast to assure that the concrete will reach a certain strength 28 days later. In practice, concrete is sampled periodically and cylinders are made as the concrete is being cast into the forms. These cylinders are tested 28 days later, affording a verification that the concrete was manufactured properly the previous month. It is thus apparent that the cylinder test is a "report card" that reveals how well a job was done 28 days before.

To assure that the concrete is of proper strength and quality at the time it is cast, rigorous measures of quality control must be instituted and maintained during its production. Over the years, simple but adequate measures have been developed which, with proper enforcement, will assure the production of good-quality concrete. These measures are unique in the industry; no other structural material is sampled for its strength and quality at the time of its final placement in the structure.

All of the desirable properties of the finished concrete, such as durability, hardness, abrasion resistance, and so on, improve as the strength increases. By controlling the strength of the concrete, its other properties may also be controlled. As a consequence, it is common practice to identify a concrete by its ultimate strength, greatly simplifying the matter of identification.

## Elasticity

The stress-strain curve for concrete can be plotted readily from the load-deflection data of a standard cylinder test. Typical stress-strain curves for various strengths of concrete are shown in Figure 1-3, where ultimate stress

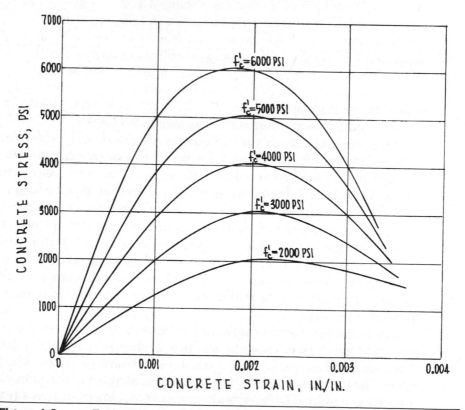

**Figure 1-3**    Typical stress-strain curves.

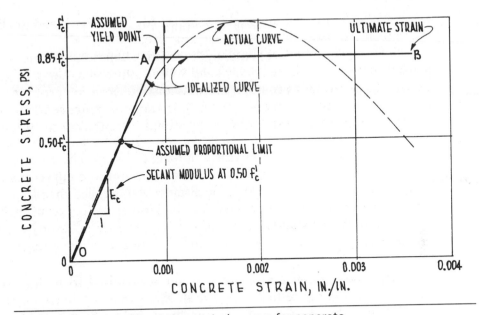

**Figure 1-4**    Idealized stress-strain curve for concrete.

is denoted $f'_c$. It should be noted that there are no distinct "breaks" in these stress-strain curves: they are continuous smooth curves.

By definition, an elastic material under load will deform along a straight line up to its yield point. Within this elastic range, it will resume its exact original configuration upon release of the load. From the curves of Figure 1-3 it is seen that concrete is not a truly elastic material under this definition, although it is reasonably elastic within the lower three-fourths of its ultimate strength. It is noted, however, that the normal range of working stresses for concrete is within the lower half of its ultimate stress. Consequently, with a normal factor of safety and under its day-to-day service loads, concrete can be expected to work at less than half of its ultimate stress and will therefore behave as an elastic material under short-term service loads.

An exaggerated stress-strain curve is shown in Figure 1-4. An approximation of the slope of the initial portion of this curve is usually defined as the modulus of elasticity; its value is the stress divided by the strain at any point. The value of the modulus of elasticity increases as the strength of the concrete increases. The modulus of elasticity, $E_c$, in pounds per square inch is computed empirically by the relationship

$$E_c = 57,000\sqrt{f_c'} \qquad\qquad (1\text{-}1)$$

where $f_c'$ is the ultimate strength of the concrete in pounds per square inch. The modulus of elasticity for each of the more common values of $f_c'$ is given in Table A-1 of the Appendix.

Since the actual stress-strain curve of Figure 1-4 has no well-defined yield point, an idealized curve is substituted for the top portion of the actual curve. The resultant idealized curve OAB forms the basis of the derivations and calculations in the following chapters. The position of line AB has been found to provide the best correlation to comprehensive test data when it is taken at $0.85f_c'$. The distance AB is called the *plastic range* of the material.

## Shrinkage and Creep

Only a relatively small amount of water is required to hydrate all the cement in the concrete mix. Any excess water that is free to migrate can evaporate as the concrete hardens, causing a slight reduction in the size of the member. This reduction in size, called *shrinkage*, can produce undesirable internal stresses and cracking.

As a very general rule of thumb, the drying shrinkage of a typical structural concrete will produce a dimensional change roughly equal to that of a temperature drop of 40°F. Unlike a temperature drop, however, which affects both the concrete and its reinforcement, shrinkage affects only the concrete. The reinforcement stays at its original dimensions while the concrete shrinks around it, thereby creating compressive stresses in the reinforcement and tensile stresses in the concrete.

Concrete also undergoes another type of reduction in size if it is subjected over a long period to sustained stresses. This reduction, called *creep* or *plastic flow*, is a permanent deformation and is only partially recovered when the load is released. Creep varies with both time and intensity of load, as shown in the typical curves of Figure 1-5. Roughly, creep strain is directly proportional to stress. Figure 1-5 gives verification that after the first few months, the strain corresponding to a stress of 600 psi is roughly double that of 300 psi.

It is also evident from Figure 1-5 that creep will occur even under very low levels of stress where the load is sustained over long periods. However, for both low and high levels of stress, it can be seen that the rate of creep diminishes sharply after about 6 months and is negligible after about 2

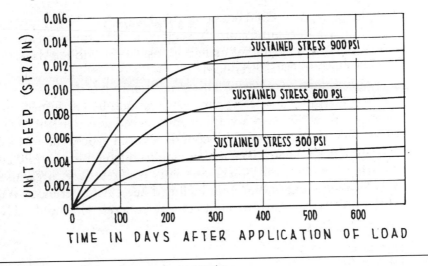

**Figure 1-5**        Rate of creep in concrete.

years. For comparison, the ultimate strength of a typical grade of structural concrete is 3000 psi; for such concrete, the stress of 300 psi in Figure 1-5 would be 10% of ultimate and 900 psi would be 30% of ultimate.

Like shrinkage, creep affects only the concrete; the reinforcement is not similarly affected. One result of creep in concrete, therefore, is to increase the compressive stresses in any adjacent reinforcement. Another result is to produce an overall increase in deformations with time, beyond those due to normal loading.

When combined, the effects of shrinkage and creep can be significant. At their maximum, deformations due to shrinkage and creep combined could be as much as the deformations that would occur due to applied loads. Obviously, these effects must be accounted for; a common means to account for shrinkage and creep is discussed in Chapter 2.

## Density and Weight

The weight of concrete varies with the density of its aggregates. An average value for unreinforced concrete is 140 pounds per cubic foot (pcf) and for reinforced concrete is 150 pcf. A median weight of 145 pcf is used throughout this text.

The weight of concrete can be reduced significantly by the use of lightweight aggregates. Concretes that weigh as low as 90 pcf are produced this way and are in common use. In such concretes, the decrease in dead load can often offset the higher materials costs, even though the strength of the lighter concrete is also reduced.

It is almost axiomatic that strength is reduced as density is reduced. Whatever causes a decrease in density, such as entrapped air or lighter aggregates, will almost always produce a reduction in strength.

There are many specialty concretes of all types: gas-formed concrete, gap-graded (popcorn) concrete, nailable concrete, fiberboard concrete, insulating concrete, and others. Specialty concretes are rarely used in structural applications and are not included here. All references to concrete in this book will mean portland cement concrete of standard weight and density.

## Thermal Coefficient

The coefficient of thermal expansion and contraction for concrete varies somewhat with the aggregates. An average value is 0.0000055 in./in. per degree Fahrenheit. Since the corresponding coefficient for reinforcing steel is quite close, 0.0000065 in./in. per degree Fahrenheit, the effects of differential thermal growth between the concrete and its reinforcement is negligible at common atmospheric temperature ranges.

## Related Properties

There are many other properties of concrete that may be highly desirable in the design of the project but which do not affect the load-carrying capabilities of the concrete. Properties such as impermeability, hardness, resistance to abrasion, resistance to freeze-thaw, resistance to chemical attack, and other desired properties could have a pronounced influence on choice of aggregates and choice of ultimate strength. Once the ultimate strength is established, however, the succeeding structural calculations are based only on that ultimate strength and the corresponding deformation properties. Other properties would have only a secondary influence on the structural design.

## Additives

Additives are chemicals added to the mix while it is plastic to impart a particular property to the concrete. The decision to use an additive should not be taken lightly: the additive cannot be withdrawn should the concrete fail at its 28-day tests.

*Water-reducing agents* (WRA) are one type of additive. They act as a lubricant on the plastic mix, allowing the W/C ratio to be reduced while keeping the workability. As the W/C ratio is reduced, all the desirable properties of concrete improve, including strength. Water-reducing agents are also used to permit a reduction in the amount of cement being used; strength and workability can thus be kept constant while shrinkage is reduced.

*Air-entraining agents* are another type of additive. They impart billions of microscopic bubbles of air, distributed uniformly throughout the mix. The presence of this air (3 to 7% by volume) makes the hardened concrete much more resistant to freeze-thaw and to chemical attack. It also reduces overall density by the same percentage and produces a corresponding slight loss of strength. The improvement in properties is reliable and dependable, but maintaining the required amount of air is sometimes quite difficult. The chemical reaction that produces air entrainment is a separate reaction; the chemical hydration of the cement is not involved.

*Accelerators*, another type of additive, serve to accelerate the setting time of the mix. In doing so, additional heat of hydration is generated. Some batching plants use accelerators indiscriminately during cold weather to keep the mix from getting cold in transit; such practices are very questionable and may often do more harm than good, particularly where the accelerator contains chlorides.

*Retardants* are additives that are used to slow the setting rate of the mix. They are used where long finishing times are required. Retardation up to 8 hours or longer is commonly produced, but the results are sometimes erratic and unpredictable.

*Plasticizers* and *superplasticizers* are additives that are used to lubricate a "stiff" mix, allowing it to be placed in intricate formwork while maintaining a low W/C ratio. The effects are temporary, usually about 30 minutes, after which time the concrete supposedly returns to its unplasticized state. Long-term performance records on concrete placed while in a "plasticized" state are scattered and incomplete.

There are other additives being sold to those who are willing to take the risk. Except for air entrainment, few additives can be recommended. Any additive that interferes with the hydration process should be used with extreme caution.

## Bond Strength

Bond strength between concrete and its reinforcement is probably the most highly variable property of concrete that must be used in design. Accordingly, bond strength is assigned a high factor of safety to account for the many uncertainties and variables. The resulting conservative values for bond strength require that reinforcements be deeply embedded in concrete to assure full development of the steel strength.

Bond strength increases as the ultimate strength of the concrete increases. It can be adversely affected by excess water, air bubbles, excessive rust on the reinforcement, improper consolidation of the concrete, accidental movement or vibration of the reinforcement after the concrete has set, and the type and number of deformations on the reinforcement. Even with identical placement conditions and practices, two embedded bars can have strikingly different pullout strengths.

Currently, the design codes no longer include smooth bars or wire for use as reinforcement. A provision of older design codes allowed half as much bond strength for smooth bars as for deformed bars, and many buildings throughout the world are still being built to that provision. The success of the many older concrete buildings still in use today attests to the validity of that provision.

## Reinforcement and Its Properties

Reinforcement is manufactured under several international specifications for a variety of yield stresses, or grades. In this text, consideration is concentrated on reinforcement grades 40 and 60. For these grades, the yield stress is 40,000 and 60,000 psi, respectively. Grade 50 steel, having a yield stress of 50,000 psi, has fallen out of common use in American practice but is still occasionally found; it is therefore included in this text.

Reinforcement is rolled in round bars, with deformed surfaces designed to improve the connection to the adjacent concrete (see Figure 1-6). In the United States, the identifying number of a reinforcing bar, up to a No. 9 bar, indicates its diameter in eighths of an inch; thus a No. 5 bar has a nominal diameter of ⅝ in. Bar sizes No. 9, No. 10, and No. 11 are also round, but their diameters are set to provide the same areas as older square bars—1 in., 1⅛ in., and 1¼ in., respectively. A No. 9 bar has a diameter of 1.128 in., a No. 10 bar has a diameter of 1.27 in., and a No. 11 bar has a diameter of 1.41 in. Bar sizes larger than No. 11 are used primarily in very heavy construction and are not included here.

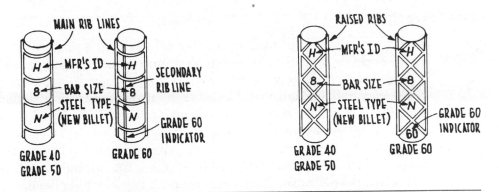

**Figure 1-6**    Deformed reinforcing bars (adapted from Ref. 11).

Although deformed bars are used almost without exception in the United States, smooth bars or drawn wire are frequently used throughout much of the rest of the world. In small diameters (6, 8, 10, and 12 mm), smooth wire can be shipped in rolls occupying a low volume at reduced shipping costs. It should be noted that labor costs are considerably higher to place a large number of small smooth wires in the forms compared to fewer larger deformed bars. Where labor is cheap and shipping is expensive, however, smooth wire will probably remain popular.

Stress-strain curves for steel are treated in detail in elementary strength-of-materials texts. For its applications in reinforced concrete, an idealized curve similar to that of concrete (Figure 1-4) is required. Such idealized curves for steel are shown in Figure 1-7 with the complete stress-strain curve shown alongside for reference. Since steel is such an ideal structural material, the idealized curves of Figure 1-7 are little different from the same portion of the actual curve.

From Table A-1 of the Appendix it is noted that the elastic modulus for steel is $29 \times 10^6$ psi. For that value of the modulus of elasticity, the strain at the yield point is 0.00138 in./in. for grade 40 steel, 0.00172 in./in. for grade 50 steel, and 0.00207 in./in. for grade 60 steel.

In addition to individual smooth wires and deformed bars, reinforcement is also fabricated into a rectangular mesh of smooth wires called *welded wire fabric* (WWF), or into a mat of deformed bars called *structural fabric*. These fabrics are made up as a convenience to construction; the structural design is the same regardless of whether the reinforcement is placed as individual bars or as a mesh.

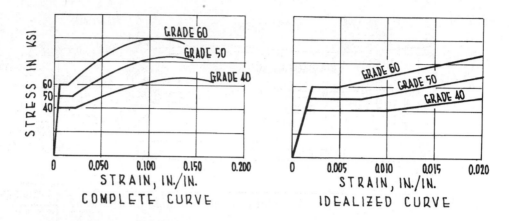

**Figure 1-7**      Idealized stress-strain curve for steel.

## Use and Placement of Reinforcement

It was noted earlier that concrete has very little tensile strength. Wherever tension occurs on a concrete structure, reinforcement is added to sustain any tensile stresses that may occur. A very large part of the design of a reinforced concrete member is therefore an analysis to find where tension will occur in the member and to provide the correct amount of reinforcement to sustain that tension.

Consider the beam shown in Figure 1-8. The moment diagram for the beam is sketched under the beam in accordance with the usual practices from elementary strength of materials, but it is common practice in the design of reinforced concrete to plot the moment diagram upside down as shown; that is, positive moment is plotted downward rather than upward. Note that under the given load conditions, the beam of Figure 1-8 will experience tension on the bottom of the section at midspan, a region of positive moment. Note further that the beam will experience tension on the top of the section in the vicinity of the overhang, a region of negative moment. In both regions, reinforcement must be provided in the correct amount and located in the correct position to sustain these tensile forces.

It is well to note at this point the reason for sketching the moment diagram upside down, that is, with positive moment being plotted downward rather than upward. It can be observed from Figure 1-8 that by being plotted this way, the moment diagram will always fall on the tension side of the member. Further, if the moment diagram is drawn accurately to

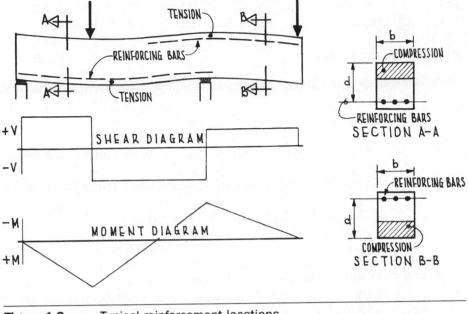

**Figure 1-8**    Typical reinforcement locations.

scale, the cutoff points for the reinforcement can actually be determined by scaling the moment diagram, a common practice in the design of reinforced concrete members.

Throughout succeeding chapters, the practice of sketching the moment diagram "on the tension side" will be followed in all design procedures. The practice not only provides a very good method for locating reinforcement but it also provides a reliable suggestion of the deflection pattern of the beam under the given loads.

## Mixing, Placing, and Curing

Volumes have been written on field practices concerning the mixing, placing, and curing of concrete. One of the most authoritative references on such matters is the *Concrete Manual* (U.S. Bureau of Reclamation, 1975). Field practices are outside the scope of this book. It is noted, however, that the tolerances, inaccuracies, and approximations used in standard field practices will have a profound effect on the final dimensions. The cumulative effect of all such deviations is accounted for in the design calculations by the strength reduction factor $\phi$, discussed at more length in Chapter 3.

## REVIEW QUESTIONS

1. List the four basic ingredients used in making portland cement concrete.
2. What does the term "calcareous" mean when applied to aggregates?
3. What is the "matrix" of a concrete mix?
4. Why is it necessary to reinforce concrete?
5. List the five types of portland cement and the distinguishing characteristic of each.
6. If a city water supply is "hard" water, could that water be safely used to make concrete?
7. What is the "ultimate strength" of concrete?
8. What are the units of ultimate strength?
9. At what age is the concrete when its ultimate strength is usually determined?
10. How long does the strength of concrete continue to increase?
11. What is the weight of a cubic foot of concrete? Of a cubic yard?
12. A concrete has a specified ultimate strength of 4000 psi at 28 days. Approximately what is its water/cement ratio? What strength can be expected at 7 days?
13. Is water/cement ratio a ratio of volumes or weights?
14. What is the primary detrimental effect of adding water to a mix, with all other ingredients unchanged?
15. A concrete has a specified ultimate strength of 3000 psi. What is its modulus of elasticity?
16. What is the coefficient of thermal expansion for concrete? For steel? Why is it important that these two coefficients be close to each other?
17. A reinforced concrete bridge spans 60 ft. From a low atmospheric temperature of $-5°F$ to a high temperature of $+115°F$, what is the overall dimensional change in its length?
18. How does creep differ from shrinkage?
19. How is creep similar to shrinkage?
20. In general terms, what is the primary cause of shrinkage?
21. At what levels of stress does creep occur?
22. What is the major objection to the use of additives?
23. What is air entrainment?
24. What is grade 50 steel?
25. What percent elongation does a typical grade 60 steel undergo in its plastic range (Figure 1-7)?
26. In older codes, what is the difference in bond strength between smooth bars and deformed bars?

# CHAPTER

# 2

# DESIGN CODES AND
# DESIGN METHODS

The design of structures in reinforced concrete is rigidly governed by codes. One may spend years learning the fine points of these codes, but the major features of design are contained in a surprisingly few paragraphs. It is these major features that are the subject of this text; the fine points are usually learned through exposure, experience, and embarrassment.

In the United States, those sections of building codes that deal with concrete design are usually based on the *Building Code Requirements for Reinforced Concrete* (American Concrete Institute, 1989). Any reference to a code in the following chapters will mean this code. It is commonly identified as ACI 318-77 or ACI 318-89, where the number following the dash is the year of the edition.

Discussions in the following chapters refer frequently to particular sections of the ACI Code. It is recommended that anyone designing concrete members obtain a current copy of the Code and become familiar with it. Where the meaning of a particular provision becomes clear only after repeated study, a marginal note will save some time several weeks or months hence when the provision is encountered again.

Where appropriate in subsequent discussions, the section number of the Code is given in parentheses following a reference to the Code. For example, in the sentence "Code (10.2.3) requires that ultimate strain in concrete be taken at 0.003 inches per inch." the referenced provision is stated in Section 10.2.3 of ACI 318-89.

## Effects of Inelastic Strains

It was noted in Chapter 1 that concrete is reasonably elastic under short-term service loads. Where loads are sustained over long periods, however, they can produce inelastic deformations due to creep. The magnitude of these deformations at any given time is unpredictable.

In addition to creep deformations, concrete is also subject to shrinkage deformations due to drying. As with creep deformations, shrinkage deformations are generally unpredictable. The concrete may regain much of the drying shrinkage loss if it is later submerged, but in dry buildings the shrinkage loss can be considered to be permanent.

The end result of long-term shrinkage and creep in concrete members is to produce an increase in the compressive strain. Such an increase occurs with no accompanying increase in load. The increase does not go on and on without limits, however. At worst, the inelastic strain due to shrinkage and creep combined can be expected to be roughly as much as the elastic strain due to applied loads. In effect, therefore, the modulus of elasticity when the long-term effects of shrinkage and creep are included can be estimated as half that due only to short-term elastic loading. For calculations, Code (B.5) does exactly that; it accounts for shrinkage and creep by taking the modulus of elasticity due to long-term loads to be half that due to short-term loads.

The response of reinforced concrete to loads is heavily dependent on the stress-strain relationships of the concrete and its reinforcement. The meaning of the idealized stress-strain curves of Chapter 1 must be fully appreciated if the analytical methods of the following sections are to be understood.

## Ultimate Strength Design (USD) Method

The ultimate strength design (USD) method for the design of individual reinforced concrete members is the method preferred by ACI. ACI has refined the name of its method to the "strength method," though the older generic name, "ultimate strength method," is probably more widely recognized.

In its crudest generic form, the ultimate strength method would consist of loading a test member to failure and recording the failure load. There is no need for mathematics, analysis, or theory; the failure load is known absolutely from the full-scale destructive test. The allowable service load on all other members that are just like this test member is then taken to be some fraction, say 60%, of this ultimate load. As long as the load on any one of these members is kept below this safe service load, then that member is known to be safe.

There are, of course, many problems in such a simplistic approach. For one thing, the ultimate strength by itself has no relationship to service conditions. A member may in fact have more than adequate strength yet deform in such a way under service conditions as to be completely unsatisfactory. A beam that is excessively "springy," for example, could have adequate strength to support its load, but its large deflections could cause breakage of glass or other brittle materials that are being supported by the beam. Obviously, requirements for performance at service levels must somehow be incorporated.

The test also applies to only one size, shape, and configuration for the test member. If the method is to extend to other sizes, shapes, and configurations, then a test loading would have to be performed for each of the new configurations. Such a program of tests would be impractical for the hundreds of thousands of concrete sections in common use in today's industry. Obviously, the approach must be relegated to some form of computations rather than tests if it is to be adopted.

As a matter of simple practicality, there are in fact two quantities that must be subject to computation if a computed solution is to be acceptable. First, the ultimate external load that the member is to sustain must be readily subject to computation. Second, the maximum resistance that a member can develop internally in order to sustain this externally applied load must also be subject to computation. With these two quantities subject to computations, a member can then be designed at its ultimate internal strength to sustain the ultimate external load.

As a part of its strength design method for concrete, ACI has included a method for computing the ultimate external load to be applied to the member. In this method, the known service loads are multiplied by "load factors" to project the service loading up to the ultimate failure load that the member must be able to sustain. A member can then be selected to resist this magnified (or ultimate) load, with the knowledge that it will actually serve its day-to-day functions at some distinctly lesser level of loading (roughly 60% of this ultimate load).

The ACI Code also prescribes a means to design a member to sustain this computed ultimate load, with the ultimate internal strength of the member also being determined by computations. In the ACI approach, the ultimate internal strength so computed is not, however, the absolute maximum collapse strength that the member can ever possibly attain. It is rather an idealized but consistent reference strength, based on near-ultimate conditions. It is this computed strength that is commonly termed the "ultimate strength" of the member in the ACI method, though, as noted, the term is applied somewhat erroneously.

The use of calculations rather than tests inherently causes the design problem to become considerably more sophisticated. Since the strength of the member must be predicted (rather than tested) for some computed ultimate load, the design of the concrete member must preclude any possibility of some small item going wrong and thereby initiating a premature failure. All potential "weak points," including those that might appear only at very high levels of load, must be eliminated from the design. The ACI Code includes numerous provisions and checks throughout the design process to eliminate such "weak points."

The end result of such an approach is a concrete member that will sustain high levels of load smoothly and predictably up to and including the ultimate load. The member will exhibit no premature breakage, no unexpected collapse, and no propagation of any local failure. The uniformity in the strength of each member so designed will ensure the safety of the structure as a whole. Insofar as strength is concerned, the final structure is safe at all levels of load up to the computed ultimate strength of the individual members.

But as noted earlier, the foregoing requirements for ultimate strength do not, in themselves, assure that the structure will perform satisfactorily down at day-to-day service levels. Measures must also be incorporated in the design process to restrict deformations to some acceptable level, to account for creep and shrinkage with time, and to assure that occasional overloads do not cause permanent deformations or damage. Such a set of requirements has been added to the Code (Chapter 9) under the title of general serviceability requirements; the limitations prescribed by Code will assure that the structure will perform satisfactorily under day-to-day service conditions.

In all of these Code requirements, however, there is no requirement that the structure behave elastically. Inherently, of course, the structure will behave elastically at all lower levels of loading, since concrete itself is elastic (except for long-term creep and shrinkage) to more than 75% of its ultimate stress. Under occasional overloads, however, it must be acknowledged that some members could possibly enter into inelastic behavior.

It is noted also that the methods of analysis used to compute the service loads under general service conditions are elastic methods. In the ACI strength method, it is the service loads computed from one of these elastic analyses that will be projected upward to predict the ultimate loads that the member must withstand. The ultimate load must therefore be projected high enough above this service level that the structure will always behave elastically throughout all service levels of loading.

All of the foregoing possibilities and probabilities can be summarized in the following statistical requirements:

1. The probability of a real load occurring in excess of the anticipated service load must be acceptably small.
2. The probability of unsatisfactory behavior under service conditions must be acceptably small.
3. The probability of a real load occurring in excess of the anticipated ultimate load must be acceptably small.
4. The probability of unsatisfactory behavior under conditions near ultimate load must be acceptably small.

The success of the ACI strength method of design over the past 30 years attests to the fact that these requirements have been met.

As a summary of these discussions of the strength method, it is observed that the strength method of design is exactly that: a method of designing a member for its strength. Code requirements for control of deformations and serviceability have been added as secondary provisions. The overriding concern in the strength method is the strength of the member at ultimate levels of load.

## Working Stress Design (WSD) Method

In the years before ACI adopted the strength method of design, concrete was designed much like other structural materials, using the working stress design (WSD) method. The working stress method was, and is, a perfectly valid method for the design of concrete members; it is still included in the ACI Code as an acceptable alternate design method. The change (begin-

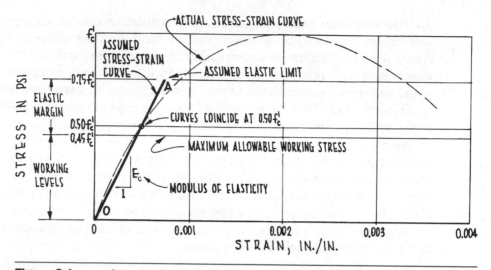

**Figure 2-1**    Assumed stress-strain curve for working stress design.

ning about 1956) from the working stress method to the ultimate strength method was spurred, at least in part, by the advent of high-strength steel reinforcement; the use of high-strength steels created questionable points in the older working stress method.

The working stress method is based on the assumption that the material is always elastic, deforming as indicated by line OA in Figure 2-1. That part of the stress-strain curve beyond the elastic limit is excluded from use. It is as if the stress-strain curves were absolutely linear up to the elastic limit, at which point rupture occurs. Rupture does not occur, of course, which leaves a considerable amount of reserve strength unused and unaccounted for in the working stress method.

A second inefficiency enters the working stress method when it is applied to concrete. Concrete is reasonably elastic for short-term loads, but even for short-term loads its elastic limit is usually taken conservatively at only 75% of its ultimate strength. If it is assumed that the concrete is allowed to work at some 60% of this elastic limit at service levels, the resulting allowable stress is then $0.6 \times 0.75 f'_c = 0.45 f'_c$, or only about 45% of the ultimate strength of concrete. This serious reduction in allowable stress makes concrete a relatively inefficient material under the working stress method.

For long-term loads, the inelastic effects of creep and shrinkage must also be accounted for. The end result of these inelastic deformations is essentially a reduction in the modulus of elasticity of concrete for long-term loads. Since 1956, the ACI Code has included corrections which approximate the long-term effects of shrinkage and creep. In essence, the modulus of elasticity for long-term loads is taken to be half that for short-term loads, a feature that must be recognized when deflections are being investigated.

In using the working stress method, one simply selects an allowable stress which is some fraction, say 60%, of the elastic limit (or the yield stress) of the material; for concrete, such a stress is shown as $0.45f'_c$ in Figure 2-1. The maximum anticipated working loads on a member are then computed. The member is then designed to sustain these maximum anticipated working loads without exceeding the given allowable stress.

The ratio of the stress at the elastic limit to the allowable stress at working levels is called the "factor of safety," designated herein as FS. Or, in more common form,

$$\frac{\text{Elastic Limit}}{\text{FS}} = \text{Allowable Working Stress}$$

The "elastic margin" shown in Figure 2-1 is closely related to the factor of safety. The elastic margin is that part of the stress-strain curve above the allowable working stress but below the elastic limit:

$$\text{Elastic Margin} = \text{Elastic Limit} - \text{Allowable Working Stress}$$

As an example in the use of these terms, the factor of safety for the concrete shown in Figure 2-1 might be taken at 1.7. The corresponding elastic margin then is 0.7, or, stated another way, the material remains elastic up to a 70% overload.

The working stress method does not require any checks for "weak points" at levels above the elastic limit of the material. Consequently, if loads should ever exceed the elastic limit (or yield point), the actual collapse mode may be quite difficult to predict. Erratic breakages and "weak points" could occur at these excessive levels of load, resulting in a rather unpredictable collapse pattern. Unlike the strength method, the working stress method does not assure smooth predictable behavior once stresses exceed the elastic limit or go into yield.

But also unlike the strength method, the working stress method does assure smooth elastic behavior of the structure at all levels of working loads. The working stress method also assures elastic behavior well above the anticipated working loads since the performance of the member remains elastic throughout the elastic margin. Further, the performance of the building under day-to-day service conditions will be quite predictable, requiring only nominal controls over its performance.

As a summary of these discussions of the working stress method, it is observed that the working stress method of design is exactly that: a method of designing a member for its performance at elastic working levels of stress. The response of the member under severe levels of overload is of secondary concern. The overriding concern in the working stress method is the elastic behavior of the member at working levels of stress.

## Design Methods for Prestressed Concrete

It is relevant at this point to compare the design methods currently being used for prestressed concrete to those being used for ordinary reinforced concrete. Prestressed concrete is a relatively recent innovation in the industry; that is, it has been in general use only since about 1950. In prestressing, a concrete member is preloaded, or prestressed, in opposition to the anticipated service loads. When the service loads are later imposed, the concrete member then becomes "unloaded" to some predetermined but acceptable stress pattern. Inherently, prestressed concrete is subjected to high levels of stress throughout its entire working life, if not from the prestressing loads then from the service loads.

The design method currently used for the design of prestressed concrete contains the best of both the strength method and the elastic (working stress) method. The member is designed first at elastic levels of stress to suit the anticipated service loads. The resulting design is then investigated and revised as necessary to assure appropriate behavior at ultimate levels of load. The end result is a member that is properly designed throughout all levels of loading, both elastic and inelastic.

The ACI Code does not yet require a similar comprehensive design procedure for ordinary reinforced concrete. To date, a reinforced concrete member may be designed either for its ultimate strength or for its elastic behavior; there is no requirement to design for both. The simplified design method presented in this textbook inherently provides such a complete design procedure for ordinary reinforced concrete, philosophically akin to that currently required for prestressed concrete.

## Simplified Full Range Design

If a concrete member is designed using the strength method, it is a somewhat clumsy and tedious procedure to review the performance of the member at service levels. Similarly, if a member is designed at elastic levels of stress, it is a somewhat clumsy and tedious procedure to review the performance of the member at ultimate load. The full range design method presented in subsequent chapters combines these two cases into one simplified operation, permitting both sets of conditions to be treated at once.

In order to be consistent with the procedures currently in use in the practice, the designs herein are performed at ultimate levels of load while incorporating all the desired properties at service levels of stress. The designs could be performed just as easily at service levels of stress, however, while incorporating all Code requirements at ultimate levels of load. Either way, the end result is a design procedure for concrete members that accounts for their elastic behavior at service levels as well as their inelastic behavior at ultimate load.

The design method used in the following chapters is completely derived from the strength design method prescribed by ACI 318-89; it is not related to the working stress design method. The two approaches are not equivalent; that is, a concrete member designed using the design methods presented in this textbook will not be the same as a member designed using the working stress method.

The full range design method utilizes the idealized stress-strain curve of Figure 2-2 for stresses in concrete at service levels. The stress in steel at service levels are those given by the idealized curves of Chapter 1, Figure 1-7. Note that for both of these sets of idealized curves, the elastic limit coincides with the yield stress. These two stress-strain curves are appropriate only for stresses at service levels. At ultimate load, the stress and strain conditions are stringently prescribed by Code; these Code requirements are rigorously observed in later derivations.

The use of the idealized stress-strain curve of Figure 2-2 in the full range design method is a distinct departure from the more complex methods of design in common use, but the end results are identical. The only difference is that using the idealized stress-strain curve permits the development of a simplified full range design method such as that developed herein.

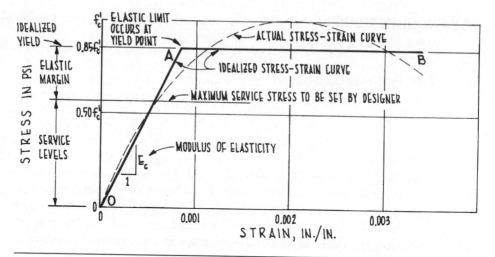

**Figure 2-2**       Idealized stress-strain curve for ultimate strength design.

## Accuracy of an Elastic Analysis

The procedures presented in this text utilize elastic methods for designing concrete at service levels of stress. The procedures also utilize the strength method of design at ultimate load. Inherently, the procedures depend on the theoretical accuracy of two independent sets of analysis, one at elastic levels and one at ultimate load.

The question immediately arises concerning the inaccuracies that accompany any analysis of concrete at elastic levels of stress. It is, in fact, these inaccuracies at elastic levels that prompted ACI to develop the ultimate strength method. While such inaccuracies do indeed occur, they are not prohibitive. One need only recall that ACI not only accepts the validity of elastic theory as an alternative design method but actually requires an elastic analysis of the most heavily stressed members in the industry, prestressed concrete members.

So while the accuracy of the elastic analysis is, in all cases, somewhat less than one may wish, an elastic analysis can certainly be used as a general indicator of the performance of the member at day-to-day working levels of load. The ideal procedure would therefore use the ultimate strength method for the actual design and it would use elastic methods as a means to control the general levels of service stress. It is exactly this type of procedure that is presented in succeeding chapters.

## REVIEW QUESTIONS

1.  State the full name of the ACI Code for reinforced concrete and the year of its latest revision.
2.  What is the long-term effect of shrinkage and creep in concrete?
3.  How do deformations due to shrinkage and creep compare in size to other deformations in reinforced concrete?
4.  How does the ACI Code account for long-term shrinkage and creep in the elastic behavior of reinforced concrete?
5.  What is a "load factor" in the ACI strength method?
6.  Why is it plausible to design each member of a structure to some "ultimate" load when it is never known exactly how the overall structure will fail?
7.  In the strength method, what assurance is there that a member will perform properly under day-to-day service loadings?
8.  In the strength method, what assurance is there that the structure as a whole will perform properly under day-to-day service conditions?
9.  What is a nominal value that one might expect for factor of safety in the working stress method of design?
10. What is the "elastic margin" in the working stress method of design?
11. How is an overload accommodated in the working stress method of design?
12. In the working stress method, what assurance is there that the structure will perform properly once the members enter yield?
13. How are the design methods for prestressed concrete markedly different from those for ordinary reinforced concrete?
14. In view of the theoretical inaccuracies that are known to exist in the elastic analysis of reinforced concrete, how can the elastic analysis be used responsibly in a design?

# 3

# CODE REQUIREMENTS

The ACI Code covers much more than just the design methods outlined in Chapter 2. Among other things, Code prescribes the maximum and minimum sizes of members that may be built of reinforced concrete, as well as minimum required amounts and sizes of reinforcement. It also prescribes limits on allowable deflections and means to predict and control those deflections. The Code also prescribes loads, load factors, and load combinations to be used in design, as well as means to allow for undercapacity in the event a concrete member is not built exactly as it was intended to be built.

In addition to design requirements, Code also prescribes construction requirements. Among the many construction items covered, Code prescribes tolerances to be held in constructing the formwork, in placing the reinforcement, and in aligning the members. It also prescribes the minimum cover that must be maintained over the reinforcement in order that the steel be adequately protected from weather, from attack by chemicals in the soil, or from corrosion by salts and deicing compounds. Code also prescribes imbedment lengths that must be maintained to assure adequate bond between the concrete and its reinforcement and it prescribes the circumstances when reinforcing bars must be placed singly or when reinforcement may be placed in bundles of several bars.

There is much to be learned about code requirements when one begins the study of reinforced concrete. A few of the more general code require-ments are introduced and discussed in this chapter; other code require-ments are introduced and discussed in later chapters when they are first encountered. The remaining requirements will be encountered by the prac-ticing designer over the years. It will be seen that these remaining require-ments are by far the largest part of the design code.

## Minimum Dimensions

The minimum size of some concrete members is set by Code and of others by limitations on forming and casting. The difficulties in forming and cast-ing very thin or very small members in the field should not be underesti-mated. Labor hours and costs required to cast such members can easily be double or triple those required to cast more conventional sizes.

The following list includes an indication of whether the minimum di-mension is fixed by Code and may not be decreased, or whether it is limited by practice and may be decreased if one is willing to pay the price.

- Bearing walls, cast in place, minimum 10 in. thick (practice)
- Bearing walls, precast, minimum 4 in. thick (Code 14.5.3.1)
- Columns, cast in place, minimum dimension 10 in. (practice)
- Foundation walls, minimum 7½ in. thick (Code 14.5.3.2)
- Nonbearing walls, cast in place, minimum 4 in. thick (Code 14.6.1)
- Slabs on grade, minimum thickness 3½ in. (practice)

## Minimum and Maximum Reinforcement

In all cases, it is assumed that concrete has no tensile strength and when-ever there is tension on the section the use of tensile reinforcement becomes necessary. In such circumstances, the concrete does in fact crack, allowing the reinforcement to undergo higher strains and correspondingly higher stresses. The "cracked section" for reinforced concrete members in flexure is the accepted state of stress.

It should not be inferred that large cracks occur in concrete beams subject to ordinary service loads. While cracks do indeed occur, they are so thin as to be essentially invisible to the naked eye. Researchers in concrete sometimes have to take rather elaborate measures to find and trace these very thin hairline cracks.

For all members subject to flexure, the ratio of tensile steel area $A_s$ to concrete area $bd$ is a widely used parameter. Called the *steel ratio* $\rho$, it is defined as

$$\rho = \frac{A_s}{bd}$$

where $A_s$, $b$, and $d$ are shown in Figure 3-1.

Reinforcement is also commonly used in the compression areas of concrete beams. Such compressive reinforcement is required when the size of the concrete section is so restricted that the area of concrete is too small to carry the entire load. The *compressive steel ratio* $\rho'$ in such members becomes a useful parameter, defined as

$$\rho' = \frac{A_s'}{bd}$$

where $A_s'$ is the area of compressive reinforcement.

It should be noted that the tension areas in the sketches of Figure 3-1 are shown as blank spaces. Concrete is assumed to have zero tensile

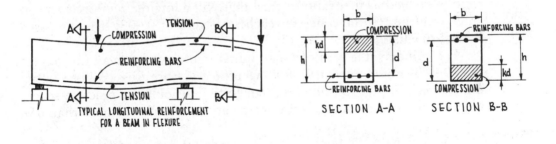

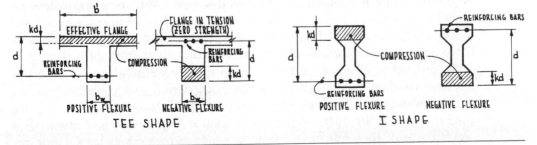

**Figure 3-1**        Typical flexure sections.

strength in all flexural calculations. Those areas of concrete in the tension zone simply cease to exist insofar as the flexural analysis is concerned, though they are still effective in carrying shear (through "shear friction").

The minimum and maximum area of flexural reinforcement given in the following paragraphs is governed by Code. These limits assure that the member will behave properly under extremes of load and that cracking in the concrete will be within acceptable sizes and will follow acceptable patterns.

For slabs having flexural reinforcement in one direction only, temperature and shrinkage reinforcement in the other direction is required. Minimum area of temperature and shrinkage reinforcement is $0.002bh$ for grades 40 and 50 steel and $0.0018bh$ for grade 60 steel, where $b$ and $h$ are gross dimensions of the section; bar spacing may not exceed 18 in. This requirement applies to all one-way slabs, including those of joist systems. This same minimum steel area also applies to the primary flexural reinforcement as well (for very lightly reinforced beams).

For flexural members, the steel ratio may not be less than $200/f_y$ nor may it be more than 75% of the balanced steel ratio. The balanced steel ratio is defined as that ratio of steel that will allow the reinforcement to enter yield just as the concrete reaches a strain of 0.003. It is discussed further in Chapter 6.

For columns, the area of longitudinal reinforcement may not be less than $0.01bh$ nor more than $0.08bh$, where $b$ and $h$ are gross cross-sectional dimensions. In practice, however, congestion at joints becomes so severe with the higher amounts of reinforcement that the use of more than 4% steel is avoided.

For bearing walls reinforced with grade 60 deformed bars no larger than No. 5, the area of reinforcement may not be less than $0.0012bh$ vertically nor less than $0.0020bh$ horizontally. These same minimums also apply to mesh reinforcement having the same bar sizes regardless of whether the bars are smooth or deformed or whether the bars are grade 40, 50, or 60 steel (smooth bars are permitted in a mesh). For all other reinforcement sizes and grades, the area of reinforcement may not be less than $0.0015bh$ vertically or $0.0025bh$ horizontally. As before, $b$ and $h$ are gross dimensions. These requirements are for total area of steel, whether placed in the middle of the wall, on one face, or on both faces.

For floor slabs on grade, Code does not require reinforcement. For crack control, however, it is common practice to meet at least Code requirements for temperature reinforcement (in both directions). For slabs more than 5 in. thick, the reinforcement mesh should be placed in two layers, at top and bottom surfaces.

## Minimum Spacing

The minimum clear distance between bars is governed by Code (7.6). The minimum limits permit the wet concrete to flow freely between bars without leaving air pockets or gaps. In earlier years, an additional provision was included to account for the nominal size of the aggregate used in making the concrete, but in more recent Codes that provision has been dropped.

A typical section of a concrete member is shown in Figure 3-2. The clear cover and clear spacing is shown in Figure 3-2a. Where bars are to be spliced, the bars being spliced will overlap each other as shown in Figure 3-2b; clear distance between adjacent sets of bars must be maintained in such splices.

The following limits on bar spacing must be maintained:

- The minimum clear spacing between single bars in a layer shall not be less than $d_b$ nor 1 inch, where $d_b$ is the diameter of the bar.
- Where parallel reinforcement is placed in two or more layers, bars in the upper layers shall be placed directly above the bars in the lower layers, with a minimum clear distance between layers of 1 inch.
- In tied columns, clear distance between longitudinal bars shall not be less than $1.5d_b$ nor less than 1½ inches.

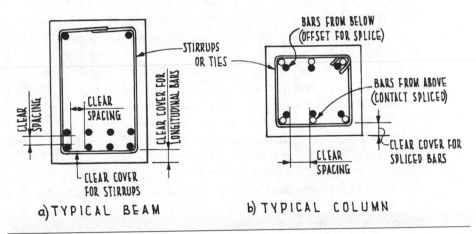

**Figure 3-2**    Cover and spacing of bars.

- Limitations on clear distance shall also apply to the clear distance between a contact lap splice and adjacent splices.
- In walls or slabs other than concrete joist construction, primary flexural reinforcement shall be spaced no farther apart than three times the wall or slab thickness, with an absolute maximum spacing of 18 inches.

As noted earlier, the foregoing limits are imposed primarily for the sake of casting wet concrete in and around the clusters of reinforcement. These spacings also have a profound effect on the formation of crack lines or "split lines," which progress from bar to bar and which disrupt bond between the bar and the concrete. Additional limitations on spacing due to such problems with bond are discussed in Chapter 12.

## Bundling of Bars

When the number of parallel bars in a concrete member are so numerous that they cause placement problems, Code (7.6.6) permits the bars to be collected into bundles of 2, 3, or 4 bars. Such bundling of bars is shown in Figure 3-3, in which the congested bar arrangement of Figure 3-3a has been changed into the more open bar arrangement of Figure 3-3b.

The following criteria apply to bundling of bars:

- Parallel reinforcing bars may be bundled in contact to act as a unit, where the maximum number of bars in a bundle is limited to four.

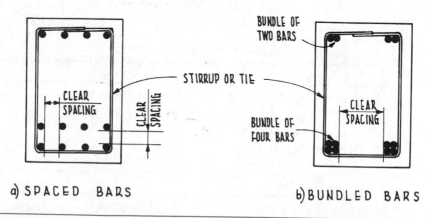

**Figure 3-3**    Bundling of bars.

- Bundled bars shall be enclosed within stirrups or ties.
- Bars larger than No. 11 shall not be bundled in beams.
- Individual bars within a bundle that are terminated within the span of flexural members shall be terminated at different points, with at least $40d_b$ staggered between cutoff points, where $d_b$ is the diameter of the bar being cut.
- Where spacing criteria or cover criteria are based on bar diameters $d_b$, a bundle of bars shall be treated as an equivalent single bar having the same total cross sectional area.

Bundling of bars is a distinct convenience in many circumstances, but the convenience does not come without a penalty. When a group of bars is bundled, it loses surface area; with the loss in surface area, the bundle loses bond strength with the adjacent concrete. This loss in bond strength is discussed further in Chapter 12, along with measures to correct for such a loss.

## Minimum Cover

The amount of clear concrete cover over the outermost reinforcement is governed by Code (7.7.1). It is this cover that provides fire protection. It also prevents the intrusion of salts or oxygen that would cause corrosion of the reinforcement. Code (7.7.5) requires that the cover be increased above Code minimums where conditions are severe.

The following criteria are taken from the ACI Code (7.7). In this summary, any dimension that is based on bar sizes is the same whether the bars are used singly or made into a mesh. Mesh reinforcement may be made of smooth or deformed bars; bars used singly must be deformed.

Minimum cover for concrete cast in place:

| | | |
|---|---|---|
| Concrete cast against earth and permanently exposed to earth | | 3 in. |
| Concrete exposed to earth or weather: | Bars larger than No. 5 | 2 in. |
| | Bars No. 5 and smaller | 1½ in. |
| Concrete not exposed to earth or weather: | | |
| Slabs, walls, and joists | Bars larger than No. 11 | 1½ in. |
| | Bars No. 11 and smaller | ¾ in. |
| Beams and columns | | 1½ in. |
| Shells and folded plates | Bars larger than No. 5 | ¾ in. |
| | Bars No. 5 and smaller | ½ in. |

Minimum cover for precast concrete elements manufactured under plant-controlled conditions:

| | | |
|---|---|---|
| Concrete exposed to earth or weather: | | |
| Wall panels | Bars larger than No. 11 | 1½ in. |
| | Bars No. 11 and smaller | ¾ in. |
| Other members | Bars larger than No. 11 | 2 in. |
| | Bars larger than No. 5 but smaller than No. 11 | 1½ in. |
| | Bars No. 5 and smaller | 1¼ in. |
| Concrete not exposed to earth or weather: | | |
| Slabs, walls, and joists | Bars larger than No. 11 | 1¼ in. |
| | Bars No. 11 and smaller | ⅝ in. |
| Beams and columns | Primary reinforcement: | |
| | one bar diameter, but not less than | ⅝ in. |
| | nor more than | 1½ in. |
| | Ties or stirrups | ⅜ in. |
| Shells and folded plates | Bars larger than No. 5 | ⅝ in. |
| | Bars No. 5 and less | ⅜ in. |

## Limits on Deflections

The following criteria concern the glass, finishes, fillers, or other non-structural elements of a building that are not themselves part of the load-carrying system but which are attached directly to a load-carrying member. Some of these elements may be damaged if the structural members supporting them undergo excessive deflections. Those nonstructural elements that would be subject to such damage are collectively termed herein the *brittle elements*.

The effects of structural deflections on the brittle elements are no different in concrete structures than in structures made of other materials. The nature and timing of the deflections themselves, however, can be quite different due to the inelastic deformations resulting from shrinkage and creep. In evaluating the effects of deflections in concrete structures, the designer must distinguish between short-term and long-term loadings and the effect that each could have on the deflections at any given time.

The following limits on deflections specified by Code (9.5) apply only to application of the short-term live load:

1. For flat roofs not supporting or attached to brittle elements, deflections are limited to $\angle/180$, where $\angle$ is the span.

2. For floors not supporting or attached to brittle elements, deflections are limited to $\angle/360$, where $\angle$ is the span.

The following limits on deflections apply only to that part of the total deflection following the placing and attaching of the nonstructural elements; earlier long-term deflections may be excluded but subsequent long-term or short-term deflections may not.

1. For roofs or floors supporting or attached to brittle elements, deflections are limited to $\angle/480$, where $\angle$ is the span.
2. For roofs or floors supporting nonstructural elements that do not include any brittle elements, deflections are limited to $\angle/240$, where $\angle$ is the span.

The computed deflections due to flexure may not exceed the foregoing limits. The computed deflections must include the elastic deflections due to short-term load, the elastic deflections due to sustained load, and any inelastic deflections due to shrinkage and creep. The elastic deflections are calculated by ordinary methods; the Code prescribes the following means to estimate the inelastic deflections due to sustained load.

Deflections due to sustained loads are known to increase with time and decrease with the amount of compressive reinforcement. Code (9.5.2.5) includes these effects in a factor $\lambda$, where

$$\lambda = \frac{\xi}{1 + 50\rho'}$$

where $\xi$ is a time factor and $\rho'$ is the ratio of compressive reinforcement $A'_s/bd$. The time factor is prescribed by Code:

$$\text{For 5 years or more: } \xi = 2.0$$
$$\text{For 12 months: } \xi = 1.4$$
$$\text{For 6 months: } \xi = 1.2$$
$$\text{For 3 months: } \xi = 1.0$$

The inelastic deflection is computed by multiplying the elastic deflection due to sustained loads times the factor $\lambda$.

Code does not place limits on any deflections other than those due to flexure. Where brittle finishes are fixed to concrete columns, such as mosaic tile, mirrors, or marble, the designer must estimate the long-term deformations by his or her own methods. The foregoing time factors offer a useful guide for estimating such long-term inelastic deformations.

## Control of Deflections

The foregoing Code criteria give the limits that must be observed for deflections of flexural members. The foregoing criteria do not, however, stipulate how the deflections are to be determined. In a separate section, Code (9.5.2) prescribes means to compute or to limit deflections in the strength method of design.

Only the deformations at service levels of loading are of interest to the designer. These are the day-to-day deformations that determine whether the building will perform satisfactorily in service. Such deformations affect cracking of plaster, "springiness" of floors, transmission of vibrations, weather seals at joints, and all other effects that arise from the building's deflections.

Code (9.5.2) permits two approaches to the control of deflections. The first approach is a "blanket" method. In this method, deflections are not computed nor is there any estimate made of their effects. The depths of all members are simply kept above prescribed minimum limits; experience has shown that buildings constructed using such member sizes will perform satisfactorily insofar as deflections are concerned.

In the "blanket" method, therefore, it is never known what the magnitude of the deflections is; it is only known that they will not be a problem. As one may suppose, this blanket approach to control of deflections is somewhat conservative. Nonetheless, the use of this method produces little or no penalty for smaller buildings. As the building size increases, however, the conservatism becomes increasingly expensive.

In most applications, only the blanket method for the control of deflections is necessary. Consequently, when the depth of a member is selected, the check for deflections consists only of a fast check against the minimum allowable depth for the member. The minimum overall depth $h$ of members is adapted from the Code (Table 9.5a) and presented in Table 3-1.

The second approach to the control of deflections is through computing the actual deflections on the member, using conventional elastic theory. For such computations, Code (9.5.2.3) permits an increase above the moment of inertia $I_{cr}$ of the cracked section. The effective moment of inertia $I_e$ then is:

$$I_e = I_{cr} + \left(\frac{M_{cr}}{M_a}\right)^3 \left(I_g - I_{cr}\right) \tag{3-1}$$

where $I_{cr}$ is the moment of inertia of the cracked section under elastic conditions;

**Table 3-1**    Minimum Overall Depth of Flexural Members[1]

| Support Condition | Overall Depth of Member | |
| --- | --- | --- |
| | Solid Slabs, One-way Flexure | Beams or Ribbed Slabs, One-way Flexure |
| Simply supported | $\angle/20$ | $\angle/16$ |
| One end continuous | $\angle/24$ | $\angle/18.5$ |
| Both ends continuous | $\angle/28$ | $\angle/21$ |
| Cantilever | $\angle/10$ | $\angle/8$ |

[1]$\angle$ is the length of span. Values are for grade 60 steel, $f_y = 60,000$ psi. For other grades of steel, these values must be multiplied by $(0.4 + f_y/100,000)$.

$I_g$   is the moment of inertia of the gross uncracked section, neglecting reinforcement;

$M_{cr}$ is the moment on the gross elastic section that produces cracking in the concrete, assumed to occur when the tensile stress in the concrete reaches $7.5\sqrt{f'_c}$;

$M_a$ is the moment acting on the section at the time deflection is being calculated;

but $I_e$ may not exceed $I_g$ in which case, at maximum $I_e$, $I_e = I_g = I_{cr}$.

Code (9.5.2.5) requires that the effects of shrinkage and creep with time be included in deflection calculations when appropriate; the time factor of the previous section would then be applicable.

A few outside calculations reveal that the second term in Eq. (3-1) is insignificant except when the steel ratio ρ is very small (less than about 0.6%). For the more common configurations of concrete sections, the second term is negligible and can be neglected. For tee beams and slabs, however, the second term can become significant and should be included in deflection calculations.

When deflections are computed, the maximum computed deflection may not exceed the limits on deflections given in the preceding section.

When deflections are being computed, it is often expedient to calculate the deflections in terms of stress rather than loads. For example, the deflection for the uniformly loaded simple beam of Figure 3-4 is given by the equation

$$\Delta = \frac{5wL^4}{384E_cI} \tag{3-2}$$

The maximum design moment on the beam is given by $M = wL^2/8$.

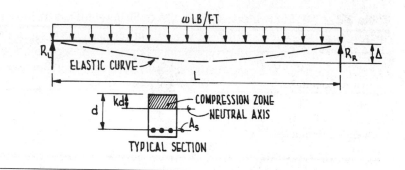

**Figure 3-4**   Uniformly loaded simple beam.

Hence, from the flexure formula, $f_c = Mkd/I$, the deflection may also be calculated as a function of stress rather than load,

$$\Delta = \frac{5f_c L^2}{48E_c kd} \qquad (3\text{-}3)$$

The use of Eq. (3-3) permits the maximum deflection to be determined for a simply supported, uniformly loaded beam when it is known only that the stress is at its maximum; it is not necessary to know the actual loads. Such a solution can sometimes simplify the calculations when loadings are uncertain.

Similarly, for a single concentrated load on a simple span, as shown in Figure 3-5, the deflection in terms of stress is given by

$$\Delta = \frac{4f_c L^2}{48E_c kd} \qquad (3\text{-}4)$$

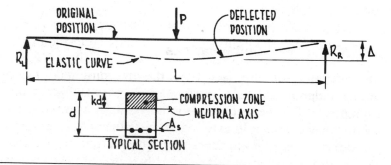

**Figure 3-5**   Concentrated load on a simple beam.

When Eq. (3-4) is compared to Eq. (3-3) it is seen that at maximum stress, the uniformly loaded span produces a higher deflection than does the concentrated load. Consequently, Eq. (3-3) can always be used for a fast comparative check of the maximum case to see if deflections are likely to be troublesome.

## Strength Reduction Factors

Construction and placement tolerances are more numerous and more generous for concrete than for other materials, presumably because concrete members are usually fabricated on site. All aspects of concrete construction have allowable tolerances, which, under normal circumstances, combine to cancel each other. If only a few of these happen to accumulate, however, a significant loss of strength could result.

Code (9.3) requires that the computed ultimate strength of concrete members be reduced somewhat to account for any accidental adverse buildup in tolerances. The amount of the reduction varies, depending on the type of loading; a higher reduction is required for columns than for beams. The following strength reduction factors, $\phi$, apply to the indicated type of loading at ultimate capacity:

| | |
|---|---|
| Beam flexure, without axial loads: | $\phi = 0.90$ |
| Axial tension, with or without flexure: | $\phi = 0.90$ |
| Axial compression, with or without flexure, square or rectangular tied columns: | $\phi = 0.70$ |
| Shear or torsion: | $\phi = 0.85$ |
| Bearing: | $\phi = 0.70$ |

Strength reduction factors are not used in the working stress method of design.

## Load Factors and Load Combinations

The strength reduction factor $\phi$ of the preceding section provides for accidental undercapacity of a section. Its purpose is to provide extra capacity in the event that the section is not built exactly as intended; it is not related to improper loading. In arriving at the value of $\phi$, it is tacitly assumed that the external loads are properly applied and that no overload occurs.

Overloading can in fact occur, however, and the chances of its occurring are completely independent of construction practices. It has already been noted that small loads may occur due to temperature or settlements that are ignored in the analysis and design of the structure. An allowance for such overload, accidental or planned, is an essential part of the design.

In allowing for overloads, it is not reasonable to design a structure for every maximum load that can be imagined during its life and to assume that all these maximum loads will occur at the same time. Maximum wind load, for example, will not likely occur during maximum earthquake load. Rather, structures are designed for reasonable combinations of loads compatible with the intended service; a reasonable margin is then provided for unpredictable circumstances.

In the working stress method this margin is provided by the *safety factor*, and in the strength method by *load factors*.

In the strength method, load factors are used to project service loads up to an "ultimate" level of load for which the concrete member is to be designed. The member is assumed to be stressed to its absolute maximum capacity under this ultimate load. Under actual service loads, the member will then work at comfortable levels of stress under much lower levels of load.

Load factors and load combinations are prescribed by Code (9.2). The following requirements are taken from the code, with $D$ representing the effects for dead load, $L$ for live load, $W$ for wind load, $T$ for settlement or temperature, and $E$ for earthquake. With this notation, the required ultimate load $U$ must be at least equal to the following combinations of loads, termed *factored* loads by ACI:

$$U = 1.4D + 1.7L, \text{ or}$$
$$U = 0.75(1.4D + 1.7L + 1.7W), \text{ or}$$
$$U = 0.9D + 1.3W, \text{ or}$$
$$U = 0.75[1.4D + 1.7L + 1.7(1.1E)], \text{ or}$$
$$U = 0.75(1.4D + 1.7L + 1.4T)$$

Where lateral loads are small, the first equation will usually yield the highest loads.

Implicit in the foregoing load combinations, Code permits an increase of 33% in overall loads where wind, earthquake, and thermal load occur in combination with dead and live loads. That provision is incorporated into the appropriate load combinations by making the design value 75% of the total load; a moment's reflection will affirm that the full load (when it occurs) will then produce a 33% overload above this design value. At this point, it is important only to recognize that Code permits an overload of some 33% above ordinary service levels for these occasionally-encountered load cases.

## Nominal Ultimate Loads and Service Loads

In the strength method for the design of concrete, there are thus two sets of factors to be used to project the loads from their actual service level up to their theoretical ultimate level. The first of these are the "load factors" and the second are the "strength reduction factors." Further, a 33% overload is allowed for certain combinations of load. These are the ultimate load conditions prescribed by Code for the design of concrete members.

In the design procedures in current use, the prescribed conditions are met through the use of a nominal ultimate load case $U_n$. The nominal ultimate load $U_n$ is found by dividing the factored load case U (given in the preceding section) by the strength reduction factor $\phi$. For example, for some of the more common load combinations, the nominal ultimate load case is given by

$$U_n = U/\phi = (1.4D + 1.7L)/\phi \qquad (3\text{-}1a)$$

$$U_n = U/\phi = 0.75(1.4D + 1.7L + 1.7W)/\phi \qquad (3\text{-}1b)$$

$$U_n = U/\phi = 0.75\left[1.4D + 1.7L + 1.7(1.1E)\right]/\phi \qquad (3\text{-}1c)$$

Or, alternatively, the nominal ultimate load case may be written in the form

$$U_n = U/\phi = 1.7\left[\left(\tfrac{1.4}{1.7}D + L\right)/\phi\right] \qquad (3\text{-}2a)$$

$$U_n = U/\phi = 1.7\left[0.75\left(\tfrac{1.4}{1.7}D + L + W\right)/\phi\right] \qquad (3\text{-}2b)$$

$$U_n = U/\phi = 1.7\left[0.75\left(\tfrac{1.4}{1.7}D + L + 1.1E\right)/\phi\right] \qquad (3\text{-}2c)$$

The combinations inside the brackets of Eq. (3-2) are designated herein to be the load cases $U_{sv}$ at service levels,

$$U_{sv} = \left(\tfrac{1.4}{1.7}D + L\right)/\phi \qquad (3\text{-}3a)$$

$$U_{sv} = 0.75\left(\tfrac{1.4}{1.7}D + L + W\right)/\phi \qquad (3\text{-}3b)$$

$$U_{sv} = 0.75\left(\tfrac{1.4}{1.7}D + L + 1.1E\right)/\phi \qquad (3\text{-}3c)$$

At these service levels of load, the reduction in factor of safety for dead loads and the increase in the factor of safety for earthquake loads is recognized simply as a feature of the ultimate strength method of design; it is implicit in the load cases given by Code (9.2).

In terms of moment, shear and axial force,

for moment,    $M_n = 1.7 M_{sv}$    with $\phi = 0.90$,    (3-4a)

for shear,    $V_n = 1.7\, V_{sv}$    with $\phi = 0.85$,    (3-4b)

for axial force,    $P_n = 1.7\, P_{sv}$    with $\phi = 0.70$,    (3-4c)

where the subscripts $_n$ and $_{sv}$ are used in the same context as before. All service loads therefore have a factor of safety of 1.7 to ultimate load, or conversely, the service loads are some 60% of the ultimate loads.

The following examples will illustrate the computation of design loads for some typical cases.

## EXAMPLE 3-1.

Computation of design loads at service levels.
Concrete member in flexure, $\phi = 0.9$ for flexure, $\phi = 0.85$ for shear.
No wind or earthquake loads occur.

**Given:**

$M_{DL} = 60$ kip-ft    $V_{DL} = 38$ kips

$M_{LL} = 44$ kip-ft    $V_{LL} = 29$ kips

**Find:**

Maximum service design moment $M_{sv}$

Maximum service design shear $V_{sv}$

**Solution:**

Since there are no lateral loads, the maximum values for shear and moment are given by the load case of Eq. (3-3a),

$$U_{sv} = \left( \tfrac{1.4}{1.7} D + L \right) / \phi$$

For moment,

$$M_{sv} = \left[ \tfrac{1.4}{1.7} (60) + 44 \right] / 0.9$$

$$M_{sv} = 104 \text{ kip-ft}$$

For shear,

$$V_{sv} = \left[ \tfrac{1.4}{1.7} (38) + 29 \right] / 0.85$$

$$V_{sv} = 71 \text{ kips}$$

These values of $M_{sv}$ and $V_{sv}$ are the values to be sustained at the maximum service level of stress in the concrete member. Note that ultimate loads $M_n$ and $V_n$ can now be found (if needed) simply by multiplying the service loads by 1.7,

$$M_n = 1.7 M_{sv} = 177 \text{ kip-ft.}$$
$$V_n = 1.7 V_{sv} = 121 \text{ kips.}$$

---

### EXAMPLE 3-2.

Computation of design loads at service levels.
Concrete member in flexure, $\phi = 0.90$ for flexure, $\phi = 0.85$ for shear.
Wind and earthquake loads included.

#### Given:

$$M_{DL} = 86 \text{ kip-ft,} \qquad V_{DL} = 40 \text{ kips}$$
$$M_{LL} = 77 \text{ kip-ft,} \qquad V_{LL} = 27 \text{ kips}$$
$$M_W = 36 \text{ kip-ft,} \qquad V_W = 12 \text{ kips}$$
$$M_E = 48 \text{ kip-ft,} \qquad V_E = 18 \text{ kips}$$

#### Find:

Maximum service design moments $M_{sv}$

Maximum service design shear $V_{sv}$

#### Solution:

The load combinations given by Eq. (3-3) will apply:

$$U_{sv} = \left( \tfrac{1.4}{1.7} D + L \right) / \phi$$
$$U_{sv} = 0.75 \left( \tfrac{1.4}{1.7} D + L + W \right) / \phi$$
$$U_{sv} = 0.75 \left( \tfrac{1.4}{1.7} D + L + 1.1E \right) / \phi$$

Substitute and solve for $M_{sv}$:

$$M_{sv} = \left[ \tfrac{1.4}{1.7} (86) + 77 \right] / 0.9 = 164 \text{ kip-ft}$$
$$M_{sv} = 0.75 \left[ \tfrac{1.4}{1.7} (86) + 77 + 36 \right] / 0.9 = 153 \text{ kip-ft}$$
$$M_{sv} = 0.75 \left[ \tfrac{1.4}{1.7} (86) + 77 + 1.1(48) \right] / 0.9 = 167 \text{ kip-ft}$$

Substitute and solve for $V_{sv}$:

$$V_{sv} = \left[\tfrac{1.4}{1.7}(40) + 27\right]/0.85 = 71 \text{ kips}$$

$$V_{sv} = 0.75\left[\tfrac{1.4}{1.7}(40) + 27 + 12\right]0.85 = 63 \text{ kips}$$

$$V_{sv} = 0.75\left[\tfrac{1.4}{1.7}(40) + 27 + 1.1(18)\right]/0.85 = 70 \text{ kips}$$

By inspection of the foregoing results, the maximum design values for $M_{sv}$ and $V_{sv}$ are selected:

Maximum $M_{sv}$ = 167 kip-ft

Maximum $V_{sv}$ = 71 kips

These values of service loads are those that must be sustained at the maximum service level of stress in the concrete member. As before, the ultimate loads $M_n$ and $V_n$ can be found (if needed) by multiplying the service loads by 1.7,

$$M_n = 1.7 M_{sv} = 284 \text{ kip-ft.}$$
$$V_n = 1.7\, V_{sv} = 121 \text{ kips.}$$

---

## EXAMPLE 3-3.

Computation of ultimate design load.
Concrete member in flexure, $\phi = 0.90$ for flexure, $\phi = 0.85$ for shear.
No wind or earthquake loads occur.

### Given:

$M_{DL} = 120$ kip-ft,     $V_{DL} = 52$ kips,     $P_{DL} = 41$ kips
$M_{LL} = 105$ kip-ft,     $V_{LL} = 37$ kips,     $P_{LL} = 22$ kips

### Find:

Ultimate design moment $M_n$
Ultimate design shear $V_n$
Ultimate design axial load $P_n$

### Solution:

With no lateral loads, the applicable load case is given by Eq. (3-1a),

$$U_n = (1.4D + 1.7L)/\phi$$

For moment,

$$M_n = (1.4 \times 120 + 1.7 \times 105)/0.90$$
$$M_n = 385 \text{ kip-ft.}$$

For shear,

$$V_n = (1.4 \times 52 + 1.7 \times 37)/0.85$$
$$V_n = 160 \text{ kips}$$

For axial load,

$$P_n = (1.4 \times 41 + 1.7 \times 22)/0.70$$
$$P_n = 135 \text{ kips}$$

Again note that the service loads $M_{sv}$, $V_{sv}$, and $P_{sv}$ may be found (if needed) simply by dividing by 1.7,

$$M_{sv} = M_n/1.7 = 226 \text{ kip-ft}$$
$$V_{sv} = V_n/1.7 = 94 \text{ kips}$$
$$P_{sv} = P_n/1.7 = 80 \text{ kips}$$

---

## EXAMPLE 3-4.

Computation of ultimate design load.
Concrete member in flexure, $\phi = 0.90$ for flexure, $\phi = 0.85$ for shear.
Wind and earthquake loads included.

**Given:**

| | | |
|---|---|---|
| $M_{DL} = 99$ kip-ft | $V_{DL} = 52$ kips | $P_{DL} = 36$ kips |
| $M_{LL} = 66$ kip-ft | $V_{LL} = 31$ kips | $P_{LL} = 26$ kips |
| $M_W = 54$ kip-ft | $V_W = 23$ kips | $P_W = 14$ kips |
| $M_E = 61$ kip-ft | $V_E = 21$ kips | $P_E = 11$ kips |

**Find:**

Ultimate design moment $M_n$

Ultimate design shear $V_n$

Ultimate axial load $P_n$

**Solution:**

The applicable load cases are those given by Eq. (3-1a, b, c),

$$U_n = (1.4D + 1.7L)/\phi$$

$$U_n = 0.75(1.4D + 1.7L + 1.7W)/\phi$$

$$U_n = 0.75 [1.4D + 1.7L + 1.7(1.1)E]/\phi$$

Substitute and solve for $M_n$,

$$M_n = (1.4 \times 99 + 1.7 \times 66)/(0.90) \qquad = 279 \text{ kip-ft}$$

$$M_n = 0.75 (1.4 \times 99 + 1.7 \times 66 + 1.7 \times 54)/0.90 = 286 \text{ kip-ft}$$

$$M_n = 0.75 (1.4 \times 99 + 1.7 \times 66$$
$$+ 1.7 \times 1.1 \times 61)/0.90 \qquad = 304 \text{ kip-ft}$$

Substitute and solve for $V_n$,

$$V_n = (1.4 \times 52 + 1.7 \times 31)/0.85 \qquad = 148 \text{ kips}$$

$$V_n = 0.75(1.4 \times 52 + 1.7 \times 31 + 1.7 \times 23)/0.85 \qquad = 145 \text{ kips}$$

$$V_n = 0.75(1.4 \times 52 + 1.7 \times 31$$
$$+ 1.7 \times 1.1 \times 21)/0.85 \qquad = 145 \text{ kips}$$

Substitute and solve for $P_n$,

$$P_n = (1.4 \times 36 + 1.7 \times 26)/0.7 \qquad = 135 \text{ kips}$$

$$P_n = 0.75(1.4 \times 36 + 1.7 \times 26 + 1.7 \times 14)/0.7 \qquad = 127 \text{ kips}$$

$$P_n = 0.75(1.4 \times 36 + 1.7 \times 26$$
$$+ 1.7 \times 1.1 \times 11)/0.7 \qquad = 123 \text{ kips}$$

By inspection of the foregoing results, the ultimate design loads are seen to be:

$$M_n = 304 \text{ kip-ft}$$

$$V_n = 148 \text{ kips}$$

$$P_n = 135 \text{ kips}$$

From these results, the service loads $M_{sv}$ and $V_{sv}$ can be found (if needed) by dividing by 1.7,

$$M_{sv} = M_n/1.7 = 179 \text{ kip-ft}$$

$$V_{sv} = V_n/1.7 = 86 \text{ kips}$$
$$P_{sv} = P_n/1.7 = 79 \text{ kips}$$

## OUTSIDE PROBLEMS

Determine the maximum design loads at service levels, then from those results compute the ultimate loads.

| Prob. No. | Axial load in kips | | | | Shear in kips | | | | Moment in kip-ft | | | |
|---|---|---|---|---|---|---|---|---|---|---|---|---|
| | $P_{DL}$ | $P_{LL}$ | $P_E$ | $P_W$ | $V_{DL}$ | $V_{LL}$ | $V_E$ | $V_W$ | $M_{DL}$ | $M_{LL}$ | $M_E$ | $W_W$ |
| 3.1 | 55 | 41 | 29 | 29 | 15 | 12 | 7 | 10 | 40 | 31 | 20 | 22 |
| 3.2 | 110 | 120 | 58 | 68 | 30 | 31 | 15 | 17 | 96 | 119 | 51 | 62 |
| 3.3 | 61 | 50 | 35 | 31 | 17 | 14 | 8 | 8 | 45 | 42 | 22 | 24 |
| 3.4 | 103 | 112 | 61 | 70 | 28 | 29 | 17 | 16 | 89 | 113 | 55 | 58 |
| 3.5 | 66 | 59 | 30 | 32 | 19 | 17 | 11 | 13 | 51 | 48 | 29 | 34 |
| 3.6 | 97 | 102 | 51 | 62 | 27 | 27 | 16 | 14 | 82 | 101 | 53 | 57 |
| 3.7 | 74 | 63 | 41 | 40 | 20 | 18 | 9 | 13 | 57 | 59 | 30 | 33 |
| 3.8 | 92 | 89 | 50 | 56 | 26 | 25 | 10 | 8 | 77 | 89 | 41 | 53 |
| 3.9 | 79 | 70 | 36 | 40 | 23 | 20 | 8 | 11 | 65 | 73 | 26 | 28 |
| 3.10 | 86 | 79 | 42 | 46 | 24 | 22 | 16 | 14 | 71 | 79 | 44 | 45 |

Determine the ultimate loads, then from those results compute the design loads at service levels.

| Prob. No. | Axial load in kips | | | | Shear in kips | | | | Moment in kip-ft | | | |
|---|---|---|---|---|---|---|---|---|---|---|---|---|
| | $P_{DL}$ | $P_{LL}$ | $P_E$ | $P_W$ | $V_{DL}$ | $V_{LL}$ | $V_E$ | $V_W$ | $M_{DL}$ | $M_{LL}$ | $M_E$ | $W_W$ |
| 3.11 | 98 | 101 | 52 | 61 | 28 | 26 | 17 | 13 | 83 | 100 | 52 | 58 |
| 3.12 | 75 | 62 | 42 | 39 | 21 | 17 | 10 | 12 | 58 | 58 | 29 | 34 |
| 3.13 | 93 | 88 | 51 | 55 | 27 | 24 | 11 | 7 | 78 | 88 | 40 | 54 |
| 3.14 | 80 | 69 | 37 | 39 | 24 | 19 | 9 | 10 | 66 | 72 | 25 | 29 |
| 3.15 | 87 | 78 | 43 | 45 | 25 | 21 | 17 | 14 | 72 | 78 | 45 | 46 |
| 3.16 | 56 | 40 | 30 | 28 | 16 | 11 | 8 | 9 | 41 | 30 | 19 | 23 |
| 3.17 | 111 | 119 | 59 | 67 | 31 | 30 | 16 | 16 | 97 | 118 | 50 | 63 |
| 3.18 | 62 | 49 | 36 | 30 | 18 | 13 | 9 | 7 | 46 | 41 | 21 | 25 |
| 3.19 | 104 | 111 | 62 | 69 | 29 | 28 | 18 | 15 | 90 | 112 | 54 | 59 |
| 3.20 | 67 | 58 | 31 | 31 | 20 | 16 | 12 | 12 | 52 | 47 | 28 | 35 |

## REVIEW QUESTIONS

1. What is the minimum dimension for a concrete column? For a cast-in-place concrete wall?
2. What is the minimum steel ratio $\rho$ for concrete members in flexure using grade 60 steel? Grade 40 steel?
3. What is the minimum required cover for No. 4 reinforcing bars in a foundation footing cast directly against soil (without forms)?
4. What is the minimum required cover for No. 6 reinforcing bars used in a second floor concrete slab in an enclosed building?
5. What is the minimum required cover for No. 6 reinforcing bars in a precast concrete "plank" which is to be used in an interior location of an enclosed building?
6. In a typical reinforced concrete beam, what is the minimum spacing that must be maintained between No. 6 bars used as flexural reinforcement?
7. In flexural reinforcement, what is the maximum number of No. 5 bars that can be collected into a bundle?
8. When more than one horizontal layer of No. 8 reinforcing bars is used for flexural reinforcement, what clear spacing must be maintained between the layers?
9. What is the maximum spacing of the bars used for flexural reinforcement in a $7\frac{1}{2}$ in. foundation wall?
10. What is the required steel ratio for reinforcement in concrete slabs on grade?
11. What is the limit on live load deflection of a typical roof slab when it supports a brittle plastered ceiling? When there is no ceiling finish?
12. A certain beam supporting a window frame has a short-term computed deflection (3 months or less) of $\frac{1}{2}$ in. The beam has no compressive reinforcement. What total allowance should be made for deflections at five years or more?
13. Why is it necessary to keep a certain amount of clear distance between reinforcing bars (or bundles of bars)?
14. A rectangular beam has a simple span of 14 feet. If deflections are not to be a problem, what is the minimum acceptable depth of the beam?
15. A second floor concrete slab is continuous over several 14'0" spans. It is reinforced with grade 40 steel. What minimum thickness must be maintained if deflections are not to be a problem?
16. Explain the need for the strength reduction factor $\phi$.
17. How is the strength reduction factor $\phi$ used in the working stress method of design?
18. What are the three most common load combinations?
19. If the moment $M_n$ is known, how is the moment $M_{sv}$ computed?
20. What is the strength reduction factor for flexural loads? For shearing loads?

21. What is the ultimate load factor for dead load? For live load? For settlements?
22. The rear face of the retaining wall in Figure 3-2 is cast against forms. Then, when the concrete has hardened, the soil is backfilled against it. How much clear concrete cover over the reinforcement is required?

# 4

# DESIGN
# PRACTICES

**A** structural member must be sized and reinforced to carry computed loads. The methods commonly used to accomplish this are called *design practices*. Once the sizes of the members are chosen, the members must be connected with reinforcement, notched for equipment mounts, and penetrated by ducts. The methods used to accomplish these and hundreds of other such details are called *detailing practices*.

General design practices suited to small projects are learned largely by exposure over a long period of time. Every designer soon develops a preferred set of practices and will adapt this consistent set of practices to each project. Even so, there are basic practices common to the industry that are used by most designers with but little variation.

Some of these basic practices are presented in the following sections. Some rules of thumb are presented and several construction practices are discussed. Foundation settlements and thermal stresses are introduced and discussed. All these topics have a bearing on design and detailing practices; their effects are illustrated in later chapters in the examples.

A major factor influencing both design and detailing practices is that of deciding whether a member is to be precast or cast in place. Significant differences apply to the two designs, to include the minimum sizes and reinforcement that must be used. The use of precast components, even those cast on the job site, is so commonplace that the following criteria include both precast and cast-in-place construction whenever appropriate.

## Accuracy of Computations

With the advent of the electronic calculator, there came a natural tendency to carry design calculations to four-, five-, or even six-figure accuracy. Such practices do no harm unless one begins to believe in this superficial accuracy. Bear in mind that the final accuracy is no better than the least accurate input, and the use of an electronic calculator does nothing to improve the accuracy of the input.

An accuracy of three significant figures will yield a final accuracy, after arithmetic manipulations, of about two significant figures or less. There is little point in performing the calculations for reinforced concrete to any higher accuracy since the loads are known only to two-place accuracy at best and the concrete can meet specifications even when it varies as much as 15% or even more. The manual 10-in. slide rule, with its three-place accuracy, has been used successfully for many years in the design of concrete structures.

The lack of numerical precision in the calculations should not be equated to the lack of sensitivity of concrete. For example, concrete is quite sensitive to torsional shears: the fact that the magnitude of the torsion is known only approximately does not decrease that sensitivity. The same degree of attention to detail must be accorded to concrete as to other materials.

## Estimated Nominal Loads

Loading on concrete buildings varies considerably for any number of reasons. Nonetheless, it is frequently necessary to estimate the total load (dead plus live) on a beam, column, or slab. For common spans and routine design, the following values offer a first guess:

| | |
|---|---|
| Floor load: | 150 psf, dead plus live load |
| Roof load: | 125 psf, dead plus live load |
| Column load: | 30 tons per story, total load |

For routine concrete buildings, the dead load should be expected to be about 60% of the total load and live load to be about 40% of the total load.

For members at the perimeter of a concrete building, the loads should be expected to be only slightly less (about 15%) than the loads at the interior. The fenestration and exterior walls commonly used in concrete buildings are so heavy that little difference should be expected.

## Rules of Thumb

Rules of thumb can be useful, particularly when trying to make an initial estimate for a beam size or a slab thickness. They can also be useful when keeping a running check on a calculation and wishing to know if the results obtained are at least reasonable. The following rules and generalities may be useful as a first guess.

The optimum column module for concrete structures can be expected to be 18 to 20 ft. This module can be extended with little penalty in cost up to 25 ft for joist systems, with many advantages in utilization of space. Spans in reinforced concrete greater than 28 ft are uncommon, although the two-way joist systems (discussed further in Chapter 11) work reasonably well with spans up to 32 ft or even more.

As a means to distinguish between girders and beams, it is assumed that girders are supported by vertical members and beams are supported by horizontal members. (Bearing walls are considered to be horizontal members.) The overall depth of a girder in reinforced concrete can be expected to be about $\angle/10$ and the overall depth of a beam can be expected to be about $\angle/12$, where $\angle$ is the span. For spans greater than about 20 ft, the depths of the members increase rapidly.

Tee beams are a commonly used shape in concrete construction; a typical tee beam arrangement is shown in Figure 4-1. Optimum spacing of the stems of tee beams is generally controlled by the floor slab they support. A floor slab 4 in. thick will span about 8 ft, a floor slab 6 in. thick will span about 12 ft, and a floor slab 7 in. thick will span about 16 ft. All of these examples apply to the usual range of floor live loads and all examples observe the common 4-ft module for building materials.

The stem of a tee beam is designed to take all the shear on the tee beam. Proportions for width and depth have been found to work well for such shear when the width of the stem $b_w$ is taken at roughly half the effective depth $d$; the dimensions $d$ and $b_w$ are shown in Figure 4-1.

Rectangular sections occur when a tee beam is placed in negative moment or when a rectangular frame is designed to receive precast floor planks. The proportions for such rectangular sections have been found to be economical when the width $b_w$ is roughly two-thirds of the effective depth $d$.

For continuous tee beams subject to both positive and negative bend-

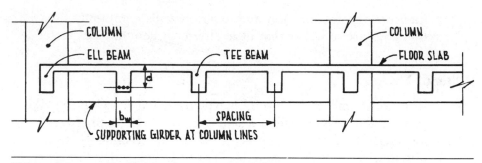

**Figure 4-1** Typical tee beam construction.

ing, the width of the stem should be somewhere between half and two-thirds of the effective depth $d$.

Concrete joist systems are floor systems that utilize narrow joists spaced closely together, supporting a thin concrete floor slab; a typical cross section of a joist system is shown in Figure 4-2. The closely spaced tee joists are significantly shallower than a regular tee-beam floor with its widely spaced tees. The overall depth $h$ of a joist system can be estimated at about two-thirds of the depth of a regular tee-beam-and-slab system.

Concrete columns subject to bending (as part of a rigid frame) can vary widely in size depending on the magnitude of the moment. As a first guess the size of a column subject only to axial load may be estimated as 12 in. on a side plus an additional 1 in. for each story above the one being considered. Thus the columns in the third story of a five-story building can be expected to be about 12 in. on a side plus 1 in. times two stories above, for a total of 14-in. square. A rectangular column for this case would have only slightly more cross-sectional area than the square column, say 10 by 20 in.

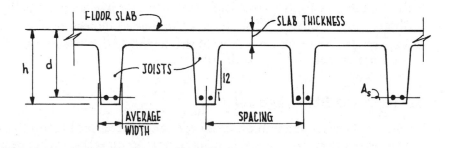

**Figure 4-2** Typical joist system.

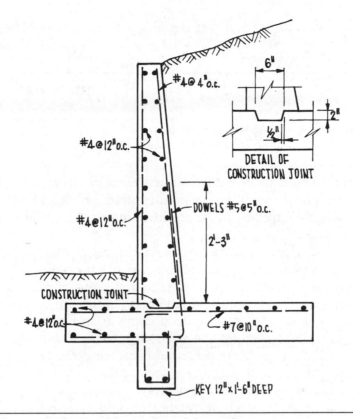

**Figure 4-3**    Typical cantilevered retaining wall.

Cantilevered retaining walls are usually tapered, being considerably thicker at the base than at the top as indicated in Figure 4-3. The thickness at the top can be expected to be about 8 in. for a wall 10 ft high and as much as 12 in. for a wall 20 ft high. The thickness at the base can be expected to be about one-ninth of the height, with a minimum thickness of 12 in.

## Temperature and Shrinkage Joints

Due to the phenomenon of shrinkage in concrete, problems of contraction and expansion of structures in concrete become somewhat more severe than in structures of other materials. The contraction due to shrinkage has

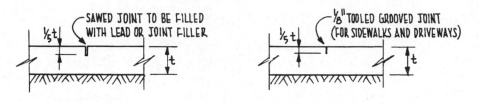

**Figure 4-4**    Typical control joints.

been likened to that due to a temperature drop of 40°F. With such an addition to overall contraction, contraction joints in concrete structures should be expected to be at closer intervals than in structures of other materials.

Code does not give a maximum distance at which a full separation joint (expansion-contraction joint) is required since such a requirement is necessarily dependent on local temperatures. As a general guide in temperate climates, when routine continuous structures are more than 100 ft long, a full separation joint starts to become desirable. At 150 ft it becomes a pressing consideration and at 200 ft it becomes essential.

Slabs on grade are protected from temperature extremes by the huge heat sink of soil that supports them. Nonetheless, slabs exposed to direct sunshine are subject to considerably higher temperatures than slabs located under shade. A common specification allows slabs located in shaded areas to be cast up to 400 ft$^2$ in area with a maximum dimension of 25 ft between joints or, for exposed locations, up to 250 ft$^2$ with a maximum dimension of 20 ft between joints. A detail of the control joints used to meet these requirements is shown in Figure 4-4.

Temperature and shrinkage deformations occur in all concrete structures. There can be hundreds of combinations and variations of these deformations that might occur even in small structures, but it is not common practice to perform calculations for such stresses. Rather, the building is conscientiously designed and detailed to prevent such stresses from occurring; no special analysis is then required. Any accidental effects due to random thermal stress is then relegated to the factor of safety.

Where thermal stresses are allowed to occur, however, they should be expected to be significant. The associated forces are quite large and can be extremely difficult to handle. Such problems are far beyond the scope of this book.

## Construction Joints

Construction joints are used to interrupt a casting; the casting can then be resumed at that joint at a later time. With few exceptions, Code permits construction joints in vertical members to be placed anywhere in the member, to include points of maximum shear and moment. A typical example may be seen at the base of a retaining wall (Figure 4-3); the construction joint in a retaining wall is almost always located at the base of the stem, where both moment and shear on the stem are highest.

Notable exceptions are drawn by Code (6.4.4) for joints in horizontal members. When joints are to be located within the span, they must be located within the middle third of the span. Additionally, the stem of a tee must be cast monolithically with the slab unless specifically designed and detailed otherwise. Similarly, haunches and drop panels must be cast monolithically unless detailed otherwise. It may be concluded that for horizontal members, the existence or absence of high flexural stress has no significant effect when placing a joint, but placing a joint in an area of high shear stress is to be avoided.

When properly designed and detailed, a construction joint does not define a plane of weakness, neither at working levels nor at ultimate strength. "Properly designed and detailed," however, when applied to joints in monolithic frames, can be quite complex. Some of these problems are treated in detail in the ACI publication *Recommendations for Design of Beam-Column Joints in Monolithic Reinforced Concrete Structures* (American Concrete Institute, 1985).

## Differential Settlements

Uniform settlement of a structure may have highly deleterious effects on sewer, water, and power connections to a building, but it has no effect on the structure. Only the differential settlement between adjacent footings or supports will affect the stress levels in the structure. The structural analysis is therefore concerned only with differential settlements between supports, not with total settlements.

If, however, the largest settlement at any footing in a group of footings is limited to 1 in., the differential settlement between any two footings in the group can be expected to be somewhat less than 1 in., say a maximum of ¾ in. It is this approach that is commonly used in practice for the design of shallow footings—that the largest settlements will be limited to about

1 in. and that differential settlements, if any occur, can then be expected to be less than about ¾ in.

As with thermal stresses, there are hundreds of combinations of differential settlements that might occur in a routine structure. If, however, the differential settlements are less than about ¾ in., the change in total stress due to any reasonable combination can be expected to be less than about 15% of the total. For this amount, a separate analysis for potential differential settlements is not usually considered to be necessary and is rarely performed in practice. Where circumstances are unusual, however, or where a key foundation is known to be subject to large settlements, a thorough study of the effects of differential settlements is necessary. For the sake of the foregoing discussion, a "routine" structure is one in which the largest column load is no more than four times the smallest column load and the shallow spread footings are between 3 ft and 8 ft on a side.

In some circumstances, a foundation system may be composed of a mixture of isolated spread footings at columns and continuous strip footings at bearing walls. Where both the spread footings and the strip footings have the same contact pressure on the soil, the strip footings can be expected to settle more than the spread footings, up to about 50% more. The use of a lesser allowable pressure under the strip footings will help equalize the settlements.

## Strengths of Steel in Common Use

The ACI Code is published in two editions: Imperial units (feet, inches, pounds, slugs), and *Systeme International* (meters, millimeters, newtons, kilograms). The two editions are not exactly identical. Some differences are due only to round-off error, but other more significant disparities occur due to the use of different standards.

An example where different standards are applied occurs in the specifications for reinforcement. In the edition containing Imperial units, the yield stress specified by ASTM A-615 for two of the grades of reinforcing steel are 40,000 and 60,000 psi, respectively. In the SI edition there are two comparable steels, but these steels have yield stresses specified by ASTM A-615M as 300 N/mm² (43,500 psi) and 400 N/mm² (58,000 psi). These differences in yield strength are enough to significantly affect a design.

Another example of disparities in standards occurs in the selection of "standard" sizes for reinforcement. Here the United States and Canada are somewhat at variance with other countries. For metric sizes, the United

**Table 4-1** U.S.-Canadian Metric Standard Steel Sizes

| Bar No. | Diameter (mm) | Area (mm$^2$) |
|---------|---------------|---------------|
| 10 | 11.3 | 100 |
| 15 | 16.0 | 200 |
| 20 | 19.5 | 300 |
| 25 | 25.2 | 500 |
| 30 | 29.9 | 700 |
| 35 | 35.7 | 1000 |
| 45 | 43.7 | 1500 |
| 55 | 56.4 | 2500 |

States and Canada specify the sizes shown in Table 4-1, which yields even increments for the cross-sectional area of the reinforcing bars rather than even increments for diameters.

The U.S.–Canadian system for metric sizes has not been widely adopted and is not widely used, even in North America. Nonetheless, these sizes are sometimes available at discount prices. The sizes of Table 4-1 are commonly provided in two grades of steel, $f_y = 300 \ N/mm^2$ and $f_y = 400 \ N/mm^2$.

In countries where the metric system (rather than SI) is in use, there is very little standardization of reinforcement sizes. Perhaps the most widely applicable standard is the UNESCO-recommended standard (UNESCO, 1971), although in any given country some of these "standard" sizes may not be available. The UNESCO standard is shown in Table 4-2. These sizes are commonly imported into the United States in various strengths of steel.

The older standard Imperial sizes are discussed in Chapter 1. Their sizes and corresponding cross-sectional areas are given in Table 4-3. These are the most widely available sizes in North America and are generally available in steel grades 40, 50, and 60.

All three sets of standard sizes shown in Tables 4-1, 4-2, and 4-3 are available in the United States. Consequently, it is somewhat difficult to determine design standards that will be widely applicable for any length of time. For the near future, however, it seems probable that the Imperial standards of Table 4-3 will continue to dominate the North American market.

Among engineers, a strong influence causing resistance to change is the data on which the Codes are based. The overwhelming bulk of those data was developed over the past 60 years using the older Imperial sizes of reinforcement, with steel having yield strengths of 40,000, 50,000, and

**Table 4-2** UNESCO-Recommended Reinforcement Sizes

| Bar Diameter (mm) | Area (mm$^2$) | Mass (kg/m) |
|---|---|---|
| 6 | 0.28 | 0.211 |
| 8 | 0.50 | 0.377 |
| 10 | 0.79 | 0.596 |
| 12 | 1.13 | 0.852 |
| 14 | 1.54 | 1.16 |
| 16 | 2.01 | 1.52 |
| 18 | 2.54 | 1.92 |
| 20 | 3.14 | 2.37 |
| 22 | 3.80 | 2.87 |
| 25 | 4.91 | 3.70 |
| 28 | 6.16 | 4.65 |
| 30 | 7.07 | 5.33 |
| 32 | 8.04 | 6.07 |
| 40 | 12.56 | 9.47 |
| 50 | 19.63 | 14.80 |
| 60 | 28.27 | 21.30 |

**Table 4-3** U.S.-Canadian Imperial Standard Steel Sizes

| Size Designation | Bar Diameter (in.) | Area (in.$^2$) |
|---|---|---|
| 3 | 0.375 | 0.11 |
| 4 | 0.500 | 0.20 |
| 5 | 0.625 | 0.31 |
| 6 | 0.750 | 0.44 |
| 7 | 0.875 | 0.60 |
| 8 | 1.000 | 0.79 |
| 9 | 1.128 | 1.00 |
| 10 | 1.270 | 1.27 |
| 11 | 1.410 | 1.56 |

60,000 psi. An understandable preference to use Code values based on the actual experimental sizes and strengths should continue to be a strong influence.

In this text, the design tables of the Appendix are based on the traditional U.S. sizes of reinforcement shown in Table 4-3, with steel grades 40, 50, and 60. Where sizes other than these are to be used, the conversions should be based on the cross-sectional areas rather than the closest equivalent diameter. The design of reinforcement will be seen later to be based on cross-sectional areas; conversions using approximate diameters (which are then squared) can reduce the accuracy of the conversion.

For the design tables of the Appendix, only the most frequently used bar sizes have been adopted. Since the Code does not permit bar sizes less than No. 3 to be used structurally, those sizes have been dropped from the tables. Similarly, for bar sizes greater than No. 11 (used in very heavy construction), the Code introduces numerous restrictions and complications, so for the sake of simplicity those sizes have also been dropped. The remaining sizes, 3 through 11, will apply to the overwhelming majority of routine structures.

## Strengths of Concrete in Common Use

In practice, concrete strengths lower than 3000 psi are rarely used for structural concrete. The design tables have been set up for only the more common concrete strengths: 3000, 4000, and 5000 psi. Beams and girders are commonly cast from concrete having an ultimate strength of 3000 or 4000 psi and columns from 4000 or 5000 psi.

## Shear and Moment Diagrams

The design of concrete beams is heavily dependent on the use of shear and moment diagrams. The size and location of reinforcement for both shear and flexure are often placed by scaling accurately-drawn shear and moment diagrams. The ability to draw shear and moment diagrams quickly and accurately is an essential skill for a designer of reinforced concrete.

Sketching of shear and moment diagrams is treated in elementary strength of materials and is not repeated here. Those lacking background in constructing these diagrams should review the subject in detail. A considerable amount of skill in sketching shear and moment diagrams is essential not only for concrete but throughout all of structural design; time spent in developing or improving such skill will indeed be time well spent.

In concrete design, the moment diagram is drawn on the "tension side" of the beam, opposite to the sign convention of basic strength of materials. The reason was given earlier: The flexural reinforcement goes on the side where the moment diagram falls. In addition, the moment diagram for this convention corresponds more closely to the elastic curve of the beam, providing a suggestion of the curvatures.

There is a particular innovation in using the shear and moment diagrams that is used more often in concrete design than it is in other materials. That innovation is in preparing the "envelope" of the shear and moment diagrams. For those who are as yet unfamiliar with such envelopes, they are briefly described in the next section.

## Shear and Moment Envelopes

The concept of shear and moment envelopes is used extensively in the design of concrete structures. The moment envelope, when accurately drawn, may be scaled to determine the cutoff points for longitudinal reinforcement. The shear envelope may be similarly scaled to determine the limiting points where shear reinforcement is required.

For the sake of simplicity in the following presentation, only uniform loads will be considered, although the same concepts apply regardless of the type of load. Only working levels of dead load and live load will be considered throughout the structural analysis. The shears and moments due to dead loads are kept separate from those due to live loads, since they must be used separately later in the design of columns and again in the design of the foundations.

As indicated in Figure 4-5, a beam may be subject to various combinations of load, any one of which may occur in the life of the structure. Each combination of load produces its own shear and moment diagrams, as indicated by the shear and moment diagrams of Figure 4-5. Depending on the circumstances, there could be numerous such combinations. It would indeed be a tedious pursuit to check every load case through every point on the beam to assure that the highest value of load has been accounted for. Rather, the shear and moment diagrams are drawn reasonably accurately to scale, and the outermost values of these diagrams at any point form the "envelope" of shears or moments that the beam must be able to sustain. It should be noted that the envelopes for Figure 4-5 are different for every span.

Some of the more common arrangements of loading are shown in Figure 4-5; maximum values of shear or moment are readily found from the diagrams. There are other arrangements of load, however, which might cause even higher values of shear or moment at some particular point. A complete analysis would therefore include all reasonable arrangements of load which could produce maximum values of shear or moment at some point along the length of the beam.

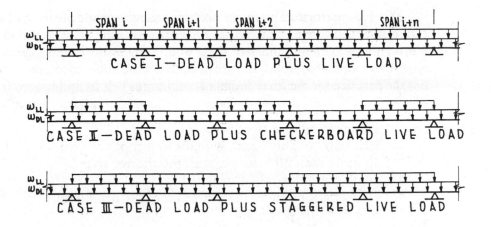

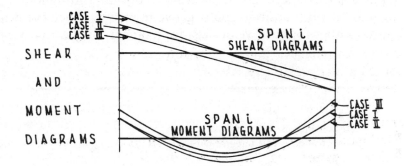

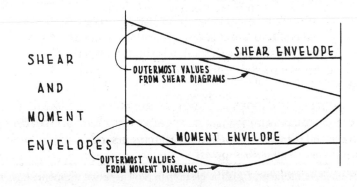

**Figure 4-5**       Shear and moment diagrams.

ACI has performed such an analysis, using a multiplicity of loading arrangements. From its analysis, ACI has arrived at the values given in Table 4-4 for the maximum values of shear and moment on beams. Termed the *ACI Approximate Method*, the use of these coefficients is accepted throughout the practice for the final design of continuous beams and braced frames. The ACI coefficients are valid where the following conditions are met:

1. There are at least two spans.
2. Spans are roughly equal, with the longer of any two adjacent spans no more than 20% longer than the shorter span.
3. Loads are uniformly distributed.
4. Live load is no more than three times the dead load.
5. Members are prismatic.

For timber or steel beams, the use of a moment envelope is rarely required since the strength of the beam is constant throughout the span and is equal for both positive and negative moments. Once the peak value is known, either for positive or negative moment, the beam can be sized for

**Table 4-4** ACI Coefficients[1]

| | |
|---|---|
| Positive moment | |
|   End spans | |
|     Discontinuous end simply supported | $wL_n^2/11$ |
|     Discontinuous end integral with support | $wL_n^2/14$ |
|   Interior spans | $wL_n^2/16$ |
| Negative moment at exterior face of first interior support | |
|   Two spans | $wL_n^2/9$ |
|   More than two spans | $wL_n^2/10$ |
| Negative moment at other faces of interior supports | $wL_n^2/11$ |
| Negative moment at face of supports for slabs with spans not exceeding 10 ft and beams where ratio of sum of column stiffnesses ($\Sigma\, I/L_n$) to beam stiffnesses ($\Sigma\, I/L_n$) exceeds eight at each end of the span | $wL_n^2/12$ |
| Negative moment at interior face of exterior support for beams or slabs built integrally with supports | |
|   Where support is a beam or girder | $wL_n^2/24$ |
|   Where support is a column | $wL_n^2/16$ |
| Shear in end members at face of first interior support | $1.15\, wL_n/2$ |
| Shear at face of all other supports | $wL_n/2$ |

[1]For positive moment, $L_n$ is the clear span. For negative moment, $L_n$ is the average of the two adjacent clear spans. For shear, $L_n$ is the clear span. In all cases, $w$ is the load per unit length.

that value and it is then known to be adequate for all other points in the span. In concrete, however, where strength of a beam varies across the span and where strength under positive moment may not be equal to strength under negative moment, the capacity of the beam at every point of the span must be examined; the shear and moment envelopes afford a graphic and easily-interpreted means to perform this examination.

It should be noted before proceeding that the use of the ACI coefficients for the final design of concrete structures is one of the more significant simplifications in all of concrete design. The ACI coefficients may be used with confidence on structures that are less than about 70 feet high and which are designed as diaphragm/shearwall structures, that is, sideways due to wind and earthquake produces no bending in the columns. Such structural systems are overwhelmingly the most common type of concrete structure in the industry; this textbook is deliberately concentrated on the design of this type of structural system.

## Constructing an Envelope from Coefficients

For simply supported beams, the moment coefficients of Table 4-4 will provide the maximum negative moment at the supports and the maximum positive moment at midspan. Given only these numerical values at a later time, the envelopes for uniform loading can then be reconstructed using very simple sketching techniques. The following observations concerning shear and moment diagrams have been found useful in making such sketches of both shear and moment envelopes. Although approximate, the resulting envelopes are accurate enough for final design of concrete members.

The shear and moment diagrams for a simply supported beam under uniform loading are shown in Figure 4-6. The maximum value of moment is $wL^2/8$ and occurs at midspan. The moment at any point $x$ is found by summing moments to the left of the point $x$:

$$M_x = \tfrac{1}{2}wLx - \tfrac{1}{2}wx^2 = \tfrac{1}{2}wx(L - x). \qquad (4\text{-}1)$$

Similarly for continuous beams, the shear and moment diagrams for one span of a continuous beam are shown in Figure 4-7. As before, the value of moment at any point $x$ is found by summing moments to the left of point $x$, assuming $M_2 > M_1$.

$$M_x = -\left[(M_2 - M_1)\frac{x}{L} + M_1\right] + \frac{1}{2}wx(L - x) \qquad (4\text{-}2)$$

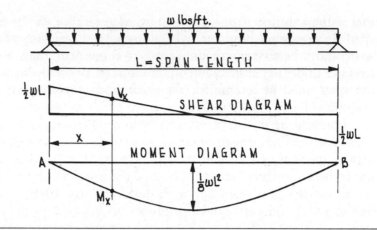

**Figure 4-6** Simply supported beam.

The second term in Eq. (4-2) is seen to be identical to the simple span moment of Eq. (4-1). The first term is seen to be the equation of a straight line drawn from $M_1$ and $M_2$ at the supports shown as $A'B'$ in Figure 4-7. The final moment diagram for this interior span of a continuous beam is thus the moment diagram of a simple beam drawn from the axis $A'B'$

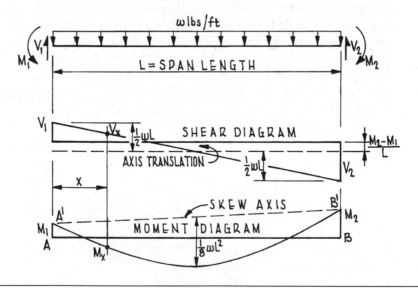

**Figure 4-7** One span of a continuous beam.

rather than the zero axis $AB$. Note that even though the horizontal axis $A'B'$ is skewed, all moments are still measured vertically.

It should also be recognized that the shear diagram of Figure 4-7 is the shear diagram of a simple span, translated up or down by the distance $(M_2 - M_1)/L$. Note that for shear diagrams the horizontal axis is only translated; it is not skewed.

The foregoing observations will be used extensively in sketching the shear and moment envelopes in subsequent discussions. The following examples demonstrate the procedure that will be used when only key values are known.

## EXAMPLE 4-1.

Sketching of shear and moment envelopes.
Key values given as shown (negative moments only).

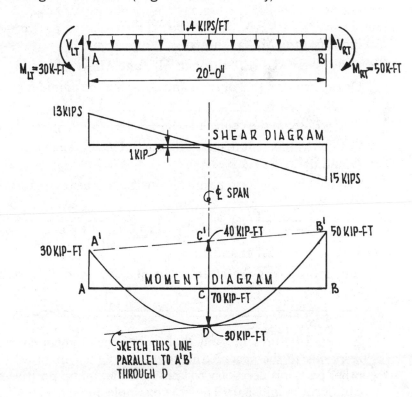

Compute end shears $V_{LT}$ and $V_{RT}$:

$\Sigma M$ about $B = 0$

$$-30 + V_{LT} \times 20 - \frac{1.4 \times 20^2}{2} + 50 = 0$$

$$V_{LT} = \frac{310 - 50}{20} = 13 \text{ kips}$$

$$V_{RT} = 1.4 \times 20 - 13 = 15 \text{ kips}$$

Shear diagram:    Plot $V_{LT}$ and $V_{RT}$ and draw shear diagram through these points as shown.

Moment diagram:  Compute the moment for a simple span:

$$M = \frac{wL^2}{8} = \frac{1.4 \times 20^2}{8}$$
$$= 70 \text{ kip-ft}$$

Plot $M_{LT}$ and $M_{RT}$ as shown.

Draw skew axis $A'B'$ through $M_{LT}$ and $M_{RT}$.

Draw the centerline of span and compute the moment at the centerline of span (at $C'$).

Plot $wL^2/8$ on the centerline, line $C'D$.

Sketch the moment diagram through $A'DB'$, tangent to a line through $D$ and parallel to $A'B'$.

To find the inflection points, write the moment equation and set it to zero:

$$M_x = -30 + 13x - 1.4x^2/2 = 0$$

Solve for $x = 2.7$ ft and $15.9$ ft.

Inflection points therefore occur at 2.7 ft and 15.9 ft from the left support.

---

Example 4-1 presents only that part of the moment envelope that corresponds to the specified negative moments. In order to sketch the other part, it is necessary to know the maximum positive moment that can occur at midspan. The next example adds this condition.

## EXAMPLE 4-2

Sketching the shear and moment envelopes.
Key values given (negative and positive moment).

As an extension to Example 4-1, the maximum positive moment at midspan is now given as 40 kip-ft.

Sketch the corresponding moment diagram, assuming that the moments at the supports occur in the same ratio as those of Example 4-1; the new part of the sketch is added to the sketch already made for Example 4-1.

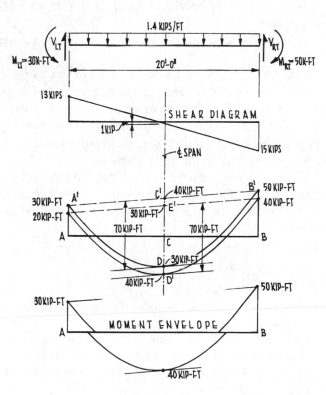

Positive moment envelope: Along centerline of span plot point $D'$, 40 kip-ft on positive side of the axis $AB$.

Move upward along centerline of span 70 kip-ft (or $wL^2/8$) and plot $E'$ as shown.

Draw an axis through $E'$ parallel to $A'B'$.

Sketch a duplicate moment diagram identical to the first but displaced downward 10 kip-ft through point $D'$.

Delete all lines except the outermost lines, leaving only the moment envelope as shown.

To find the inflection points, write the moment equation and set it to zero,

$$M_x = -20 + 13x - 1.4x^2/2 = 0$$

Solve for $x$ = 1.7 ft and 16.9 ft.

Inflection points therefore occur at 1.7 ft and 16.9 ft from the left support.

---

The moment envelope of Example 4-2 is seen to be a complete moment envelope, constructed using only the key values of negative moment at the supports and the positive moment at midspan. Note that the maximum vertical ordinate that can occur in either case is the simple span moment of 70 kip-ft. Note also that the shear diagram is valid for either case and is the final design shear diagram for this case.

The point to be realized is that the final diagrams for shear and moment are obtained simply by shifting the simple beam shear and moment diagrams up or down and, in the case of the moment diagram, skewing the horizontal axis.

## OUTSIDE PROBLEMS

4.1    Using the ACI coefficients, draw the shear and moment envelopes for the end span of a multispan continuous slab 6 in. thick where the slab is built integrally with all its supporting beams and all spans are 12′6″ clear. Uniform live load is 120 lb/ft². Use working loads; that is, do not use load factors.

4.2    Draw the shear and moment envelopes for the first interior span of the slab of Problem 1.

4.3    Using the ACI coefficients, draw the shear and moment envelopes for the end span of a fivespan continuous girder built integrally with the wall supports at the ends of the girder and being built integrally with the column supports elsewhere. Include the location of inflection points, both for the part due to negative moments and the part due to positive moments. Load is uniform, 4250 lb/ft, including the dead load. End spans are 16′6″ clear and all other spans are 18′6″ clear. Use working loads; that is, do not use load factors.

**4.4** Draw the shear and moment envelopes for the center span of the beam of Problem 3.

## REVIEW QUESTIONS

1. What is the difference between designing and detailing?
2. About how much load can be expected on a footing that supports an interior column in a three-story building?
3. About how many pounds of structural material are required to carry 2 lb of live load in an average concrete building?
4. A continuous floor slab is to be supported by beams spaced at 8 ft on center. What thickness of slab can be expected?
5. What size can be expected for a concrete column at the bottom floor of a five-story building?
6. A rectangular concrete beam is to be built to span 24 ft. About what depth can be expected for such a beam?
7. Using a 60/40 ratio of dead load to live load, derive the overall load factor of 1.52 for the total load on concrete buildings when no lateral loads are present.
8. At what length of continuous structure does a full expansion/contraction joint become a consideration?
9. What is the controlling factor in concrete design that restricts the accuracy of all calculations?
10. Why isn't a thermal analysis commonly made for routine concrete structures?
11. What range of bar sizes is most commonly used in reinforced concrete construction in American practice?
12. A particular structural member spans between two columns. Is it classified as a beam or a girder?
13. A particular structural member spans between two girders. Is it classified as a beam or a girder?
14. A particular structural member is supported at its ends by masonry bearing walls. Is it classified as a beam or a girder?
15. An elevated floor slab is to be 4 in. thick. About how closely spaced should the stems of the supporting tee beams be placed?
16. A cantilevered retaining wall must retain a backfill height of 9 ft. What thickness might be expected at the top and at the bottom of the wall?
17. Why aren't differential settlements included with routine stress calculations for buildings?
18. At equal contact pressures, which would settle more: a square footing supporting a column or a strip footing supporting a masonry wall?

19. What strengths of reinforcing steel are most commonly used in American practice? What are their grade numbers?
20. What strengths of concrete are commonly used to cast concrete columns? Beams?
21. Why are shear and moment envelopes so necessary in the design of reinforced concrete but are rarely used to design steel or timber members?
22. For what types of loads are the ACI coefficients valid?
23. In using the ACI coefficients to determine negative moment at an interior support, what is the length $L_n$?
24. In using the ACI coefficients to determine positive moment, what is the length $L_n$?
25. In using the ACI coefficients to determine positive moment, where is this maximum positive moment assumed to occur?
26. When constructing a moment envelope from the ACI coefficients, why are the moments measured vertically even when the horizontal axis is skewed?

CHAPTER

# 5

# ELASTIC
# FLEXURE

The ultimate strength design method prescribed by ACI 318-83 specifies only the ultimate load condition for the design of concrete members; there are no required conditions of stress or load prescribed by Code for the day-to-day service conditions. There are, however, requirements for serviceability, deflection, and crack control that must be imposed at working levels regardless of what the ultimate load requirements may be. At this point in the development, however, it is emphasized that there is no specified limiting stress that must be observed when concrete members are subjected to the service moment $M_{sv}$.

Methods of investigating and designing concrete flexural members for elastic stress are developed in this chapter. As noted in Chapter 2, concrete has been found to be elastic to more than 75% of its ultimate stress, or, within acceptable error, up to its idealized yield stress of $0.85f'_c$.

## The Elastic Flexure Formula

The flexure formula is well known from elementary strength of materials. With $f_c$ denoting the stress on the compression side of the beam, the flexure formula is

$$f_c = \frac{Mc}{I} \qquad (5\text{-}1a)$$

where  $M$  = moment acting on the section

$I$   = moment of inertia about the neutral axis

$c$   = distance from the neutral axis to the level where $f_c$ is computed.

Typical sections in steel, timber, and concrete are shown in Figure 5-1, with the distance $c$ given for the various configurations.

In an alternative form, the flexure formula can also be used to compute the required magnitude of the section modulus $S_c$, where $S_c = I/c$:

$$\frac{M}{f_c} = S_c \qquad\qquad (5\text{-}1b)$$

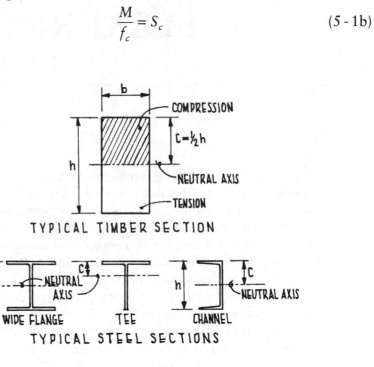

TYPICAL TIMBER SECTION

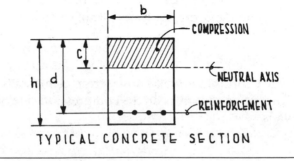

TYPICAL STEEL SECTIONS

TYPICAL CONCRETE SECTION

**Figure 5-1**    Typical flexure sections.

The symbol $S_c$ is used to denote the section modulus, taken to the outermost compression fiber. The section modulus is seen to be the ratio between the moment acting on the section and the stress which that moment produces at the outermost fibers. It is a property of the section; its value depends only on the size and shape of the cross section. Its units are moment per unit stress (lb – in./psi or $N$ – mm/$MPa$) which reduce to in.$^3$ or mm.$^3$

For design in steel, the section modulus of a structural steel member is rarely computed; it is almost always found by looking it up in a table in the steel manual. For American practice, the *Steel Construction Manual* (American Institute of Steel Construction, 1989) lists the section modulus for each of the several hundred steel sections commonly manufactured in the United States. To try to design a steel structure without such a section modulus table would be impractical.

For design in timber, the section modulus may be computed by formula ($S = bh^2/6 = 0.1667bh^2$) or, again, it may be looked up in a standard section modulus table for timber sections. It must be remembered that a timber section nominally designated 6 in. × 12 in. is actually $5\frac{1}{4}$ in. × $11\frac{1}{4}$ in. and that the actual dimensions must be used in the flexure formula. It is usually more convenient to use a table to find a suitable section modulus in timber than to try to remember the rules for finding the actual dimensions.

For design in concrete, the procedure used herein is no different from that for steel or timber. The section modulus is found from a section modulus table and used in the same way. Admittedly, the existence of the larger number of variables in a concrete section makes the tables somewhat longer than for other materials, but there is no difference in concept. Even with a large number of variables, however, it is a relatively simple matter to develop a table of values for the section modulus of a rectangular concrete section. The derivation of such a table follows.

## Derivation of the Elastic Section Modulus

The flexure formula itself can be adapted readily to beams of two materials, such as reinforced concrete beams. To do so, it is only necessary to derive the moment of inertia and the section modulus for such a beam. In the following derivations, the section modulus is always taken to the compression side of the cross section.

Concrete is assumed to have no tensile strength. Wherever concrete is placed in tension, it is assumed to be cracked and the net tensile stress is zero. Reinforcing steel is provided to take all of the tensile load.

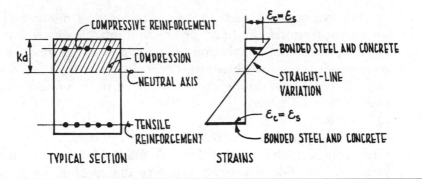

**Figure 5-2**      Strains in a reinforced section.

Regardless whether the concrete is cracked or uncracked, when concrete and its reinforcing steel are bonded together in a beam, their strains at any point are equal, as shown in Figure 5-2.

The relationship between the stresses is fixed by the equality of strains:

$$\varepsilon_c = \varepsilon_s = \frac{f_{cA}}{E_c} = \frac{f_{sA}}{E_s} \qquad (5\text{-}2)$$

where   $\varepsilon_c, \varepsilon_s$   = strains in adjacent concrete and steel, respectively
$\quad\quad\ f_{cA}, f_{sA}$   = stresses in adjacent concrete and steel, respectively
$\quad\quad\ E_c, E_s$   = moduli of elasticity of concrete and steel, respectively

Eq. (5-2) is solved readily for stress in steel,

$$f_{sA} = n f_{cA} \qquad (5\text{-}3)$$

where $n = E_s/E_c$. It was observed in Chapter 2 that the modulus of elasticity of concrete for long-term compressive loads is taken to be half that for short-term loads. For long-term loading, therefore,

$$f_{sA} = 2 n f_{cA} \text{ for steel in compression} \qquad (5\text{-}4a)$$

$$f_{sA} = n f_{cA} \text{ for steel in tension} \qquad (5\text{-}4b)$$

A cross section of a concrete beam is shown in Figure 5-3 along with its stress diagram. It should be noted that the section contains reinforcement on the compression side of the beam. Such compressive reinforcement is commonly used where the size of the beam is so restricted that the concrete alone is not adequate to carry the compressive load within the specified limits for deflections. Alternatively, it is far less expensive (but not always

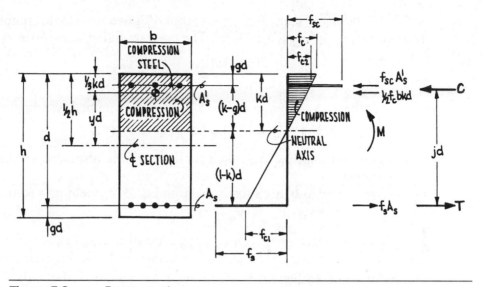

**Figure 5-3**    Beam section.

possible) to make the concrete section bigger and avoid the use of compression steel.

By ratio, the stress in the concrete adjacent to the compressive steel is found to be

$$f_{c2} = \frac{f_c(k-g)}{k} \tag{5-5a}$$

The stress in the compressive steel is, then, from Eq. (5-4a):

$$f_{sc} = 2nf_{c2} = \frac{2nf_c(k-g)}{k} \tag{5-5b}$$

The stress in the tensile steel is found similarly:

$$f_s = nf_{ci} = \frac{nf_c(1-k)}{k} \tag{5-5c}$$

It is observed that the resisting moment of the section can be shown as an internal couple formed by the forces $C$ and $T$ in Figure 5-3. An average value for the distance between the forces, designated $jd$ in Figure 5-3, will be shown later to be about $0.90d$. The average distance $jd$ is affected but little by the existence of compressive steel.

Refer again to Figure 5-3. The symbol $A_s'$ is used to denote compressive steel and $A_s$ to denote tensile steel. The corresponding steel ratios are then

$$\rho' = \frac{A_s'}{bd} \text{ and } \rho = \frac{A_s}{bd} \qquad (5\text{-}6a, b)$$

Horizontal forces shown in Figure 5-3 are now summed, yielding

$$\Sigma H = 0 = \tfrac{1}{2}f_c bkd - f_{c2}A_s' + f_{sc}A_s' - f_s A_s \qquad (5\text{-}7)$$

The second term in Eq. (5-7) deducts the concrete displaced by the compressive steel.

Eq. (5-5) and (5-6) are substituted into Eq. (5-7), yielding a solution for $k$:

$$k = \sqrt{\left[n\rho + (2n-1)\rho'\right]^2 + 2\left[n\rho + (2n-1)g\rho'\right]} - \left[n\rho + (2n-1)\rho'\right] \qquad (5\text{-}8)$$

Refer again to Figure 5-3. Moments are summed about the neutral axis, yielding

$$M = \tfrac{1}{2}f_c bkd\,\frac{2kd}{3} - f_{c2}A_s'(k-g)d + f_{sc}A_s'(k-g)d + f_s A_s(1-k)d \qquad (5\text{-}9)$$

Eq. (5-5) and (5-6) are substituted into Eq. (5-9) and the result solved for the section modulus $S_c$ to the outermost compression fiber,

$$\frac{M}{f_c} = S_c \qquad (5\text{-}10)$$

$$\text{where } S_c = \left[\frac{k^2}{3} + (2n-1)\rho'\frac{(k-g)^2}{k} + n\rho\frac{(1-k)^2}{k}\right]bd^2$$

and where the value of $k$ is given by Eq. (5-8).

Eq. (5-8) and (5-10) would be rather formidable expressions to solve manually in order to find the elastic (or service) section modulus $S_c$. Fortunately, it is not necessary to solve them manually. They can be solved readily by a small computer and the results tabulated for ready reference. Such a tabulation is presented in the Appendix in Tables A-5 through A-7.

With the section modulus known, however computed, the moment of inertia of the cracked section $I_{cr}$ is readily computed from the definition of the section modulus:

$$S_c = \frac{I}{c} \quad \text{hence} \quad I_{cr} = S_c kd \qquad (5\text{-}11)$$

For computing the moment of inertia $I_{cr}$ for short-term deflections, the long-term creep and shrinkage effects should not be included. For such computations, the term $(2n - 1)$ in Eq. (5-10) becomes $(n - 1)$; the value $(n - 1)$ was used in calculating $I_{cr}$ in Tables A-5 through A-7. The coefficient for the moment of inertia $I$ for long-term deflections can be found, if needed, simply by multiplying the coefficient of $S_c$ given in the tables by the value of $k_{sv}$; the long-term moment of inertia $I$ is then this result times $bd^3$.

The distance $yd$ from the centerline of section to the center of force on the concrete is shown in Figure 5-4; this distance will also be needed later.

The distance $yd$ taken from the geometry of the section:

$$yd = \frac{d + gd}{2} - \frac{kd}{3} = \left( \frac{1+g}{2} - \frac{k}{3} \right) d \qquad (5\text{-}12)$$

Since the factor $y$ in Eq. (5-12) is based on the effective depth $d$ rather than the total depth $h$, the value of $y$ may sometimes be larger than 0.5.

It will also be necessary to know the ratio of stresses $f_s/f_c$. A diagram showing these stresses is given in Figure 5-5.

The ratio $f_s/f_c$ is found by similar triangles from the stress diagram:

$$\frac{f_s}{f_c} = \frac{n(1 - k)}{k} \qquad (5\text{-}13)$$

A table of the foregoing section properties is presented in Tables A-5 through A-7 of the Appendix. The tables are entered with the allowable

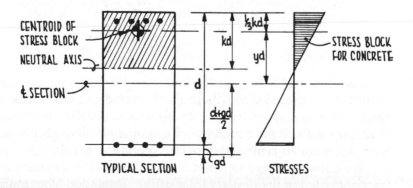

**Figure 5-4**    Centroid of compressive stress block.

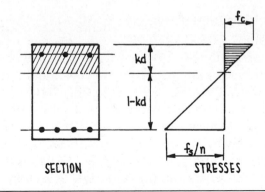

**Figure 5-5**    Elastic stresses on a concrete section.

values of $f_s/f_c$ and the ultimate strength of the concrete. The columns in the tables with the heading "At Service Loads" reflect the tabulated values of Eq. (5-8) and (5-10) through (5-13). To distinguish the values of $y$ and $k$ from those of the ultimate load analysis (presented later), a subscript $_{sv}$ has been added to $y$ and $k$ in the tables.

Rather than being computed from allowable values of $f_s$ and $f_c$, the ratio $f_s/f_c$ can be deliberately selected to provide exactly the same factors of safety (to yield) for both steel and concrete. Such a feature produces optimum efficiency throughout the elastic range of the materials, since neither material can then enter yield ahead of the other. For such a "balanced" stress condition,

$$\text{for steel,} \qquad f_s = f_y / FS \qquad\qquad (5\text{-}14)$$

$$\text{for concrete,} \quad f_c = 0.85 f_c' / FS, \qquad\qquad (5\text{-}15)$$

which provides a ratio of $f_s / f_c$ of

$$\frac{f_s}{f_c} = \frac{f_y}{0.85 f_c'} \qquad\qquad (5\text{-}16)$$

The tables may therefore be entered with the fixed ratio $f_y/0.85f_c'$. The section so chosen will have exactly the same factors of safety and elastic margins for both the steel and the concrete over the entire elastic range of both materials. Both materials will enter their yield ranges at exactly the same time, an efficient and desirable feature in the design.

For convenience, a separate table for the balanced stress condition has been included in the design tables of the Appendix. The table, Table A-8, has been prepared utilizing exact values for $f_y/0.85f_c'$; these exact values eliminate the need for interpolation that can occur when using the more general design tables, Tables A-5, A-6, and A-7. In developing the table, the fixed stress ratio $f_y/0.85f_c'$ was substituted into Eq. (5-13) and the result solved for $k$. The value of $k$ thus obtained was used to find the steel ratio $\rho$ from Eq. (5-8). With both $k$ and $\rho$ known, all the remaining section constants were readily computed.

As an alternative means to enter the design tables, the tables may be entered with known values of $\rho$ and $\rho'$. Such a case will occur when the size and reinforcement of a section are already known, and the section is being investigated to find its capacity for moment.

## Applications at Elastic Levels of Stress

Some examples will illustrate the use of the tables of elastic section constants.

### EXAMPLE 5-1.

Determination of elastic stress due to moment.

**Given:**

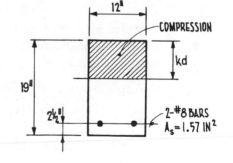

Section as shown
Elastic moment = 46 kip-ft
Grade 50 steel, $f_c'$ = 4000 psi

**Find:**

Maximum stress in concrete
Maximum stress in reinforcement

**Solution:**

Calculate the steel ratio $\rho$ for the given section:

$$\rho = \frac{A_s}{bd} = \frac{1.57}{12 \times 16.5} = 0.008$$

Enter Table A-6 with this steel ratio and find the section constants:

$$\rho = 0.008, \ S_c = 0.135bd^2, \ f_s/f_c = 18.76$$

Solve for the stress in concrete:

$$f_c = \frac{M}{S_c} = \frac{46,000 \times 12}{0.135 \times 12 \times 16.5 \times 16.5} = 1250 \text{ psi}$$

Solve for the stress in reinforcement:

$$\frac{f_s}{f_c} = 18.76$$

$$f_s = 18.76 f_c = 18.76 \times 1250 = 23,450 \text{ psi}$$

Stress in concrete: 1250 psi (compression)
Stress in reinforcement: 23,500 psi (tension)

## EXAMPLE 5-2.

Determination of a suitable section.

### Given:

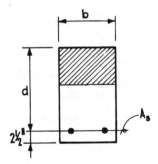

Section as shown
Grade 60 steel, $f'_c$ = 4000 psi
Allowable concrete stress = 1800 psi
Allowable steel stress = 24,000 psi
Applied moment = 76 kip-ft.

### Find:

Required size of section and reinforcement

### Solution:

Desired ratio for steel and concrete stresses:

$f_s/f_c$ = 24,000/1800 = 13.3

Enter Table A-6 with this ratio of stresses and find the section constants:

$\rho$ = 0.014, $S_c$ = 0.164$bd^2$, $f_s/f_c$ = 13.44
(close enough to 13.3 to be acceptable)

Calculate the required magnitude of the section modulus:

$$S_c = \frac{M}{f_c} = \frac{76000 \times 12}{1800} = 507 \text{ in.}^3$$

Solve for the required size of section, assuming $b = 0.6d$:

$$S_c = 0.164bd^2; \; 507 = 0.164 \times 0.6d \times d^2$$

Solve for $d = 17.27$ in.
Solve for $b = 0.6d = 10.36$ in.

Solve for the required steel area:

$$A_s = \rho bd = 0.014 \times 10.36 \times 17.27 = 2.50 \text{ in.}^2$$

From Tables A-3 and A-4, select 2 No. 10 bars, $A_s$ furnished = 2.53 in.$^2$

Round dimensions up, to nearest $\frac{1}{2}$ in.:

Use $b = 10.5$ in., $d = 17.5$ in., 2 No. 10 bars, as shown at the beginning of the example.

---

## EXAMPLE 5-3.

Determination of capacity of a known section.

### Given:

Section and reinforcement as shown
Grade 50 steel, $f'_c = 4000$ psi
Allowable stress in steel = 20,000 psi

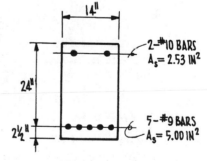

### Find:

Allowable moment on section
Whether stress in the section is balanced

### Solution:

Calculate the steel ratio for the given section:

$$\rho = \frac{A_s}{bd} = \frac{5.00}{14 \times 24} = 0.015, \quad \rho' = 0.5\rho(\pm)$$

Enter Table A-6 with $\rho = 0.015$, $\rho' = 0.5\rho$, and find the section constants:

$$\rho = 0.015, \quad S_c = 0.211bd^2, f_s/f_c = 15.94$$

Determine the allowable stress in the concrete:

$$\frac{f_s}{f_c} = 15.94; \qquad f_c = \frac{f_s}{15.94} = \frac{20,000}{15.94} = 1254 \text{ psi}$$

Determine the allowable moment on the section:

$$M = f_c S_c = 1254 \times 0.211 \times 14 \times 24 \times 24$$
$$= 2134 \text{ kip-in.} = 178 \text{ kip-ft}$$

Determine the stress ratio for balanced stress condition:

At balanced stress,
$f_s/f_c = f_y/0.85f'_c = 50,000/0.85 \times 4000 = 14.7$

Actual stress ratio = 15.94 ≠ 14.7; conclude that the stresses are not balanced.

Final results:

Allowable moment = 178 kip-ft; section is not balanced for stresses.

The elastic section constants may also be used to design members and to compute deflections under short-term loads. Some examples will illustrate.

---

## EXAMPLE 5-4.

Design of a beam and determination of its deflections.

**Given:**

Uniformly loaded beam, simple span of 20'0", exterior exposure
Dead load = 2.2k/ft, live load 1.8k/ft
Grade 60 steel, $f'_c = 3000$ psi
Service stress = 1500 psi for balanced design

**Find:**

Service moment $M_{sv}$

Required size of member with balanced stresses

Required reinforcement

Maximum deflection at midspan

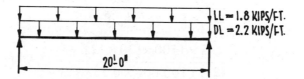

**Solution:**

Calculate the dead load and live load moments:

$$M_{DL} = \frac{wL^2}{8} = \frac{2.2 \times 20^2}{8} = 110 \text{ kip-ft}$$

$$M_{LL} = \frac{wL^2}{8} = \frac{1.8 \times 20^2}{8} = 90 \text{ kip-ft}$$

Calculate the service moment:

$$M_{sv} = \left(\tfrac{1.4}{1.7} M_{DL} + M_{LL}\right)\phi = \left(\tfrac{1.4}{1.7} \times 110 + 90\right)/0.9$$
$$= 201 \text{ kip-ft}$$

Solve for the required magnitude of the section modulus:

$$S_c = \frac{M_{sv}}{f_{sv}} = \frac{201,000 \times 12}{1500} = 1608 \text{ in.}^3$$

For the balanced stress condition, enter Table A-8, select section constants for a beam with no compressive reinforcement:

$$\rho = 0.0060, \quad S_c = 0.128bd^2, \quad k_{sv} = 0.283, \quad f_s/f_c = 23.53$$

Determine the required sizes and reinforcement, assuming $b = 0.6d$:

$$S_c = 0.128\,bd^2; \quad 1608 = 0.128 \times 0.6d \times d^2$$

Solve for $d = 27.6$ in., $b = 16.5$ in.

Steel area $= \rho bd = 0.0060 \times 16.5 \times 27.6 = 2.73$ in.$^2$

From Tables A-3 and A-4, select 2 No. 8 and 2 No. 7 bars.
As furnished = 2.77 in.²

Use $b$ = 16.5 in., $d$ = 28.0 in., 2 No. 8 and 2 No. 7 bars.

Determine the maximum deflection at midspan [Eq. (3-3)], using modulus of elasticity $E_c$ as given in Table A-1:

$$\Delta_{max} = \frac{5f_cL^2}{48E_ckd} = \frac{5f_{sv}L^2}{48E_ck_{sv}d}$$

$$= \frac{5 \times 1500 \times (20 \times 12)^2}{48 \times 3,100,000 \times 0.283 \times 28}$$

$\Delta_{max}$ = 0.37 in. at midspan

The chosen section is shown in the following sketch.

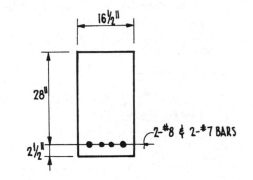

---

## EXAMPLE 5-5.

Design of a beam and determination of its deflections.

### Given:

Simple span 36'0", exterior exposure
Dead load 1.6 kips/ft
Live load is a single moving load, 48 kips
Grade 50 steel, $f'_c$ = 5000 psi
Service stress = 2600 psi for balanced design
Deflection limited by brittle elements to $L/360$ for live load

**Find:**

Service moment $M_{sv}$

Required size of member with balanced stresses

Required reinforcement

Check deflections

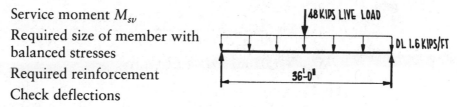

**Solution:**

Calculate the dead load and live load moments:

$$M_{DL} = \frac{wL^2}{8} = \frac{1.6 \times 36^2}{8} = 259 \text{ kip-ft}$$

$$M_{LL} = \frac{PL}{4} = \frac{48 \times 36}{4} = 432 \text{ kip-ft (maximum)}$$

Determine the service moment $M_{sv}$:

$$M_{sv} = \left(\tfrac{1.4}{1.7} M_{DL} + M_{LL}\right)/\phi = \left(\tfrac{1.4}{1.7} \times 259 + 432\right)/0.9$$
$$= 717 \text{ kip-ft}$$

Solve for the required magnitude of the section modulus:

$$S_c = \frac{M_{sv}}{f_{sv}} = \frac{717,000 \times 12}{2600} = 3310$$

For the balanced stress condition, enter Table A-8, select section constants for a beam with no compressive reinforcement:

$$\rho = 0.0161, \quad S_c = 0.166bd^2, \quad k_{sv} = 0.379, \quad f_s/f_c = 11.76$$

Determine the required sizes and reinforcement. Assume $b = 0.6d$ and substitute into the section modulus:

$S_c = 0.166bd^2$;  $3310 = 0.166 \times 0.6d \times d^2$

Solve for $d = 32.2$ in., $b = 19.3$ in.

Steel area $= \rho bd = 0.0161 \times 19.3 \times 32.2 = 10$ in.$^2$

From Tables A-3 and A-4 select 10 No. 9 bars, 2 layers of 5 bars ea. $A_s$ furnished $= 10.0$ in.$^2$

Use $b = 20$ in., $d = 33$ in., 10 No. 9 bars in 2 layers of 5 ea.

Determine the maximum deflection due to dead load plus live load.

At maximum stress $f_{sv}$, for dead plus live load:

$$\Delta_{max} = \frac{5f_{sv}L^2}{48E_c k_{sv}d} = \frac{5 \times 2600 \times (36 \times 12)^2}{48 \times 4,000,000 \times 0.379 \times 33}$$

$$\Delta_{max} = 1.01 \text{ in.} (DL + LL)$$

This $\Delta_{max}$ must be less than $L/240$,

Allowable $\Delta = L/240 = (36 \times 12)/240 = 1.8 > 1.01$ (O.K.)

Determine the deflections due only to live load.

By ratios:

$$\Delta_{LL} = \frac{M_{LL}/\phi}{M_{sv}}\Delta_{TOTAL} = \frac{432/0.9}{717} \times 1.01$$

$$\Delta_{LL} = 0.67 \text{ in.}$$

This $\Delta_{LL}$ must be less than $L/360$.

Allowable $\Delta = L/360 = 360 \times 12/360 = 1.2 > 0.67$ (O.K.)

The final section is shown in the following sketch.

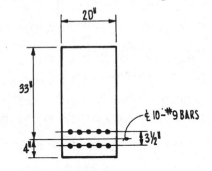

## Common Arrangements of Reinforcement

Some of the more common arrangements of flexural reinforcement were used in the preceding examples. There are other types of reinforcement that occur in beams, however, for which other arrangements become necessary. One such additional type of reinforcement is that for shear, as shown in Figure 5-6. This type of shear reinforcement is called a "stirrup." Stirrups are placed at a spacing of $\frac{1}{2}d$ or $\frac{1}{2}d$ along the length of the beam where shears are high.

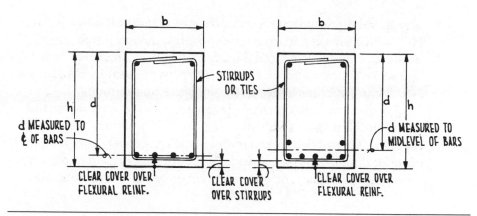

**Figure 5-6**    Typical reinforced sections.

Reinforcement for shear is presented in detail in Chapter 10. At this point, it is necessary only to consider the effects that the existence of shear reinforcement will have on the overall size of the section. Minimum cover must be provided for the shear reinforcement as well as for the flexural reinforcement; the cover over both types of reinforcement must therefore be considered when one is establishing the required depth of section $h$.

Requirements for minimum cover given in Chapter 3 apply also to shear reinforcement, or stirrups. A dimension sketch is shown in Figure 5-7 for typical steel arrangements that include shear reinforcement. For flex-

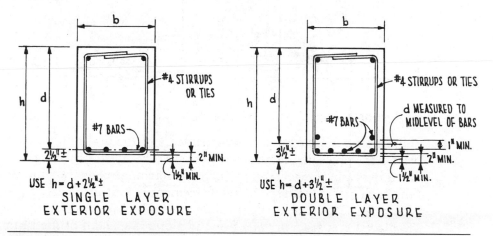

**Figure 5-7**    Overall depth of section with stirrups.

ural members in interior exposures, the minimum cover is ¾ in. for bars No. 11 and smaller. In exterior exposures, the minimum cover is 1½ in. for bars No. 5 and smaller and 2 in. for bars larger than No. 5, as indicated in Figure 5-7. As a matter of interest, stirrups are usually made from No. 4 or No. 5 bars; stirrups larger than No. 5 are uncommon.

For flexural reinforcement larger than No. 8 bars or for interior exposures, the dimensions shown in Figure 5-7 will vary somewhat. Such arrangements must be checked individually for proper cover as appropriate. In general, dimensions that include cover are rounded up to the nearest ½ in., with the allowances for minimum cover being rather generous.

## Concept of the Internal Couple

One last item concerning the stresses on an elastic section can now be developed. In Figure 5-8, a resultant couple is shown on the stress diagram. The couple is composed of the resultant $T$ of all tensile forces and the resultant $C$ of all compressive forces, with the arm between the two forces being designated $jd$. The moment on the section may be computed alternatively by

$$M = Tjd = f_s A_s jd \tag{5-17}$$

In most concrete beams, the factor $j$ varies between very close limits, with an average value of 0.9 being a good approximation at elastic levels of

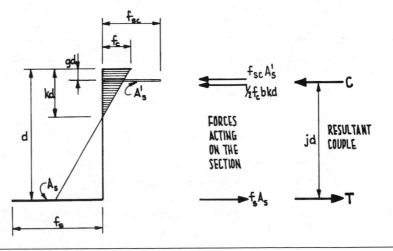

**Figure 5-8**    Resultant couple.

stress. This feature of the couple can sometimes allow an approximate design to be made.

For example, given a section having fixed dimensions—that is, $b$ and $d$ are fixed—and given a reasonably consistent value for stress in steel, $f_s$, the allowable moment on the section is governed by the corresponding steel ratio $\rho_1$,

$$M_1 = f_s(\rho_1 bd)(j_1 d) \qquad (5\text{-}18)$$

Similarly, for any other steel ratio $\rho_2$ proposed for use in this section, the moment $M_2$ corresponding to this second steel ratio is given by

$$M_2 = f_s(\rho_2 bd)(j_2 d) \qquad (5\text{-}19)$$

Eq. (5-16a) is divided by Eq. (5-16b) and the result solved for $\rho_2$, yielding

$$\rho_2 = \frac{M_2}{M_1}\rho_1 \text{ (approximately)} \qquad (5\text{-}20)$$

where it is assumed that $j_1$ is roughly equal to $j_2$. Consequently, given a moment $M_1$ and its corresponding steel ratio $\rho_1$ for a particular section, the steel ratio $\rho_2$ required to sustain some other moment $M_2$ on the section may be found by the use of Eq. (5-20).

## OUTSIDE PROBLEMS

5.1  A rectangular concrete beam, $f'_c = 3000$ psi, has a width of 14 in., a total height of 28 in., and an effective depth of 25.5 in.; reinforcement consists of 4 No. 7 bars, grade 40 steel. Sketch the cross section and compute the stress in the steel when the moment is 106 kip-ft.

5.2  A rectangular concrete beam, $f'_c = 5000$ psi, carries a moment of 84 kip-ft. The width of the beam is 12 in. and the effective depth is 26 in. Reinforcement on the tension side is 4 No. 8 bars and on the compression side is 2 No. 8 bars, all grade 60 steel. Cover over all reinforcement is $1\frac{1}{2}$ in. Sketch the cross section and determine the maximum stress in the concrete.

5.3  A rectangular concrete beam, $f'_c = 4000$ psi, is loaded in flexure in the laboratory. During the test, the elastic stress in the concrete is measured and found to be 1357 psi; at the same time, the stress in the steel is found to be 30,100 psi. If no compression steel was used, what was the tensile steel ratio used in building this beam?

5.4  A rectangular concrete beam having $f'_c = 4000$ psi has a tensile steel ratio $\rho = 0.012$ and a compression steel ratio of 0.0048. Sketch the stress diagram and determine the stress in both the tensile steel and the compressive steel when the maximum stress in the concrete is 1150 psi. $g = 0.125$

For the following conditions, select a suitable size for a rectangular concrete beam and select the required reinforcement:

| Prob. No. | Concrete Strength $f'_c$ | Allowable Stress in Steel, psi | Allowable Stress in Concrete, psi | Applied Moment in kip-ft |
|---|---|---|---|---|
| 5.5 | 3000 | 20,000 | 900 | 60 |
| 5.6 | 3000 | 24,000 | 1200 | 70 |
| 5.7 | 4000 | 20,000 | 1200 | 80 |
| 5.8 | 4000 | 24,000 | 1600 | 90 |
| 5.9 | 5000 | 22,000 | 1600 | 100 |
| 5.10 | 5000 | 22,000 | 2000 | 110 |
| 5.11 | 3000 | 30,000 | 1100 | 60 |
| 5.12 | 3000 | 36,000 | 1350 | 70 |
| 5.13 | 4000 | 30,000 | 1500 | 80 |
| 5.14 | 4000 | 36,000 | 1800 | 90 |
| 5.15 | 5000 | 33,000 | 2100 | 100 |
| 5.16 | 5000 | 33,000 | 2400 | 110 |

**5.17** A rectangular concrete beam, $f'_c$ = 4000 psi, is limited to an overall depth of 16 in. The beam must sustain a moment of 102 kip-ft. Stress in the concrete is not to exceed 1500 psi and stress in the steel is not to exceed 22,000 psi. Clear cover over the reinforcement must not be less than 1.5 in. Select a suitable beam and its reinforcement.

**5.18** For the conditions of Problem 5.17, select a suitable beam if, in addition, the width must not exceed 20 in.

Select a suitable rectangular concrete section and its reinforcement under service loads at balanced stress conditions for the given simply supported beams. Then determine the maximum deflection that will occur under the service load conditions.

| Prob. No. | Uniform Load kips/ft | | Span ft | Concrete $f'_c$ psi | Allow. $f'_c$ psi | Steel Grade |
|---|---|---|---|---|---|---|
| | $w_{DL}$ | $w_{LL}$ | | | | |
| 5.19 | 1.9 | 1.8 | 26 | 3000 | 1200 | 40 |
| 5.20 | 2.2 | 2.8 | 18 | 5000 | 2200 | 60 |
| 5.21 | 1.8 | 2.1 | 24 | 4000 | 1800 | 50 |
| 5.22 | 2.1 | 2.6 | 18 | 4000 | 1600 | 40 |
| 5.23 | 2.0 | 2.3 | 22 | 5000 | 2000 | 60 |
| 5.24 | 1.9 | 2.4 | 20 | 3000 | 1350 | 50 |

## REVIEW QUESTIONS

1. Define the elastic section modulus of a concrete beam in flexure.
2. Under what conditions might compressive reinforcement be required in a concrete beam?
3. Given the elastic compressive stress $f_c$ in concrete and the elastic stress $f_s$ in the tensile steel, how can the stress in the compressive reinforcement be computed?
4. Describe the "balanced stress condition" at elastic levels of stress in a concrete beam.
5. Given the section modulus $S_c$ and the distance to the neutral axis $kd$, how is the moment of inertia I computed?
6. In flexural design, how is the effect of creep and shrinkage accounted for?
7. Define the steel ratios $\rho$ and $\rho'$.
8. What is the difference between margin of safety and factor of safety in elastic design methods?
9. When the flexural reinforcement in a beam is placed in two levels, to what level is the effective depth $d$ measured?
10. In a particular rectangular beam subject to exterior exposures, the flexural reinforcement consists of No. 5 bars and the shear reinforcement consists of stirrups made from No. 4 bars. What clear cover is required for the flexural reinforcement? For the stirrups? Sketch your result and show the clear cover finally selected.
11. What is the effect of creep and shrinkage on the stress in the tensile reinforcement of a concrete beam?
12. What is the effect of creep and shrinkage on the stress in the compressive reinforcement of a concrete beam?

# 6

# ULTIMATE
# FLEXURE

Traditionally, most structural materials have been designed at elastic levels of stress rather than at ultimate load. The reason for such a limitation is that once the stresses enter yield and the strains become inelastic, the deformations become so large that the member is no longer considered to be usable; stresses are therefore held within elastic levels in order to limit large deformations. Too, once the strains become inelastic, the beam reactions begin to change, sometimes drastically, and the reactions and moments derived from the elastic analysis are no longer valid.

There is, however, a very large reserve of strength in a member after it enters into inelastic strains. That reserve of strength is ignored when the design is limited to the elastic range. It should be remembered that the factor of safety used in elastic theory is based on a maximum yield stress, not on collapse load. If the factor of safety in elastic theory were to be recomputed and based on collapse load, it would increase considerably.

With the 1956 Code, ACI recognized that this large reserve of strength does indeed exist in concrete members and that the design of concrete members, even at elastic levels, can take into account this reserve of strength; the ACI strength method of design in current use has been developed from that viewpoint. This chapter presents the ACI strength method of design.

## The Plastic Section Modulus

The ACI Code does not prescribe the methods to be used in achieving an end result; it prescribes only the end result that must be achieved. For concrete members in flexure, for example, Code (10.2) prescribes only the pattern of strains that must exist in a member when it is at its ultimate capacity. The methods for selecting the sizes and reinforcement of a member so that the prescribed pattern of strains does in fact occur are not prescribed.

There are numerous approaches to designing a concrete member so that the member will sustain a prescribed ultimate load at a prescribed pattern of strains. Some of the methods are slow but simple, others are faster but more complex. Some are suited to manual solutions, others only to computer applications. The use of a "plastic section modulus," as used in this textbook, is regarded as one of the slower but simpler manual methods. It has the distinct advantage to the intermittent user, however, in that it is much easier to remember and apply over the years than are the faster but more complex methods.

The plastic section modulus is directly analogous to the elastic section modulus used in Chapter 5. Recall that the elastic section modulus $S_c$ is the ratio of moment $M$ to concrete stress $f_c$ where all deformations are elastic:

$$S_c = \frac{M}{f_c} \tag{5-1b}$$

By analogy, the plastic section modulus $Z_c$ is the ratio of nominal ultimate moment $M_n$ to the idealized yield stress $0.85f'_c$, with deformations at specified inelastic levels:

$$Z_c = \frac{M_n}{0.85f'_c} \tag{6-1}$$

Since the nominal ultimate moment $M_n$ is readily calculated from the ACI load factors and the idealized yield stress $0.85f'_c$ is known, the required magnitude of the plastic section modulus $Z_c$ can be easily computed from Eq. (6-1). It remains only to find some way to express $Z_c$ in terms of the member's cross section so that the size of the member can be easily computed. Such a means to express $Z_c$ is developed next.

## Derivation of the Plastic Section Modulus

Chapter 5 dealt with the design and investigation of members using the elastic method of analysis. The following sections deal with the same problems using the strength method. Although the final concern of the strength method is the ultimate load (rather than stress), the analysis must start with the usual stress distribution and develop that into the corresponding ultimate load.

The analysis for the ultimate moment on a concrete section is stringently prescribed by Code (10.2). To begin, it is required by Code (10.2.3) that the maximum strain in the concrete be 0.003 at ultimate moment, which is presumed to include all effects of shrinkage and creep. To accompany this prescribed strain in the concrete, the strain in the steel must take whatever value is required to produce equilibrium.

Code (10.2.2) also requires that the variation in strains across the section shall be assumed to follow a straight-line variation. Such a straight-line variation in strain at ultimate load conditions is shown graphically in Figure 6-1. Strain in steel at yield is shown as $\varepsilon_{sy}$. The strain in concrete is 0.003 in./in., as required by Code (10.2.3).

Strains falling in stress range $I$ of Figure 6-1 are those for a heavily reinforced section, where the neutral axis is low and the stress in the compressive reinforcement is in yield. Strains falling in stress range $II$ are those for lightly reinforced sections, for which the neutral axis is much higher on the section and the compressive reinforcement is elastic. Line $OA$ indicates the strains when the tensile steel is just at yield, $OB$ when the compressive

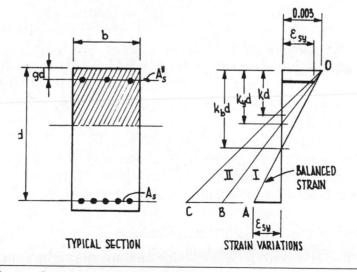

**Figure 6-1**    Strains at ultimate load.

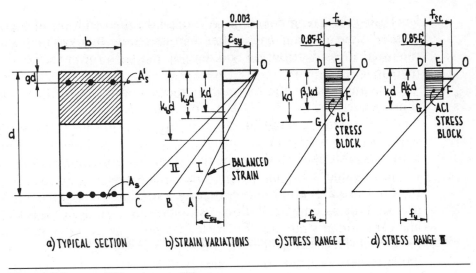

**Figure 6-2**     Ultimate conditions.

steel is just at yield, and $OC$ at some arbitrarily chosen minimum level of reinforcement.

The variations in stress that accompany the strain diagrams of Figure 6-1 are shown in Figure 6-2. In Figure 6-2c, the compressive steel is still in yield; in Figure 6-2d it has emerged from yield and is in its elastic range. In both cases, the tensile steel remains in yield, denoted $f_y$.

The variations in stress should theoretically follow the stress-strain curve of Figure 2-2, but it does not matter since Code (10.2.7) also specifies an empirically derived "stress block" to be used in computing the compressive force. In all succeeding derivations it is assumed that both the concrete and the tensile steel are well into yield, but that compressive steel may or may not be in yield.

The value of $k$ that occurs when the compressive steel is just at yield, designated $k_y$, can be computed by similar triangles from the line $OB$ in the strain diagram of Figure 6-2b:

$$\frac{0.003}{k_y d} = \frac{\varepsilon_{sy}}{k_y d - gd} \quad \text{where } \varepsilon_{sy} = \frac{f_y}{E_s} \tag{6-2}$$

The solution for $k_y$ is, with $E_s = 29,000,000$ psi,

$$k_y = \frac{0.003 E_s g}{0.003 E_s - f_y} = \frac{87,000 g}{87,000 - f_y} \quad f_y \text{ in psi} \tag{6-3}$$

For all values of $k$ greater than $k_y$ the stress diagram of Figure 6-2c applies; for values of $k$ less than $k_y$, the stress diagram of Figure 6-2d applies. Taken together, these two stress diagrams define the state of stress in concrete beams throughout the range of interest. The ultimate section modulus (plastic section modulus) can be developed readily from these diagrams.

The value of $\beta_1$ shown in the stress diagrams for the ACI stress block is prescribed by Code (10.2.7.3). Although the Code permits approaches other than this, the Code value of $\beta_1$ is known to provide a close correlation to test data and is accepted here. When the prescribed value of $\beta_1$ is used, the force in the compression stress block is found to be $0.85f_c'b\beta_1kd$.

In the form of an equation, the factor $\beta_1$ is given by Code (10.2.7.3):

$$\beta_1 = 0.85 - 0.05(f_c' - 4), \; f_c' \text{ in ksi} \qquad (6\text{-}4)$$

In addition, $\beta_1$ may not be greater than 0.85 nor less than 0.65.

From Figure 6-3, the sum of horizontal forces yields an expression for $\beta_1k$. For the first case, take $k > k_y$, placing the compressive steel in yield, that is, $f_{sc} = f_y$:

$$\Sigma H = 0 = 0.85f_c'b\beta_1kd + f_y\rho'bd - f_y\rho bd \qquad (6\text{-}5)$$

The solution for $\beta_1k$ is, where $k > k_y$ and compressive steel is in yield, $f_{sc} = f_y$

$$\beta_1k = \frac{f_y(\rho - \rho')}{0.85f_c'} \qquad (6\text{-}6)$$

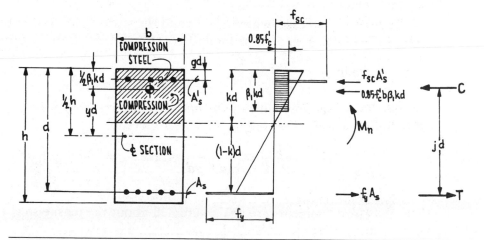

**Figure 6-3**    Stresses at ultimate load.

Note that since $\beta_1$ is governed by Code, it is not appropriate to deduct the compressive steel area from the stress block.

Refer again to the stress diagram of Figure 6-3. The sum of moments about tensile steel yields

$$M_n = 0.85f'_c b\beta_1 kd(d - \tfrac{1}{2}\beta_1 kd) + f_y A'_s(d - gd) \qquad (6\text{-}7)$$

Solve Eq. (6-7) for $M_n/0.85f'_c$, substitute $f_y$ from Eq. (6-6), and solve for the plastic section modulus $Z_c$. With $k > k_y$, the compressive steel is in yield and the plastic section modulus $Z_c$ is:

$$\frac{M_n}{0.85f'_c} = Z_c \qquad (6\text{-}8)$$

$$\text{where } Z_c = \beta_1 k\left(1 - \frac{\beta_1 k}{2} + \rho'\frac{1-g}{\rho - \rho'}\right)bd^2$$

and where $\beta_1 k$ is the value given by Eq. (6-6).

Eq. (6-6) and (6-8) are the values of $\beta_1 k$ and $Z_c$ when the compressive steel is in yield. When the compressive steel emerges from yield and is in the elastic range, its elastic strain $\varepsilon_s$ can be found from Figure 6-1d by ratio, where $E_s = 29,000,000$ psi,

$$\frac{\varepsilon_s}{0.003} = \frac{kd - gd}{kd} \qquad (6\text{-}9a, \text{ b})$$

$$\text{hence } f_{sc} = \varepsilon_s E_s = 87,000\frac{k-g}{k}$$

For the second case, where $k < k_y$ and the stress $f_{sc}$ in the compressive steel is elastic, the sum of horizontal forces shown in Figure 6-3 again yields an expression for $k$:

$$\Sigma H = 0 = 0.85f'_c b\beta_1 kd + f_{sc}\rho'bd - f_y\rho bd \qquad (6\text{-}10)$$

Eq. (6-9b) is now substituted and the result solved for $\beta_1 k$, where $k < k_y$ and stress in the compression steel is elastic,

$$\beta_1 k = \sqrt{\left[\frac{f_y\rho - 87,000\rho'}{2 \times 0.85f'_c}\right]^2 + \frac{87,000g\rho'\beta_1}{0.85f'_c}} + \frac{f_y\rho - 87,000\rho'}{2 \times 0.85f'_c} \qquad (6\text{-}11)$$

As before, moments are again summed about the tensile steel (Figure 6-3), with $f_{sc}$ elastic,

$$M_n = 0.85f_c'b\beta_1kd(d - \tfrac{1}{2}\beta_1kd) + f_{sc}\rho'bd(d - gd) \qquad (6\text{-}12)$$

Eq. (6-9b) is substituted into Eq. (6-12) and the result solved for $M_n/0.85f_c'$. For $k < k_y$, the stress in the compression steel is elastic and the plastic section modulus $Z_c$ is:

$$\frac{M_n}{0.85f_c'} = Z_c \qquad (6\text{-}13)$$

where $Z_c = \left[\beta_1k\left(1 - \frac{\beta_1k}{2}\right) + \frac{87{,}000\rho'}{0.85f_c'}(1 - g)\frac{k - g}{k}\right]bd^2$

and where $\beta_1k$ is the value given by Eq. (6-11).

Eq. (6-6), (6-8), (6-11), and (6-13) provide the plastic section modulus for a concrete beam throughout its range of interest. As with the elastic analysis, it is not necessary to manually solve these rather lengthy equations. They can be solved by computer and the results tabulated for reference. Such a tabulation is given in Tables A-5, A-6, and A-7 of the Appendix, with $k$ becoming $k_n$ to distinguish it from $k_{sv}$ in the elastic constants.

As in the elastic analysis, the distance $yd$ from the centerline of the section to the center of the compressive force will be needed; the distance is shown in Figure 6-4.

The distance $yd$ is found from the geometry of Figure 6-4,

$$yd = \frac{h}{2} - \frac{\beta_1kd}{2} = \tfrac{1}{2}\left(1 + g - \beta_1k\right)d \qquad (6\text{-}14)$$

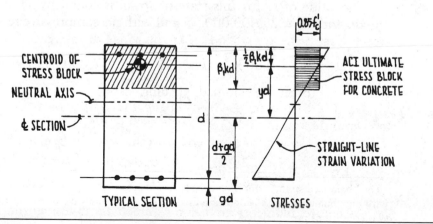

**Figure 6-4**    Centroid of ultimate load stress block.

As observed earlier, this factor $y$ may sometimes be greater than 0.5 since it applies to $d$ rather than $h$.

Tables A-5 through A-7 of the Appendix include the values of $Z_c$, $\beta_1 k$, and $y$ given above; a subscript $_n$ has been added both to $k$ and to $y$ to distinguish them from the comparable terms in the elastic analysis. Similar to the elastic analysis, the tables may be entered with the values of $\rho$ and $\rho'$, the ultimate strength of concrete and the grade of steel. The columns headed "at ultimate strength" give the tabulated values of Eq. (6-8), (6-13), and (6-14).

The value of $k_n$ at ultimate load is not often of interest, but its value can be obtained from the listed value of $\beta_1 k_n$ simply by dividing by $\beta_1$; the value of $\beta_1$ is given by Eq. (6-4) and is also given at the top of each beam table.

For ultimate load design, the Code places limits on both the minimum amount of steel that may be used and the maximum amount as well. Code (10.5.1) gives the minimum steel ratio by a simple formula:

$$\text{minimum } \rho = \frac{200}{f_y} \tag{6-15}$$

The maximum steel ratio is somewhat more involved. When there is no compressive reinforcement in the section, the maximum steel ratio is given by Code (10.3.3) as 75% of that required to produce the balanced strain* condition (line $OA$ of Figure 6-2b). At the balanced strain condition, the concrete reaches its ultimate strain of 0.003 in./in. just as the steel enters yield. This state of strain is shown again in Figure 6-5, along with the corresponding stresses.

The value of $k_b$ for this state of strain is found by similar triangles, where, with $E_s = 29,000,000$ psi and with no compressive reinforcement,

$$\frac{f_y / E_s}{0.003} = \frac{d - k_b d}{k_b d}; \quad k_b = \frac{87,000}{87,000 + f_y} \tag{6-16}$$

Horizontal forces are summed, yielding

$$0 = 0.85 f_c' b \beta_1 k_b d - f_y \rho_b b d \tag{6-17}$$

This result is solved for $\rho_b$ where, with the value of $k_b$ given by Eq. (6-16),

---

*The "balanced strain" condition used here and prescribed by Code should not be confused with the "balanced stress" condition used earlier as a means to control elastic margins. The two conditions are unrelated.

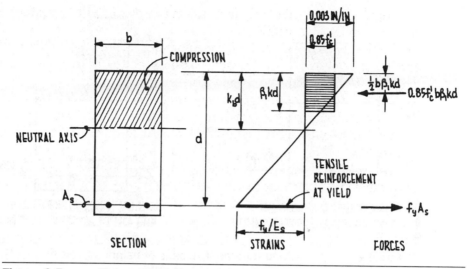

**Figure 6-5**      Balanced strain condition.

$$\rho_b = \beta_1 \frac{0.85f_c'}{f_y}\left(\frac{87,000}{87,000+f_y}\right), \quad f_y \text{ and } f_c' \text{ in psi} \qquad (6\text{-}18)$$

and maximum allowable

$$\rho = 0.75\rho_b \qquad (6\text{-}19)$$

The limitation thus imposed on the area of steel assures that the tensile steel will always enter yield before the concrete reaches its ultimate strain, thereby avoiding the chance of brittle failure in the concrete.

Code (10.3.4) permits the area of tensile reinforcement to exceed this level of $0.75\rho_b$, provided that the excess area of tensile reinforcement is matched by a corresponding addition of compressive reinforcement. The extra steel areas are shown schematically in the sketch of Figure 6-6.

Code does not require the excess area of tensile reinforcement to be reduced by the factor 0.75. The total amount of tensile reinforcement permitted by Code is then, when compressive reinforcement is used,

$$\text{maximum } \rho = 0.75\rho_b + \rho' \qquad (6\text{-}20)$$

As a practical matter, large amounts of reinforcement can produce serious problems with congestion in placing and holding the steel in the

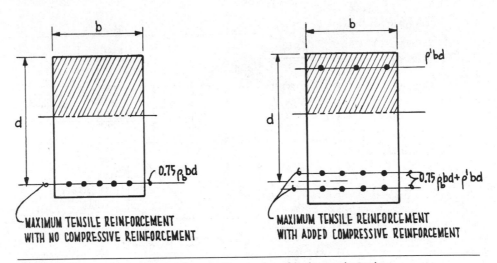

**Figure 6-6**    Tensile steel area in excess of balanced strain.

forms. A practical limit on the total area of tensile and compressive rein-forcement has been found to be about 8% of gross area; steel areas in excess of this amount produce unacceptable problems in placement.

Tables A-5 through A-7 of the Appendix reflect the foregoing maxi-mum and minimum values of $\rho$ permitted by Code. Values greater than that given by Eq. (6-19) are not entered. Values less than that given by Eq. (6-15) are those above the interior line in the tables; the reason for includ-ing these smaller values is given later with the discussion of concrete tee beams.

Under the ACI strength method, there are no further rules for choosing the steel ratio $\rho$; it may fall anywhere between the maximum and minimum allowable values. The more economical sections, however, are those where $\rho$ is chosen about midway between the maximum and minimum allowable values. Where appropriate in the following examples, the steel ratio $\rho$ is therefore chosen at about the midpoint in the range of allowable values.

## Applications at Ultimate Load

Some examples will illustrate the use of the plastic section modulus in the design of concrete sections.

**EXAMPLE 6-1.**

Strength method of design.
Determination of moment capacity.

**Given:**

Section as shown
Grade 50 steel, $f'_c$ = 4000 psi

**Find:**

Ultimate moment $M_n$ the section
will sustain.

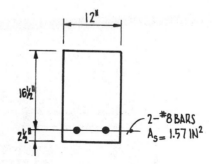

**Solution:**

Calculate the steel ratio ρ for the given section,

$$\rho = \frac{A_s}{bd} = \frac{1.57}{12 \times 16.5} = 0.008$$

Enter Table A-6, find section constants,

$\rho = 0.0008, \ Z_c = 0.111bd^2$

Solve capacity for moment

$M_n = 0.85f'_c Z_c = 0.85 \times 4000 \times 0.111 \times 12 \times 16.5 \times 16.5$

$M_n$ = 1233 kip-in. = 103 kip-ft

The given section will fail at a nominal ultimate moment of 103 kip-ft.

---

**EXAMPLE 6-2.**

Strength method of design.
Determination of moment capacity.

**Given:**

Section as shown
Grade 60 steel, $f'_c$ = 3000 psi, $\phi$ = 0.90
No wind or earthquake
Dead load moment = 121 kip-ft

**Find:**

Maximum allowable live load moment $M_{LL}$

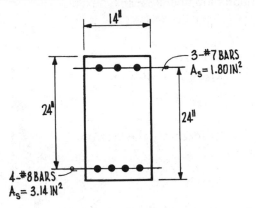

**Solution:**

Calculate the steel ratios $\rho$ and $\rho'$

$$\rho = \frac{A_s}{bd} = \frac{3.14}{14 \times 24} = 0.0094$$

$$\rho' = \frac{A'_s}{bd} = \frac{1.80}{14 \times 24} = 0.0054$$

$$\frac{\rho'}{\rho} = 0.57, \quad \rho' = 0.57\rho; \quad \text{use } \rho' = 0.60\rho.$$

Enter Table A-5, find section constants (by interpolation),

$$\rho = 0.0094, \quad \rho' = 0.6\rho, \quad Z_c = 0.199bd^2$$

Determine the nominal ultimate moment $M_n$

$$M_n = 0.85f'_c Z_c = 0.85 \times 3000 \times 0.199 \times 14 \times 24 \times 24$$
$$= 4092 \text{ kip-in.} = 341 \text{ kip-ft}$$

Solve for the live load moment

$$M_n = (1.4M_{DL} + 1.7M_{LL})/\phi$$
$$341 = (1.4 \times 121 + 1.7 \times M_{LL})/0.9$$
$$M_{LL} = 81 \text{ kip-ft}$$

The section will sustain a live load moment of 81 kip-ft

## EXAMPLE 6-3

Strength method of design.
Design of a rectangular section in flexure.
No limitations on dimensions.

### Given:

$M_{DL} = 88$ kip-ft; $M_{LL} = 68$ kip-ft
Grade 60 steel, $f'_c = 4000$ psi, $\phi = 0.90$

### Find:

Suitable section to sustain the load

### Solution:

Calculate the ultimate load moment $M_n$. With no wind or earthquake, the load case given by Eq. (3-1a) applies.

$$M_n = (1.4M_{DL} + 1.7\,M_{LL})/\phi$$
$$= (1.4 \times 88 + 1.7 \times 68)/0.90$$
$$= 265 \text{ kip-ft}$$

Determine the required magnitude of the section modulus $Z_c$,

$$Z_c = \frac{M_n}{0.85f'_c} = \frac{265,000 \times 12}{0.85 \times 4000}$$
$$Z_c = 935 \text{ in.}^3$$

Enter Table A-6, select a section about midway between minimum and maximum steel ratios with no compressive reinforcement,

$$\rho = 0.0080,\ Z_c = 0.131bd^2$$

Solve for minimum required dimensions,

$$Z_c = 935 = 0.131bd^2, \text{ assume } b = 0.6d,$$
$$d = 22.8 \text{ in., } b = 13.7 \text{ in.}$$

Solve for steel area $A_s$

$$A_s = \rho bd = 0.0080 \times 13.7 \times 22.8 = 2.50 \text{ in.}^2$$

From Tables A-3 and A-4,

Use 6 No. 6 bars, $A_s$ furnished = 2.65 in.$^2$

Round up to nearest inch or half-inch,

Use $b$ = 14 in., $d$ = 23 in., 6 No. 6 bars.

The selected section is shown in the following sketch.

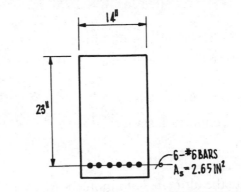

Often, the effective depth of a section must be restricted. Such circumstances commonly occur around air conditioning ductwork, pipe and cable trays, and wall openings. Such a case is considered in the next example.

## EXAMPLE 6-4

Strength method of design.
Design of a rectangular section in flexure.
Depth of section limited.

### Given:

$M_{DL}$ = 88 kip-ft; $M_{LL}$ = 68 kip-ft

Grade 60 steel, $f'_c$ = 4000 psi, $\phi$ = 0.90

Effective depth $d$ limited to 20 in.

### Find:

Suitable section to sustain the load

### Solution:

Calculate the ultimate design moment $M_n$. With no wind or earthquake,

$$M_n = \left(1.4M_{DL} + 1.7M_{LL}\right)/\phi$$
$$= \left(1.4 \times 88 + 1.7 \times 68\right)/0.90$$
$$= 265 \text{ kip-ft.}$$

Determine the required magnitude of the section modulus $Z_c$,

$$Z_c = \frac{M_n}{0.85f_c'} = \frac{265,000 \times 12}{0.85 \times 4000}$$
$$= 935 \text{ in.}^3$$

Enter Table A-6, select a section about midway between minimum and maximum steel ratios, with no compressive reinforcement,

$$\rho = 0.0080, \quad Z_c = 0.131bd^2$$

Solve for minimum required dimensions,

$Z_c = 935 = 0.131bd^2$, use $d = 20$ in.,
$b = 17.8$ in., $d = 20$ in.

Solve for steel area $A_s$

$A_s = \rho bd = 0.008 \times 17.8 \times 20 = 2.85$ in.²

From Tables A-3 and A-4

Use 5 No. 7 bars, $A_s$ furnished = 3.01 in.²

Round up to nearest inch or half inch,

Use $b = 18$ in., $d = 20$ in., 5 No. 7 bars

Final section is shown in the following sketch.

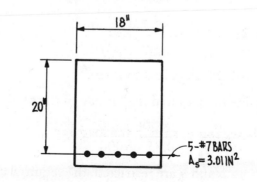

It is pointed out that Example 6-4 has the identical set of design conditions as Example 6-3, except that the effective depth has been limited. The resulting section in this second case is somewhat wider than desirable, but it still was not found to be necessary to go to heavier amounts of reinforcement. When both the width $b$ and the depth $d$ are restricted, however, the use of heavier levels of reinforcement may become unavoidable. Such circumstances occur in the next example.

## EXAMPLE 6-5.

Strength method of design.
Design of a rectangular section in flexure.
Both width and depth are limited.

### Given:

$M_{DL}$ = 88 kip-ft; $M_{LL}$ = 68 kip-ft
Grade 60 steel, $f'_c$ = 4000 psi, $\phi$ = 0.90
Width $b$ limited to 12 in.
Effective depth $d$ limited to 18 in.

### Find:

Suitable section to sustain the load

### Solution:

Calculate the ultimate design moment $M_n$
With no wind or earthquake,

$$
\begin{aligned}
M_n &= \left(1.4M_{DL} + 1.7M_{LL}\right)/\phi \\
&= \left(1.4 \times 88 + 1.7 \times 68\right)/0.90 \\
&= 265 \text{ kip-ft}
\end{aligned}
$$

Determine the required magnitude of the section modulus $Z_c$,

$$
Z_c = \frac{M_n}{0.85f'_c} = \frac{265,000 \times 12}{0.85 \times 4000} = 935 \text{ in.}^3
$$

Since both $b$ and $d$ are restricted, the required coefficient of $Z_c$ may be computed,

$Z_c = 935 = $ coeff. $\times b \times d^2$, $b = 12$ in., $d = 18$ in.

coeff. $= 0.240$

From Table A-6, with no compressive reinforcement, select the required steel ratio. (It should be noted that the section selected under these conditions must be chosen near the maximum value for $\rho$ rather than being about midway between the minimum and maximum values.)

$\rho = 0.0160$, $Z_c = 0.242bd^2$

Solve for steel area $A_s$

$A_s = \rho bd = 0.0160 \times 12 \times 18 = 3.46$ in.$^2$

From Tables A-3 and A-4

Use 3 No. 10 bars, $A_s$ furnished $= 3.80$ in.$^2$

The final section is shown in the following sketch.

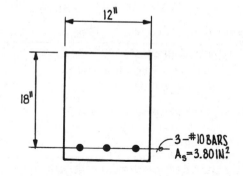

In Example 6-5, the required amount of reinforcement is considerably greater than that in Example 6-3. Such is the penalty to be paid when sizes are restricted. Such restrictions, however, are often unavoidable, arising typically when mechanical equipment must be accommodated in tightly enclosed spaces. In many cases, the cost of additional reinforcement is the most economical alternative.

## Use of Compressive Reinforcement

The results of Example 6-5 suggest that sometimes it may be worthwhile to add compressive reinforcement to a section to increase its capacity in moment. Unfortunately, when tensile steel ratios are below those at the balanced strain condition [given by Eq. (6-17)], the addition of compression steel adds but little to the overall moment capacity of the section. The greatest benefits are to reduce the compressive stress in the concrete at service levels of load and to reduce the long-term deflections.

As noted earlier, however, it is permitted by Code (10.3.3) to exceed the maximum steel ratio for tensile reinforcement given by Eq. (6-17). Such an increase can increase the moment capacity appreciably, but the increase can only be done by adding both tensile and compressive reinforcement. The tables of the Appendix have been extended to include such increased levels of reinforcement. The same procedures that were used earlier to design beams at lower levels of reinforcement remain valid at these higher levels.

As an example of the use of compressive reinforcement, the beam in Example 6-5 will be redesigned, in this case using compressive reinforcement.

### EXAMPLE 6-6

Strength method of design.
Design of a rectangular section in flexure.
Both width $b$ and depth $d$ limited.

**Given:**

Beam of Example 6-5

**Find:**

Suitable section using compressive reinforcement

**Solution:**

From the solution of Example 6-5,

Design moment $M_n$ = 265 kip-ft.

Required magnitude of $Z_c$ = 935 in.

Required coefficient of $Z_c$ :

coeff. = 0.240

From Table A-6 it is now possible to choose a section having its steel ratio about midway between the minimum and maximum allowable values of $\rho$. For a coefficient of 0.240, it is seen that a section having 60% compressive reinforcement suits this choice. Select

$$\rho = 0.0152, \quad \rho' = 0.60\rho, \quad Z_c = 0.240bd^2$$

Solve for steel areas $A_s$ and $A'_s$

$$A_s = 0.0152 \times 12 \times 18 = 3.28 \text{ in.}^2$$
$$A'_s = 0.60A_s = 1.97 \text{ in.}^2$$

For $A_s$ use 3 No. 8 and 2 No. 7 bars.

For $A'_s$ use 2 No. 9 bars.

The final section with its compressive reinforcement is shown in the following sketch, along with the solution from Example 6-5 for comparison.

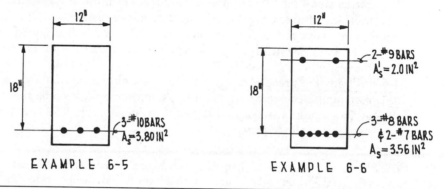

values of $\rho$. For a coefficient of 0.240, it is seen that a section having

From the sketch of the solutions of Examples 6-5 and 6-6, it is immediately apparent that the section with no compressive reinforcement has a total of 3.80 in.$^2$ of reinforcement while the section with compressive reinforcement has 5.56 in.$^2$ of reinforcement. Significantly more steel seems to be required when compressive steel is used. One must wonder whether it is practical to use compressive reinforcement at all. The reason for this large difference in this case is that the section having compressive reinforcement was selected about midway between the minimum and maximum allowable values of $\rho$, while the section having no compressive reinforcement was selected at the upper extremes of $\rho$, where $\rho$ was approaching its maximum allowable value. The two sections are therefore quite different in the way they respond to load and should not be compared. It will be shown in a later chapter that the section having no compressive reinforcement will develop much higher service stresses (roughly 30% higher) and will deform considerably more under long-term service loads (roughly 50% more).

From such results, one might suspect that there is more involved in the use of compressive reinforcement than just strength. Such suspicions are justified; there is indeed a great deal more involved. Some of the more prominent of these additional factors are treated in depth in Chapter 9, where the investigation of known sections is examined.

## OUTSIDE PROBLEMS

6.1 A rectangular concrete beam, $f'_c$ = 3000 psi, has a width of 14 in., a total depth of 28 in., and an effective depth of 25.5 in. Tensile reinforcement consists of 4 No. 7 bars, grade 40 steel. Sketch the cross section and determine the ultimate moment $M_n$ the section can sustain.

6.2 Compressive reinforcement consisting of 2 No. 7 bars is added to the beam of Prob. 6.1. Sketch the cross section and determine the ultimate moment $M_n$ the section can sustain.

Using the strength method for the following conditions, select a suitable rectangular concrete section and its reinforcement, keeping the steel ratio ρ at about half the maximum allowable value. Provide 1.5 in. minimum cover over the reinforcement.

| Prob. No. | Concrete Strength $f'_c$ psi | Steel Grade | Limits on Width $b$ | Limits on Height $h$ | Dead Load Moment kip-ft | Live Load Moment kip-ft | E'quake Moment kip-ft | Wind Moment kip-ft |
|---|---|---|---|---|---|---|---|---|
| 6.3 | 3000 | 40 | None | None | 30 | 21 | 13 | 10 |
| 6.4 | 3000 | 50 | None | None | 59 | 72 | 40 | 44 |
| 6.5 | 3000 | 60 | None | None | 32 | 25 | 15 | 13 |
| 6.6 | 4000 | 40 | None | None | 56 | 65 | 37 | 38 |
| 6.7 | 4000 | 50 | None | 14 in. | 35 | 28 | 15 | 21 |
| 6.8 | 4000 | 60 | None | 14 in. | 52 | 58 | 27 | 30 |
| 6.9 | 5000 | 40 | None | 14 in. | 38 | 32 | 17 | 18 |
| 6.10 | 5000 | 50 | None | 14 in. | 49 | 54 | 33 | 32 |
| 6.11 | 5000 | 60 | 10 in. | 14 in. | 40 | 37 | 20 | 24 |
| 6.12 | 4000 | 40 | 12 in. | 14 in. | 47 | 49 | 33 | 30 |
| 6.13 | 4000 | 50 | 10 in. | 14 in. | 43 | 41 | 24 | 28 |
| 6.14 | 4000 | 60 | 12 in. | 14 in. | 45 | 45 | 25 | 28 |

Using the strength method for the following conditions, select a suitable rectangular concrete section and its reinforcement, using tensile steel ratios above $0.75\rho_b$ where necessary. Provide 1.5 in. minimum cover over the reinforcement.

| Prob. No. | Concrete Strength $f_c'$ psi | Steel Grade | Limits on Width $b$ | Limits on Height $h$ | Dead Load Moment kip-ft | Live Load Moment kip-ft | E'quake Moment kip-ft | Wind Moment kip-ft |
|---|---|---|---|---|---|---|---|---|
| 6.15 | 3000 | 60 | 18 in. | 24 in. | 168 | 192 | 0 | 0 |
| 6.16 | 3000 | 40 | 18 in. | 27 in. | 154 | 181 | 96 | 106 |
| 6.17 | 3000 | 60 | 24 in. | 24 in. | 194 | 206 | 0 | 0 |
| 6.18 | 4000 | 40 | 24 in. | 26 in. | 212 | 221 | 121 | 141 |
| 6.19 | 4000 | 50 | 16 in. | 25 in. | 216 | 204 | 0 | 0 |
| 6.20 | 4000 | 50 | 20 in. | 22 in. | 155 | 161 | 106 | 116 |
| 6.21 | 5000 | 60 | 16 in. | 23 in. | 201 | 216 | 0 | 0 |
| 6.22 | 5000 | 40 | 24 in. | 22 in. | 198 | 171 | 99 | 89 |

## REVIEW QUESTIONS

1. Define the plastic section modulus of a concrete beam in flexure.
2. What is the maximum value of strain that is assumed to occur in concrete at ultimate load?
3. What is the maximum value of strain that is assumed to occur in the reinforcing steel at ultimate load?
4. In terms of strains, how is the ultimate resistance of a section defined by Code?
5. At ultimate load, how does strain vary across the section?
6. How is the value of the ACI factor $\beta_1$ determined?
7. What happens to stress in reinforcing steel once the steel enters yield?
8. How is the minimum stress ratio $\rho$ determined?
9. Verify the algebraic derivation of Eq. (6-18).
10. For low steel ratios, that is, for steel ratios less than the maximum given by Eq. (6-17), what is the primary benefit to be realized in the use of compressive reinforcement?
11. A concrete section is made from concrete having $f_c'$ = 3000 psi and grade 60 steel. The steel ratio for this section was chosen to be 0.0120 [well under the maximum allowable of $0.75\rho_b$ as given by Eq. (6-17)], with no compressive steel being used. If compressive steel were to be added, however, increasing the total steel area by 60%, what would be the percent increase in moment capacity of this section? Based on your results, what conclusion can be drawn concerning the use of compressive steel when the tensile steel ratio is kept less than $0.75\rho_b$?
12. For the beam in Question 11, compare the long-term deflections of the beam with no compressive reinforcement to the deflections of the beam with

100% compressive reinforcement. (Use the deflection factor $\lambda$ presented in Chapter 3.)

13. Under what conditions can the maximum tensile steel ratio $0.75\rho_b$ be exceeded?

14. If the maximum tensile steel ratio $0.75\rho_b$ is exceeded, why must the excess tensile steel area be compensated by an equal area of compressive steel?

# 7

# BALANCED FLEXURAL DESIGN

To this point, the discussions of flexure in concrete have been centered around two independent concepts: elastic stress at service levels of load and inelastic forces at ultimate levels of load. There has been no effort so far to link together these two concepts and to find how one influences the other. They are, of course, intimately interrelated; their interrelationship is one topic to be examined in this chapter.

A second topic to be examined in this chapter is the design of members to achieve something approaching optimum efficiency, both at elastic levels of stress and at ultimate levels of load, when the interrelationships between the two are considered. It will be seen that optimum efficiency can be achieved only in certain types of beams; in others one must simply live with inefficient stress levels in the materials.

## Balanced Stress Condition

By definition of the plastic section modulus $Z_c$, the nominal ultimate moment $M_n$ is computed by

$$M_n = 0.85 f_c' Z_c \qquad (7\text{-}1)$$

Similarly, by definition of the elastic section modulus $S_c$, the service moment $M_{sv}$ is computed by

$$M_{sv} = f_{sv}S_c \qquad (7\text{-}2)$$

where $M_{sv} = M_n/1.7$ and $f_{sv}$ is the service stress corresponding to this value of $M_{sv}$.

Eq. (7-1) is divided by Eq. (7-2), yielding

$$\frac{M_n}{M_{sv}} = \frac{0.85f_c'Z_c}{f_{sv}S_c} \qquad (7\text{-}3)$$

It is recognized that $M_n/M_{sv} = 1.7$, hence

$$\text{service } f_{sv} = 0.5f_c'\frac{Z_c}{S_c} \qquad (7\text{-}4)$$

It is noted immediately that the service stress $f_{sv}$ defined by Eq. (7-4) is an invariant property of the section, dependent only on section properties and materials properties. It is the elastic stress that will occur in the concrete at 60% of the ultimate moment (or, more exactly, at $M_n/1.7$). The service stress $f_{sv}$ thus provides the link between service levels of stress and ultimate levels of load. For reference, the service stress $f_{sv}$ for various steel ratios, steel grades, and concrete strengths has been computed and included in Tables A-5, A-6, and A-7 of the Appendix.

With this addition of $f_{sv}$, the derivation of Tables A-5, A-6, and A-7 is now complete. Each line of these tables defines a valid and usable section under the Code. The tables may be entered with a known value for the ratio $f_s/f_c$ or with a desired value for the service stress $f_{sv}$. All other properties, both elastic and plastic, may then be read from the line so selected.

The stress ratio $f_s/f_c$ is now deliberately chosen at the balanced stress condition presented earlier; that is,

$$\frac{f_s}{f_c} = \frac{f_y}{0.85f_c'} \qquad (7\text{-}5)$$

The elastic section constants $S_c$, $k_{sv}$, and $y_{sv}$ corresponding to this balanced stress condition have already been introduced; these elastic constants have been computed and tabulated in a special table, Table A-8, of the Appendix. To this special table, the plastic section constants $Z_c$, $k_n$, and $y_n$ can now be added, where the steel ratio $\rho$ is that found earlier with the elastic constants. With both $Z_c$ and $S_c$ thus known, the service stress $f_{sv}$ for the balanced stress condition has been found from Eq. (7-4) and listed, thereby completing the special table, Table A-8.

## Transition from Elastic to Plastic Flexure

For the balanced stress condition, the transition from elastic behavior to plastic behavior can now be traced. As indicated in Figure 7-1 a and b, the stresses remain elastic throughout all the lower values of moment, up to the transition moment $M_{TR}$. At the transition point, the concrete and the steel are simultaneously on the verge of yield. At this point, the maximum elastic resistance to moment has been developed; any further rotation will produce inelastic strains.

As rotations increase further, the inelastic strains begin; both concrete and steel enter yield. At that point the arm $jd$ begins to increase slightly, which provides a very small increase in moment capacity. Thereafter, a small but steady increase in $jd$ (and moment capacity) will accompany any further inelastic rotations. The inelastic rotations end when the strain in the concrete reaches 0.003. At that point, as indicated in Figure 7-1c, the ultimate moment $M_n$ has been reached and the arm $jd$ is at its maximum, with $M_n = f_y A_s jd$.

Alternatively, the transition may be traced by plotting the rotation of

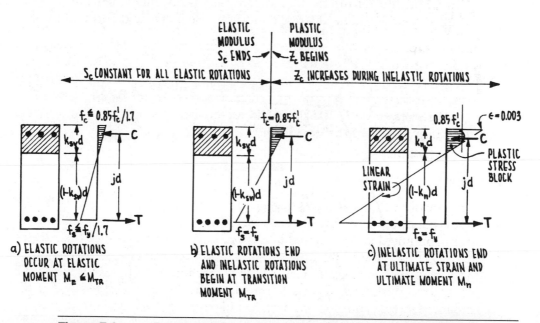

**Figure 7-1**     Rotations of a section under balanced stress conditions.

the section against the corresponding moment. As shown in Figure 7-2, rotations of the section remain elastic up to the transition point. As rotations progress past the transition point, both concrete and steel simultaneously enter yield. As yield in the concrete continues to increase, the arm $jd$ also increases slightly, adding a small increase in moment capacity. Moment capacity continues to increase slightly until the maximum concrete strain of 0.003 in./in. is reached.

In Figure 7-2, a comparison between the transition moment $M_{TR}$ and the ultimate moment $M_n$ can be drawn using the relative values of Table A-8. The maximum elastic moment is $M_{TR}$, at which point the stress in both steel and concrete is just at yield, for which case, $M_{TR} = 0.85f'_c S_c$. For comparison, the ultimate moment is $M_n$, for which case $M_n = 0.85f'_c Z_c$. The difference between these two moments is a direct consequence of the difference between $Z_c$ and $S_c$. From Table A-8, it is seen that for the balanced stress condition, $Z_c$ and $S_c$ are always very close to each other; their difference, in fact, is always less than 4%. As indicated in Figure 7-2, therefore, the transition moment $M_{TR}$ will always be within 4% of the ultimate moment $M_n$.

It is concluded from the foregoing observations that when a member is designed for the balanced stress condition, the inelastic deformations will

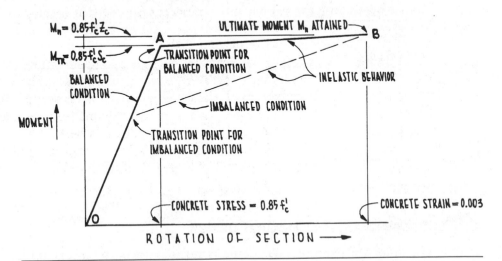

**Figure 7-2**    Moment versus rotation at the balanced stress condition.

begin just as the concrete and steel simultaneously reach their yield stresses. At that point, more than 96% of the resistance to flexural load will have been attained. A member designed for the balanced stress condition will therefore behave elastically at all levels of load up to some 96% of its ultimate load, at which point the member will (idealistically) begin its final inelastic deformations. Thereafter, rotations will increase rapidly and inelastically to the final collapse of the member.

It is observed also that the use of the balanced stress condition in combination with the ACI strength method of design will incorporate all the best features of both ultimate load design and elastic stress design. A member so designed will perform smoothly and elastically all the way through its working levels of load up to its elastic limits (or yield stresses), at which point it will undergo inelastic deformations smoothly and predictably to ultimate collapse. There is no early failure, no premature inelastic behavior, and there will be less than 4% wasted residual strength.

If, however, the service stress $f_{sv}$ is chosen at any other level than at this balanced stress condition, there will be an "imbalance" between the elastic margins for the steel and the concrete. One material will necessarily enter yield before the other. As loads are increased above this level, the member will undergo some sort of nonlinear rotations as it progresses on up to ultimate load; at that point the elastic values used to compute the design moment $M_n$ will no longer be valid. It is important to note, however, that *Code does not require a balanced state of stress*, and that insofar as Code is concerned, the only disadvantage to using such an imbalance in the design is in the inefficient use of materials.

A final observation can be drawn from Table A-8 concerning the service stress $f_{sv}$. It is noted that the service stress is, in all cases, just slightly more than $0.5f'_c$; the maximum variation from this norm is seen to be less than 4%. With only a small error, therefore, the nominal service stress $f_{sv}$ under balanced stress conditions can be thought of as being 60% of the idealized yield stress of concrete, or $f_{sv} = 0.6(0.85f'_c)$. Similarly, the corresponding service stress in the steel will also be roughly 60% of yield, or $f_s = 0.6f_y$.

## Applications of Balanced Flexural Design

The following examples will demonstrate the use of the special table, Table A-8, in the design of concrete members.

### EXAMPLE 7-1.

Design of a rectangular section in flexure. No limitation on dimensions.

**Given:**

$$M_{DL} = 36 \text{ kip-ft}, M_{LL} = 42 \text{ kip-ft}$$
$$f_c' = 3000 \text{ psi, grade 60 steel}$$

**Find:**

Suitable section to sustain the load

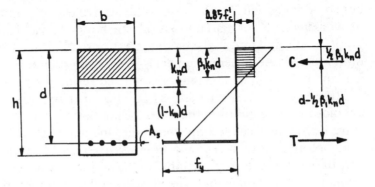

**Solution:**

Calculate the design moment at ultimate load

$$M_n = (1.4M_{DL} + 1.7M_{LL})/\phi = (1.4 \times 36 + 1.7 \times 42)/0.9$$
$$= 135 \text{ kip-ft}$$

Determine the magnitude of the required plastic section modulus $Z_c$,

$$Z_c = \frac{M_n}{0.85f_c'} = \frac{135,000 \times 12}{0.85 \times 3000} = 635 \text{ in.}^3$$

Select the service stress and section constants for the balanced stress condition (both the steel and the concrete will then have the same elastic margin to yield).

From Table A-8, no compressive reinforcement,

use $\rho = 0.0060$, $f_{sv} = 1539$ psi, $Z_c = 0.132bd^2$, $f_s/f_c = 23.53$

Solve for minimum required sizes,

$Z_c = 635 = 0.132bd^2$

Assume that $b = 0.6d$, then

$635 = 0.132 \times 0.6d^3$,

Solve for $d = 20.0$ in., $b = 12.0$ in.

Steel area $= \rho bd = 0.006bd = 0.006 \times 12.0 \times 20.0 = 1.44$ in.$^2$

From Tables A-3 and A-4,

use 5 No. 5 bars, as furnished $= 1.53$ in.$^2$

Dimensions are now rounded up to even inches or half inches,

Use $b = 12.0$ in., $d = 20.0$ in., $A_s = 5$ No. 5 bars.

The final section is shown in the following sketch.

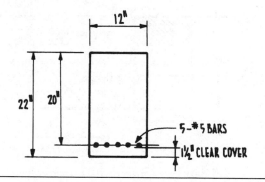

Without making any further investigations, it is immediately known that the beam section selected in Example 7-1 will develop a concrete stress of 1539 psi when subjected to its service moment of $M_n/1.7$ or 79.4 kip-ft. The stress in the grade 60 steel under that moment will be $1539(f_s/f_c)$ or 36,200 psi. Further, the beam will have the same margins to elastic yield for both the steel and the concrete. And finally, the beam will perform elastically throughout its entire load-carrying range from zero moment all the way up to 96% of its ultimate moment. Inherently, when a section is selected at its balanced stress condition, there is no question what its service performance will be.

On occasion, it becomes necessary to limit the overall depth of a beam. When there is no corresponding limit on the width $b$, one way to design the member is to make it wider and flatter. Such a case is considered in the next example.

### EXAMPLE 7-2.

Design of a rectangular section in flexure. Depth of section limited.

**Given:**

$M_{DL} = 37$ kip-ft, $M_{LL} = 25$ kip-ft

$f'_c = 4000$ psi, grade 50 steel

Depth $h$ limited to $16\frac{1}{2}$ in.

Interior exposure

**Find:**

Suitable section to sustain the load

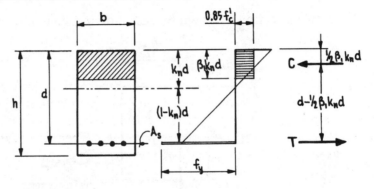

With overall depth $h$ limited, it is necessary to estimate the maximum allowable value for $d$, as shown in the following sketch.

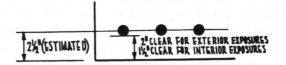

**Solution:**

Estimate maximum allowable $d = h - 2\frac{1}{2}$ in. = 14 in.

Calculate the design moment at ultimate load

$$M_n = (1.4M_{DL} + 1.7M_{LL})/\phi = (1.4 \times 37 + 1.7 \times 25)/0.9$$
$$= 105 \text{ kip-ft}$$

Determine the magnitude of the required plastic section modulus $Z_c$,

$$Z_c = \frac{M_n}{0.85f'_c} = \frac{105,000 \times 12}{0.85 \times 4000} = 371 \text{ in.}^3$$

Select the service stress and section constants to provide a balanced stress condition.

From Table A-8, No Compressive Reinforcement,

Try $\rho = 0.0120$, $f_{sv} = 2067$ psi, $Z_c = 0.161bd^2$, $f_s/f_c = 14.71$

Solve for required minimum sizes

$Z_c = 371 = 0.161bd^2$

For $d = 14$ in., solve for $b = 11.8$ in.

Steel area $= \rho bd = 0.0120 \times 11.8 \times 14 = 1.98$ in.$^2$

From Tables A-3 and A-4,

Use 2 No. 9 bars, $A_s$ furnished $= 2.0$ in.$^2$

Use $b = 12.0$ in., $d = 14$ in., $h = 16\frac{1}{2}$ in., 2 No. 9 bars.

Note that $b$ and $d$ are used at their exact values for computing $A_s$, but are rounded off to the nearest inch or $\frac{1}{2}$ inch after all calculations are complete. Note also that this section is wider than would ordinarily be desirable, but the restriction on $d$ will not permit a narrower section to be used unless compressive reinforcement is added.

---

It is again reassuring to note that due to the use of the balanced stress condition, the concrete in the beam of Example 7-2 will work at an elastic stress of 2067 psi when the moment on the section is 105 kip-ft/1.7 or 61.8 kip-ft. The stress in the steel at that point will be $f_{sv}(f_s/f_c)$ or 30,400 psi. Further, both the steel and the concrete will be elastic all the way to 96% of the ultimate moment of 105 kip-ft.

It is emphasized that in selecting section constants from the tables of the Appendix, all sections listed in Tables A-5, A-6, and A-7 are usable and are within all requirements of the Code. It is simply good practice to avoid the use of compressive steel (where possible) and to keep the section balanced for its elastic margins. Nonetheless, in Example 7-2 a section having a steel ratio of 2.75% (see Table A-6) and its accompanying theoretical service stress of 3200 psi, or $0.80f'_c$ is an acceptable section under the Code.

On occasion, both the depth $d$ and width $b$ of a concrete section may be limited by outside constraints. Such a case is considered in the next example.

### EXAMPLE 7-3.

Design of a rectangular section in flexure; both depth $d$ and width $b$ limited.

**Given:**

$M_{DL} = 62$ kip-ft, $M_{LL} = 35$ kip-ft

$f'_c = 3000$ psi, grade 40 steel

Overall $h$ limited to 20 in.

Width $b$ limited to 12 in.

Exterior exposure, ½ in. ties

**Find:**

Suitable section to sustain the load

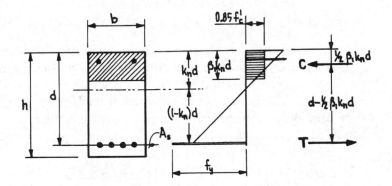

With overall height $h$ limited to 20 in., it becomes necessary to estimate the maximum effective depth $d$, as shown in the following sketch.

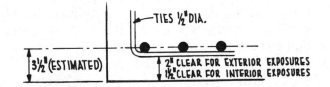

**Solution:**

The maximum allowable value for $d$ is estimated at 20 in. − 3½ in. = 16½ in.

Calculate the design moment at ultimate load.

$$M_{sv} = (1.4M_{LL} + 1.7M_{LL})\phi = (1.4 \times 62 + 1.7 \times 35)/0.90$$
$$= 163 \text{ kip-ft}$$

Determine the magnitude of the required plastic section modulus $Z_c$,

$$Z_c = \frac{M_n}{0.85f_c'} = \frac{163,000 \times 12}{0.85 \times 3000} = 767 \text{ in.}^3$$

Since both $b$ and $d$ are known in this case, the coefficient of the plastic section modulus can be computed directly, where $Z_c = \text{coeff.} \times bd^2$,

$$Z_c = 767 = \text{coeff.} \times 12 \times 16.5^2$$
$$\text{coeff.} = 0.235$$

It is now required to find in Table A-8 a section having a coefficient of $Z_c$ at least 0.235. It is seen immediately that a section without compressive reinforcement will not work; the coefficient for that case is only 0.169. Further, providing compressive steel in the amount of 20% of tensile steel is still inadequate; its coefficient is only 0.197. But for a compressive steel ratio of 40% of tensile steel, the section constants are

$$\rho = 0.0169, \ Z_c = 0.238bd^2, \ f_{sv} = 1541 \text{ psi}, \ f_s/f_c = 15.69$$

The result is verified:

$$M_n = 0.85f_c'Z_c = 0.85 \times 3000 \times 0.238 \times 12 \times 16.5^2/12$$
$$= 165 > 164 \text{ kip-ft (O.K.)}$$

The area of tensile reinforcement is selected:

$A_s = \rho bd = 0.0169 \times 12 \times 16.5 = 3.35 \text{ in.}^2$,

Use 4 No. 9 bars in tension

$A_s' = 0.40 \ A_s = 0.40 \times 3.35 = 1.34 \text{ in.}^2$,

Use 2 No. 8 bars in compression.

Use $b = 12$ in., $d = 16.5$ in., $h = 20$ in.

4 No. 9 bars in tension

2 No. 8 bars in compression

The final cross section is similar to that shown at the beginning of this example.

If it had not been a requirement to design for the balanced stress condition, the section of Example 7-3 could have been designed to sustain the moment $M_n$ with no compressive reinforcement at all. For such a design, the required tensile steel ratio is found from Table A-5 to be 0.0174, producing a required steel area of 3.45 in². The earlier choice of 4 No. 9 bars for the balanced stress condition is thus seen to be adequate even without the addition of compressive reinforcement. The conclusion is drawn that the inclusion of the two No. 8 bars in compression adds very little to the moment capacity of the section.

It is further noted, however, that the addition of the 2 No. 8 bars in Example 7-3 reduces the service stress in the concrete from 1920 psi (or $0.64f'_c$) when no compressive reinforcement is used, down to 1541 psi (or $0.51f'_c$) if compressive reinforcement is used. Further, the long-term deflections of the beam will be reduced by some 33% if compressive reinforcement is added (see factor $\lambda$, Chapter 3). As noted earlier in Chapter 6, it is again observed that the addition of compressive steel adds little to the moment capacity when the tensile steel ratio is less than $0.75\rho_b$; the primary benefit derived from such reinforcement is in the reduction of the day-to-day service stresses in concrete and in the reduction of long-term deflections.

When a section is being selected from the tables of the Appendix, the selection is made (usually) on the basis of the conditions at ultimate load. The section constants at service levels are immediately available in the tables, however, which permits routine checks on deflections to be made quite readily. Some examples will illustrate such calculations.

**EXAMPLE 7-4.**

Design of a rectangular section in flexure. No limitations on dimensions. Deflections are to be computed.

**Given:**

$M_{DL}$ = 48 kip-ft, $M_{LL}$ = 56 kip-ft

Uniformly loaded simple span, $L$ = 24′ 0″

$f'_c$ = 3000 psi, grade 60 steel

Deflections limited to $L/240$ for $DL + LL$

Deflections limited to $L/360$ for $LL$ only.

**Find:**

Suitable section to sustain the load

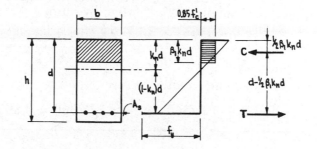

**Solution:**

Calculate the design moment at ultimate load

$$M_n = (1.4M_{DL} + 1.7M_{LL})/\phi = (1.4 \times 48 + 1.7 \times 56)/0.9$$
$$= 180 \text{ kip-ft}$$

Determine the magnitude of the required plastic section modulus $Z_c$,

$$Z_c = \frac{M_n}{0.85f_c'} = \frac{180,000 \times 12}{0.85 \times 3000} = 847 \text{ in.}^3$$

Select the service stress and design constants for the balanced stress condition

From Table A-8, No Compressive Reinforcement,

Use $\rho = 0.0060$, $Z_c = 0.132bd^2$, $f_{sv} = 1539$ psi, $k_{sv} = 0.283$

Solve for minimum required sizes

$Z_c = 847 = 0.132bd^2$

Assume $b = 0.6d$, then $d = 22.03$ in., $b = 13.22$ in.

Steel area $= 0.006 \times 13.22 \times 22.03 = 1.75$ in.$^2$

From Tables A-3 and A-4 use 3 No. 7 bars.

Check deflections for simple span [see Ch. 3, Eq. (3-3)] at maximum level of service loads,

$$\Delta_{sv} = \frac{5f_{sv}L^2}{48E_c k_{sv}d} = \frac{5 \times 1539 \times (24 \times 12)^2}{48 \times 3,100,000 \times 0.283 \times 22.03}$$

$$= 0.69 \text{ in.}$$

At full service load, $DL + LL$, allowable $\Delta_{sv}$ is:

Allow $\Delta_{sv} = L/240 = (24 \times 12)/240 = 1.20 > 0.69$ in. (O.K.)

Under live load only, $\Delta_{LL}$ is found by ratio,

$$\Delta_{LL} = \frac{1.7 M_{LL}}{1.4 M_{DL} + 1.7 M_{LL}} \times \Delta_{sv} = \frac{1.7 \times 56}{1.4 \times 48 + 1.7 \times 56} \times 0.69 = 0.40 \text{ in.}$$

For live load only, allowable $\Delta$ is

Allow $\Delta_{LL} = L/360 = (24 \times 12)/360 = 0.80 > 0.40$ in. (O.K.)

---

## EXAMPLE 7-5.

Verification of deflections.

### Given:

Beam design of Example 7-3
Uniformly loaded simple beam, span = 20′ 0″

### Find:

Verify that deflection is less than $L/240$ for combined $DL + LL$ loading

### Solution:

The section constants are, for $\rho = 0.0169$

$\rho' = 0.40\rho$, $Z_c = 0.238bd^2$, $f_{sv} = 1541$ psi
$k_{sv} = 0.372$, $f_s/f_c = 15.69$

The maximum service deflection is, for $d = 16.5$ in.,

$$\Delta_{sv} = \frac{5 f_{sv} L^2}{48 E_c k_{sv} d} = \frac{5 \times 1541 \times (20 \times 12)^2}{48 \times 3,100,000 \times 0.372 \times 16.5}$$

$$= 0.486 \text{ in.}$$

Maximum allowable deflection is given by $L/240$,

Allow $\Delta = L/240 = (20 \times 12)/240 = 1.0 > 0.486$ (O.K.)

It is concluded that the computed deflection is within allowable limits.

## OUTSIDE PROBLEMS

For the following conditions, select a suitable rectangular concrete section and its reinforcement at the balanced stress condition. Provide 1.5-in. cover over the reinforcement. For the chosen section, indicate the stress in the concrete and steel under service conditions.

| Prob. No. | Concrete Strength $f'_c$ psi | Steel Grade | Limits on Width $b$ | Limits on Height $h$ | Dead Load Moment kip-ft | Live Load Moment kip-ft | E'quake Moment kip-ft | Wind Moment kip-ft |
|---|---|---|---|---|---|---|---|---|
| 7.1 | 3000 | 40 | None | None | 30 | 21 | 13 | 10 |
| 7.2 | 3000 | 50 | None | None | 59 | 72 | 40 | 44 |
| 7.3 | 3000 | 60 | None | None | 32 | 25 | 15 | 13 |
| 7.4 | 4000 | 40 | None | None | 56 | 65 | 37 | 38 |
| 7.5 | 4000 | 50 | None | 14 in. | 35 | 28 | 15 | 21 |
| 7.6 | 4000 | 60 | None | 14 in. | 52 | 58 | 27 | 30 |
| 7.7 | 5000 | 40 | None | 14 in. | 38 | 32 | 17 | 18 |
| 7.8 | 5000 | 50 | None | 14 in. | 49 | 54 | 33 | 32 |
| 7.9 | 5000 | 60 | 10 in. | 16 in. | 40 | 37 | 20 | 24 |
| 7.10 | 4000 | 40 | 14 in. | 16 in. | 47 | 49 | 33 | 30 |
| 7.11 | 4000 | 50 | 10 in. | 16 in. | 43 | 41 | 24 | 28 |
| 7.12 | 4000 | 60 | 12 in. | 16 in. | 45 | 45 | 25 | 28 |

Select a suitable rectangular concrete section under balanced stress conditions for the given simply supported beams, to include required reinforcement. Then check the maximum deflection (by computation) that will occur under service load conditions. Allowable deflection = $L/240$.

| Prob. No. | Uniform Load kips/ft | | Span ft | Concrete $f'_c$ psi | Steel Grade |
|---|---|---|---|---|---|
| | $w_{DL}$ | $w_{LL}$ | | | |
| 7.13 | 1.9 | 1.8 | 26 | 3000 | 40 |
| 7.14 | 2.2 | 2.8 | 18 | 4000 | 50 |
| 7.15 | 1.8 | 2.1 | 24 | 5000 | 60 |
| 7.16 | 2.1 | 2.6 | 18 | 3000 | 60 |
| 7.17 | 2.0 | 2.3 | 22 | 4000 | 40 |
| 7.18 | 1.9 | 2.4 | 20 | 5000 | 50 |

## Underreinforced and Overreinforced Sections

The balanced stress condition provides the line of separation between two distinct cases of reinforcement. A very lightly reinforced section will have a steel ratio ρ lower than the steel ratio at the balanced stress condition; such sections are classified here as underreinforced. At the other extreme, a heavily reinforced section will have a steel ratio ρ higher than the steel ratio at the balanced stress condition; such sections are classified here as over-reinforced. These two cases of reinforcement merit further examination; they exhibit very different patterns of performance under yield conditions.

In an underreinforced section, the steel will enter yield before the concrete does. Typical strain conditions for an underreinforced section are shown in Figure 7-3. At service levels of loading, the strains are those indicated by line $AE$, falling somewhere between $00'$ and $BF$, with the position of the neutral axis remaining fixed at the position indicated by $k_{sv}$.

As loads increase, the service strains indicated by line $AE$ in Figure 7-3 rotate up to line $BF$. As the steel approaches yield, line $BF$, the stress in the concrete can be quite low, as low as $0.75 f'_c$ or lower. As loads continue to increase, the steel enters yield. Once the steel enters yield, rotations of the section proceed more rapidly but the stress in the steel remains constant at $f_y$, regardless of how much the rotations increase. As loading continues further, the steel continues to yield and the strains rotate past line $BF$ on up to the line $CG$, with the neutral axis shifting upward to the position indicated as $k_y$. When strains reach line $CG$, the stress in the concrete has also

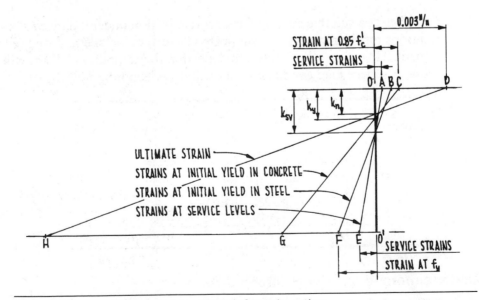

**Figure 7-3**    Strains in an underreinforced section.

reached yield. As loads increase and rotations increase even further, the neutral axis continues to shift upward until the strain in the concrete eventually reaches its ultimate value of 0.003 in./in., at which point the neutral axis is at the position indicated $k_n$. At that point, the section has reached its ultimate capacity.

In an underreinforced section, then, it is seen that the increase in rotations of the section beyond initial yield will occur primarily due to the unrestricted deformations of the steel. As the rotations increase beyond the initial yield in the steel, the value of $k$ decreases and the neutral axis shifts upward. As loads continue to increase, the ever-increasing rotations produce an increase in stress (and strain) in the concrete until the concrete eventually reaches its ultimate strain of 0.003 in./in. At that point, the section is at its ultimate capacity.

An overreinforced section will also undergo inelastic rotations similar to those just described for an underreinforced section, except that in an overreinforced section it is the strain in the concrete that reaches yield first. Strains in a typical overreinforced section are shown in Figure 7-4.

At service levels of loading, the strains in an overreinforced section are those indicated in Figure 7-4 by line $AE$, falling somewhere between line $00'$ and line $BF$, with the neutral axis being fixed at the position indicated

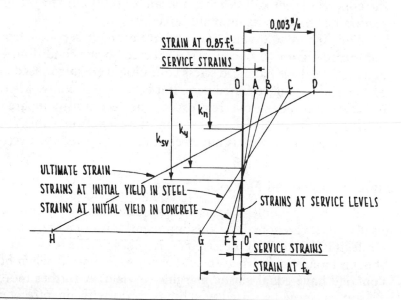

**Figure 7-4**    Strains in an overreinforced section.

by $k_{sv}$. As loads increase, the strains rotate up to line *BF*, at which point the stress in the concrete reaches yield but the stress in the steel remains below yield. As load increases further and strains increase beyond the line *BF*, the concrete enters yield and rotations of the section thereafter proceed more rapidly. In this range of rotations, the location of the neutral axis may actually shift downward slightly as the stress in the steel continues to climb toward yield. As load continues to increase and the concrete continues to yield, rotations eventually reach the position indicated by line *CG*, at which point the steel also enters yield. Thereafter, the rotations proceed even more rapidly to the final strain condition indicated by line *DF*. At this point, the section has reached its ultimate capacity. (It is the limit on the maximum allowable area of tensile reinforcement that assures that the steel will enter yield before the concrete reaches its ultimate strain of 0.003 in./in.)

In an overreinforced section, then, it is seen that the increase in rotations of the section beyond initial yield will occur primarily due to the unrestricted deformations in the concrete. As loads continue to increase, rotations increase further until the stress in the steel also reaches yield, with the position of the neutral axis shifting to the equilibrium position. Thereafter, rotations proceed much like those in an underreinforced section until the concrete eventually reaches a strain of 0.003 in./in. At that point, the section has reached its ultimate capacity.

The line that divides the underreinforced sections from the overreinforced sections is of course the balanced stress condition. The rotations of a section at the balanced stress condition were introduced and discussed earlier and are not repeated here. If it is desired to know whether a section is underreinforced or overreinforced, one need only compare its steel ratio with that of the balanced stress condition given in Table A-8. If its steel ratio is less, the section is underreinforced; if more, it is overreinforced.

## Discussion of Balanced Flexural Design

In all of the examples of this chapter, the balanced stress condition was used as if it were somehow a requirement. It is not a requirement under the Code; it is, however, a preferred design choice. The reason given earlier for this preference concerned the efficiency of materials when both steel and concrete have equal elastic margins to yield. A further more compelling reason can now be stated which has to do with the effects of premature yield on the structural analysis.

Methods of structural analysis in current use often utilize the elastic rotations of the concrete cross section as a means to compute service moments at a support. (For those already familiar with structural analysis, such methods include slope-deflection, moment distribution, three-moment equation, etc.) It is the moments calculated from these elastic methods of analysis that will be projected upward to find the ultimate load for which the section will be designed; their accuracy must be assured if the design moments are to be valid.

It was noted in the discussions involving Figure 6-1, however, that when the stress in either the steel or the concrete enters yield, the rotation is no longer linear; even a small increase in moment can produce inordinately large rotations. To preserve the accuracy of the analysis, it is therefore necessary to prevent any premature entry into yield by either of the materials. The use of the balanced stress condition does indeed preclude any possibility of such premature yield since both materials enter yield at exactly the same time. The balanced stress condition is therefore a recommended choice as a means to prevent such premature yield.

In addition to such theoretical considerations, it is of practical importance to choose concrete sections that fall generally in the more economically-sized ranges. At one extreme, the use of large, lightly reinforced sections will produce inordinately heavy and expensive structures with excessively high story heights. At the other extreme, the use of small, heavily reinforced sections can produce members that are comparatively expensive and so small that deflections become troublesome. Over the years, it has been found that there are certain ranges of sizes between these extremes that produce economical, serviceable designs. With their efficient proportions of materials, concrete members selected at their balanced stress condition have been found to fall within these economical ranges of sizes.

A further reason to use the full range design method concerns, very simply, designer preference. It has been noted repeatedly that ACI does not require any direct control over elastic stresses in its strength method of design. Even so, it should be noted that at the higher steel ratios, the day-to-day service stresses listed in Tables A-5 through A-7 of the Appendix can approach the yield stress of the concrete. A designer who is inclined to worry may begin to feel uncomfortable with a structure in which stresses are allowed to approach $0.85f'_c$ under day-to-day service loads and such a designer may wish to control these stresses. The author is one such worrier who chooses to control these service stresses.

## REVIEW QUESTIONS

1. How can the service stress $f_{sv}$ be called the "link" between ultimate levels of load and elastic levels of stress?

2. A section is designed without compressive reinforcement at the balanced stress condition; $f'_c$ = 4000 psi, steel is grade 60. What is the day-to-day service stress in the concrete?

3. A section without compressive reinforcement is designed for ultimate strength but not for balanced stress conditions; $f'_c$ = 3000 psi and steel is grade 60. A steel ratio of 0.016 is selected. What is the day-to-day service stress in the concrete? What percentage of ultimate strength is this service stress? What can be concluded about the use of the higher values of steel ratios?

4. At the balanced stress condition, about how much margin of strength is left when the steel and concrete first enter yield?

5. At the balanced condition, what is the nominal day-to-day service stress in the concrete, expressed as a percentage of $f'_c$.

6. At the balanced stress condition, what is the nominal day-to-day service stress in the tensile reinforcement, expressed as a percentage of the yield stress $f_y$?

7. When the steel ratio $\rho$ of a particular beam is less than that at the balanced stress condition, which material will enter yield first, the concrete or the steel? How is it that the beam does not collapse completely at this point?

8. When the steel ratio of a particular beam is more than that at the balanced stress condition, what happens to the stress in the steel after the concrete enters yield?

9. For overreinforced beams, what feature in the design assures that the tensile steel will reach yield before the concrete reaches a strain of 0.003 in./in.?

10. Why is the balanced stress configuration inherently an economical configuration?

11. What purpose is served in using compressive reinforcement at the balanced stress condition?

12. Under what conditions should the use of compressive steel be considered when the tensile steel ratio is less than $0.75\rho_b$?

13. Under what conditions should the use of tensile steel in excess of $0.75\rho_b$ be considered?

14. What penalty is encountered in using heavily loaded compressive reinforcement in beams?

# 8

# ELEVATED FLOOR
# AND ROOF SLABS

Elevated concrete floor slabs and roof slabs are so common in today's industry that it would be difficult to imagine modern construction without them. There is another type of concrete floor slab, the slab on grade, which is supported directly by the underlying soil; slabs on grade are discussed in Chapter 14 along with foundations.

Elevated slabs act as wide flat beams. They are analyzed and designed as rectangular beams, one-foot wide. Steel is selected for this typical one-foot strip and is then used at regular spacing throughout the width of the slab, as shown in Figure 8-1.

Where negative moments occur over supports (producing tension on top), the reinforcement is placed at the top of the slab. At midspan, where moments are positive (producing tension on bottom), the reinforcement is placed at the bottom of the slab. A typical longitudinal section is included in Figure 8-1, showing typical locations for the positive and negative reinforcement.

Reinforcing bars are placed at a fixed spacing across the width of the slab. For the most commonly used spacings and bar sizes, steel areas per foot of width are tabulated in Table A-2. Bars may be no closer together than one bar diameter (clear distance), and no farther apart than three times the slab thickness $h$, with an absolute minimum clear distance of 1 in. and an absolute maximum spacing of 18 in.

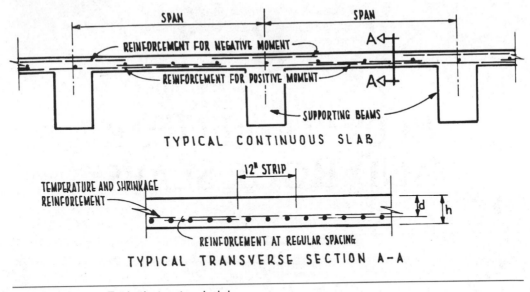

**Figure 8-1** Typical structural slab.

In American practice, slab thicknesses are varied in ½-in. increments. Forms, screeds, and accessories are manufactured in these dimensions and the tradition is so strong that change seems unlikely. In SI units, increments of 10 mm is common; reinforcement cannot reasonably be placed to closer tolerances than 10 mm.

Because the thickness of slabs is small, the reinforcement in slabs is similarly limited to smaller sizes. Bar sizes larger than No. 6 are not generally used in slabs; No. 4 is probably the most commonly used size, with No. 6 a close second. Bar sizes less than No. 4 are not used as structural reinforcement but may be used as temperature steel.

Slabs are almost never reinforced for compression. If a slab is made so thin that compressive reinforcement becomes necessary, the problems with deflections become almost insurmountable. It is far easier, cheaper, and more practical to add ½ in. to the depth.

Where the shape of a slab is square, or nearly so, considerable savings in both concrete and steel can be effected by designing the slab for flexure in two directions. In such "two-way" slabs, part of the load is assumed to be carried in one direction, the remaining part in the other direction. The ACI Code includes a special section on the design of two-way slabs.

**Table 8-1**   Minimum Steel Ratios in Slabs for Temperature and Shrinkage Reinforcement

| Steel Grade | $\rho$ |
|---|---|
| 40 | .0020 |
| 50 | .0020 |
| 60 | .0018 |

Very often, the area of steel in a slab is governed by minimum requirements for temperature and shrinkage reinforcement. Such minimum steel requirements are discussed in Chapter 3; they are presented again in tabular form in Table 8-1 for immediate reference.

In addition to requiring that minimum areas of reinforcement be provided in the primary direction of stress, Code also requires that the same minimum area of reinforcement be provided transverse to the primary direction; such reinforcement is required for temperature/shrinkage stresses regardless whether any flexural stress exists. The minimum steel ratios given in Table 8-1 are therefore applicable in both primary and transverse directions.

The following sections present the design of elevated slabs under simple support conditions. Except for very short spans, the thickness of elevated slabs will be governed by deflections rather than by strength; the minimum thicknesses given in Table 3-1 will govern in such cases. Whenever the thickness of a slab is governed by deflections, the compressive stress in the concrete is invariably low and the balanced stress condition cannot be achieved. For such slabs, one simply accepts the inefficiencies inherent in designing to meet limits on deflections.

## One-way Slabs, Simply Supported

When an elevated slab has simple supports on two edges only, as shown in Figure 8-2, the slab undergoes bending in only one direction and is called a one-way slab. Since every foot of width is identical to every other foot, the slab may be designed for a one-foot width between supports, then the design may be extended laterally as far as support conditions remain unchanged.

An example will illustrate the design of such a simply supported slab.

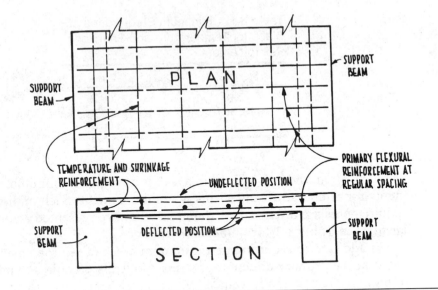

**Figure 8-2**    One-way slab.

## EXAMPLE 8-1.

Design of a concrete floor slab. Simple supports, exposed to weather.

**Given:**

Slab as shown, live load 100 psf

Grade 60 steel, $f'_c = 4000$ psi

Normal-weight concrete, exterior exposure

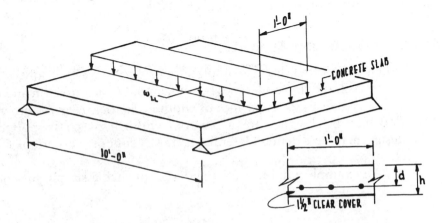

**Find:**

Suitable section to sustain the load

**Solution:**

Calculate the live-load moment. For a 1-ft strip, the uniform live load becomes a distributed load of 100 plf, and

$$M_{LL} = \frac{w_{LL}L^2}{8} = \frac{100 \times 10 \times 10}{8} = 1250 \text{ lb. - ft/ft}$$

For control of deflections, overall slab thickness is limited (see Table 3-1). For this slab, overall thickness is limited to $L/20$ or 6 in.

Estimate thickness: $h$ = 6 in. ±, $d$ = 4 in. ±

Determine the dead load of slab per foot of width:

Dead load  = (150 pcf)(⁶⁄₁₂)

= 75 psf

Find the estimated dead-load moment

$$M_{DL} = \frac{w_{DL}L^2}{8} = \frac{75 \times 10 \times 10}{8} = 938 \text{ lb - ft/ft}$$

Determine the moment at ultimate load

$$M_n = (1.4M_{DL} + 1.7M_{LL})/\phi = (1.4 \times 938 + 1.7 \times 1250)/0.9$$
$$= 3820 \text{ lb-ft} = 45.8 \text{ kip-in.}$$

Solve for the magnitude of the required plastic section modulus,

$$Z_c = \frac{M_n}{0.85f_c'} = \frac{45,800}{0.85 \times 4000} = 13.50$$

Select the service stress $f_c$ and design constants for balanced stress conditions

From Table A-8, with no compressive reinforcement,

$\rho$ = 0.0089, $f_c$ = 2000, $Z_c$ = $0.140bd^2$

Solve for required depth, where $b$ = 12 in.

$Z_c$ = 13.50 = 0.140 × 12 × $d^2$

$d$ = 2.83 in.

The value of $d$ is much less than the 4 in. estimated earlier, indicating that deflection conditions rather than stress conditions will control this design. Obviously, a balanced stress condition will not be possible so a new steel ratio will be selected.

For slabs it can be assumed that, as a general rule, the design condition will be imbalanced and the methods of Example 6-5 can be used to determine the steel area. For such a case, the solution for the section modulus becomes, for $b$ = 12 in. and $d$ = 4 in.,

$$\frac{M_n}{0.85f_c'} = Z_c = \text{coeff} \times bd^2,$$

$$\frac{45,800}{0.85 \times 4000} = \text{coeff} \times 12 \times 4^2$$

$$\text{coeff} = 0.0702$$

For this coefficient of $Z_c$, the steel ratio is found by interpolation in Table A-6 to be $\rho$ = 0.042. For this value of $\rho$,

$$\rho = 0.042, \ f_c = 1352 \text{ psi}, \ Z_c = 0.071bd^2, \ f_s/f_c = 27.29$$

Verify the ultimate moment $M_n$ for this section,

$$M_n = 0.85f_c'Z_c = 0.85 \times 4000 \times 0.071 \times 12 \times 4^2$$
$$= 46,300 > 45,800 \text{ lb-in. (O.K.)}$$

Select reinforcement,

Steel $A_s = \rho bd = 0.0042 \times 12 \times 4 = 0.202$ in.$^2$/ft

From Table A-2, use No. 4 bars @ 10 in. o.c.

Minimum steel ratio for temperature and shrinkage reinforcement is found from Table 8-1,

Min $\rho$ = 0.0018 $bh$ = 0.0018 $\times$ 12 $\times$ 6.0
= 0.13 in$^2$/ft

The primary reinforcement provides 0.42 in$^2$/ft so no additional reinforcement will be necessary in the primary direction. In the transverse direction where there is no primary reinforcement, add No. 4 bars @ 18" o.c., to be placed directly on top of the other reinforcement.

Use $h$ = 6 in., $d$ = 4 in., $A_s$ = No. 4 bars @ 10 in. o.c.

Use No. 4 bars @ 18″ o.c. in the transverse direction, placed on top of the primary reinforcement.

For this slab, the concrete is working at very low stress levels, about 40% of idealized yield, indicating poor efficiency in materials. The design is controlled by deflections, however, and in such cases the reduced efficiency in stresses is simply accepted.

___

The investigation of a known section to find an allowable load is the reverse of the design procedure. The following example will illustrate.

### EXAMPLE 8-2.

Investigation of a section to determine the allowable live load on the given slab. Simple supports, no exposure to weather.

**Given:**

Slab as shown
Grade 50 steel
$f'_c$ = 3000 psi
Simple span, 12 ft

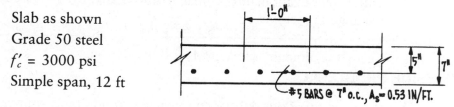

**Find:**

Allowable uniform Live load $w_{LL}$

**Solution:**

Calculate ρ and find the section modulus from tables.

$$\rho = \frac{A_s}{bd} = \frac{0.53}{12 \times 5} = 0.0088; \text{ use } 0.009$$

Select the design constants from Table A-5

$$f_c = 1627 \text{ psi}, \ Z_c = 0.161bd^2$$

Determine the ultimate moment the section can sustain,

$$M_n = 0.85f'_cZ_c = 0.85 \times 3000 \times 0.161 \times 12 \times 5^2$$
$$M_n = 123 \text{ kip-in.} = 10.3 \text{ kip-ft}$$

Determine the dead load and the dead load moment for a simple span:

$$w_{DL} = 150 \text{ pcf} \times \frac{7}{12} = 87.5 \text{ psf}$$

$$M_{DL} = \frac{w_{DL}L^2}{8} = \frac{87.5 \times 12^2}{8} = 1575 \text{ lb-ft}$$

Solve for the live load moment

$$M_n = (1.4M_{DL} + 1.7M_{LL})/\phi$$

$$10300 = (1.4 \times 1575 + 1.7 \times M_{LL})/0.9$$

$$M_{LL} = \frac{0.9 \times 10,300 - 1.4 \times 1575}{1.7} = 4160 \text{ lb-ft}$$

Solve for the uniform live load

$$M_{LL} = \frac{w_{LL}L^2}{8}; \quad 4160 = \frac{w_{LL} \times 12^2}{8}$$

$$w_{LL} = 230 \text{ lb/ft}^2$$

Other investigations are similar to those already discussed with the rectangular sections.

## OUTSIDE PROBLEMS

Design a one-way floor slab for the given loads and simply supported spans in an interior exposure.

| Prob. No. | Concrete Strength psi | Steel Grade | Live Load kips/ft$^2$ | Clear Span Feet |
|---|---|---|---|---|
| 8.1 | 3000 | 40 | 60 | 12 |
| 8.2 | 3000 | 50 | 100 | 12 |
| 8.3 | 3000 | 60 | 150 | 12 |
| 8.4 | 4000 | 40 | 75 | 14 |
| 8.5 | 4000 | 50 | 125 | 14 |
| 8.6 | 4000 | 60 | 200 | 14 |
| 8.7 | 5000 | 40 | 100 | 16 |
| 8.8 | 5000 | 50 | 150 | 16 |
| 8.9 | 5000 | 60 | 200 | 16 |
| 8.10 | 5000 | 40 | 75 | 16 |
| 8.11 | 4000 | 50 | 150 | 14 |
| 8.12 | 3000 | 60 | 300 | 12 |

Determine the amount of live load the given slabs can carry over the given simply supported span.

| Prob. No. | Concrete Strength $f'_c$ psi | Steel Grade | Height $h$ in. | Effective Depth $d$ in. | Reinf. | Span Feet |
|---|---|---|---|---|---|---|
| 8.13 | 3000 | 40 | 6 | 4½ | #5@12 | 10 |
| 8.14 | 4000 | 40 | 6½ | 5 | #6@10 | 12 |
| 8.15 | 5000 | 50 | 7 | 6 | #4@ 8 | 14 |
| 8.16 | 3000 | 50 | 8½ | 7 | #5@10 | 15 |
| 8.17 | 4000 | 60 | 7 | 6 | #4@ 6 | 14 |
| 8.18 | 5000 | 60 | 9 | 7½ | #6@12 | 16 |

## One-way Slabs Continuous Over Several Supports

In many if not most cases in modern construction, an elevated concrete slab will be continuous over three or more supports, as shown in Figure 8-3. Such slabs may be designed as continuous beams, with the design shears and moments being computed from the ACI coefficients presented in Chapter 4. Where negative moments occur at the supports, producing tension on the top face of the slab, the tensile reinforcement is placed at the top face, as shown in Figure 8-3. Similarly, toward the center of the span where the moments are positive, tension occurs at the bottom face of the slab and the reinforcement is placed at the bottom face.

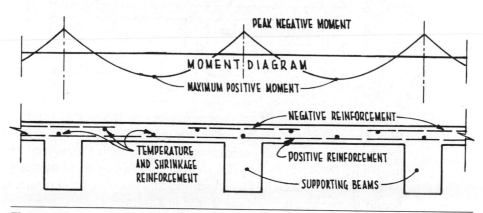

**Figure 8-3**    Continuous one-way slab.

In flexural members, Code permits the reinforcing bars to be cut off when they are no longer needed. There are stringent Code requirements for establishing cutoff points, however, which will assure that the full strength of the bars will be available wherever the bar comes under stress. These anchorage requirements are discussed more fully in Chapter 12, but a few basic requirements are introduced here.

An example will illustrate the design of a continuous one-way slab.

### EXAMPLE 8-3.

Design of a continuous slab.

**Given:**

Continuous slab as shown

Grade 60 steel, $f_c' = 4000$ psi

Normal weight concrete, exterior exposure

Live load 100 psf, clear span 15' 0"

**Find:**

Suitable section in reinforced concrete

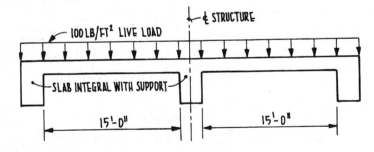

The slab reinforcement will obviously be symmetrical about the center-line. The design can therefore be limited to only one side and repeated for the other side. The coefficients for the design shears and moments are found from the ACI coefficients given in Chapter 4; they are shown in the following sketch. The envelopes and the inflection points are also found using the methods presented in Chapter 4.

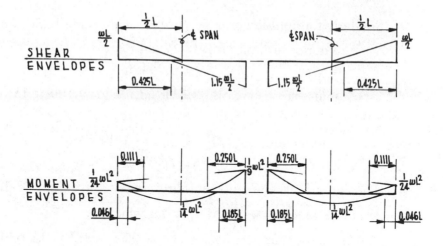

In order to establish the dead load of the slab, a trial thickness of the slab is obtained from the rules of thumb in Chapter 4. A trial thickness of about 7 in. is estimated.

A check is also made for the minimum thickness required to control deflections. From Table 3-1 the minimum thickness is found to be $L/24$ where $L$ is the length of span of a one-way slab. For a span of 15′0″, the minimum thickness is then 7½ in.

### Solution:

From the foregoing estimates, an initial trial thickness of 7½ in. is adopted, for which the dead load is $(7.5/12)(145\#/\text{ft}^3)$ or 90.7 lb/ft². The effective depth $d$ is taken to be 5.5 in.

The maximum design moments are then,

For dead load on a 1-ft strip,

$$\text{At exterior support, } M_{DL} = \tfrac{1}{24}\, wL^2 = \tfrac{1}{24} \times 90.7 \times 15^2$$
$$= 0.85 \text{ k-ft}$$

$$\text{At midspan, } \qquad M_{DL} = \tfrac{1}{14}\, wL^2 = \tfrac{1}{14} \times 90.7 \times 15^2$$
$$= 1.46 \text{ k-ft}$$

$$\text{At interior support, } M_{DL} = \tfrac{1}{9}\, wL^2 = \tfrac{1}{9} \times 90.7 \times 15^2$$
$$= 2.27 \text{ k-ft}$$

Similarly for live load on a 1-ft strip,

At exterior support, $M_{LL} = \frac{1}{24}\, wL^2 = \frac{1}{24} \times 100 \times 15^2$
$$= 0.94 \text{ k-ft}$$

At midspan, $M_{LL} = \frac{1}{14}\, wL^2 = \frac{1}{14} \times 100 \times 15^2$
$$= 1.61 \text{ k-ft}$$

At interior support, $M_{LL} = \frac{1}{9}\, wL^2 = \frac{1}{9} \times 100 \times 15^2$
$$= 2.50 \text{ k-ft}$$

The nominal ultimate moments are computed from Eq. (3-1a),

$M_n = (1.4 M_{DL} + 1.7\, M_{LL})/\phi$, for which,

At exterior support, $M_n = (1.4 \times 0.85 + 1.7 \times 0.94)/0.9$
$$= 3.09 \text{ k-ft}$$

At midspan, $M_n = (1.4 \times 1.46 + 1.7 \times 1.61)/0.9$
$$= 5.31 \text{ k-ft}$$

At interior support, $M_n = (1.4 \times 2.27 + 1.7 \times 2.50)/0.9$
$$= 8.25 \text{ k-ft}$$

Since both the effective depth $d$ and the width $b$ have been established (or estimated), the coefficient of $Z_c$ can be found:

At the exterior support,

$$Z_c = \frac{M_n}{.85 f_c'} = \frac{3090 \times 12}{.85 \times 4000} = 10.90$$

$Z_c = \text{coeff} \times bd^2 = 10.90 = \text{coeff} \times 12 \times 5.5^2$
$\text{coeff} = .030$

From Table A-6, $\rho = .0017$,
$A_s = .0016\ A_s = .0017 \times 12 \times 5.5 = 0.112 \text{ in}^2$

At midspan,

$$Z_c = \frac{M_n}{.85 f_c'} = \frac{5310 \times 12}{.85 \times 4000} = 18.74$$

$Z_c = \text{coeff} \times bd^2; \ 18.74 = \text{coeff} \times 12 \times 5.5^2$
$\text{coeff} = .0516$

From Table A-6, $\rho$ = .0030,

$A_s = \rho bd = .0030 \times 12 \times 5.5 = 0.198$ in²/ft

At the interior support,

$$Z_c = \frac{M_n}{.85 f_c'} = \frac{8250 \times 12}{.85 \times 4000} = 29.12$$

$Z_c = \text{coeff} \times bd^2 \quad 29.12 = \text{coeff} \times 12 \times 5.5^2$
    coeff = 0.0802

From Table A-6, $\rho$ = 0.00475

$A_s = .00475 \quad A_s = .00475 \times 12 \times 5.5 = 0.313$ in.²/ft

Check minimum area requirements for temperature and shrinkage reinforcement:

Min $A_s$ = .0018 bh = .0018 × 12 × 7.5
    = 0.162 in.²/ft.

This minimum $A_s$ is compared to the values of $A_s$ just computed for strength. It is noted that the value of $A_s$ = 0.112 in.²/ft at the exterior support does not meet these minimum requirements. That value of $A_s$ is therefore revised upward to 0.14 in.²/ft.

Transverse reinforcement is set at its minimum value of 0.14 in.²/ft.

The final results are then, from Table A-2,

At exterior support,   $A_s$ = 0.140 in²/ft, use No. 4 @ 16 in. o.c.

At midspan,       $A_s$ = 0.198 in²/ft, use No. 4 @ 12 in. o.c.

At interior support,   $A_s$ = 0.313 in²/ft, use No. 4 @ 7½ in. o.c.

Verify that the thickness is adequate for shear. Maximum shear force occurs at the interior support, for which $V = 1.15\ wL/2$,

$V_{DL}$ = 1.15 × 90.7 × 15/2 = 782 lb/ft

$V_{LL}$ = 1.15 × 100 × 15/2 = 863 lb/ft

$V_n$  = (1.4 $V_{DL}$ + 1.7 $V_{LL}$)/$\phi$ = (1.4 × 782 + 1.7 × 863)/0.85
          = 3.01 k/ft.

The capacity of the section in shear is, with no shear reinforcement,

$$V_n = V_c = 2\sqrt{f_c'}bd = 2\sqrt{4000} \times 12 \times 5.5$$
$$= 8.4 \text{ k/ft} > 3.01 \text{ k/ft load (O.K.)}$$

It is concluded that the section is adequate in shear.

A great deal of latitude may be exercised in selecting and arranging the reinforcement. One solution is given in the following sketch. (Temperature and shrinkage reinforcement is not shown.)

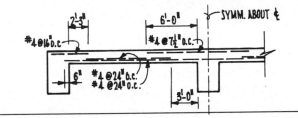

The cutoff points for the reinforcement in Example 8-3 are governed by anchorage requirements. One of those anchorage requirements is that at least ¼ of the positive reinforcement shall extend into the supports by a distance not less than 6 in.; the remaining positive reinforcement shall extend past the point of inflection by 6 in. or by a distance of 12 bar diameters. Another anchorage requirement is that the negative reinforcement shall extend beyond the point of inflection by a distance $d$, 12 bar diameters, or span/16, whichever is larger. These requirements for cutoff points are met in the layout shown in the sketch. Such anchorage requirements are treated in further detail in Chapter 12.

## OUTSIDE PROBLEMS

Design a suitable continuous concrete slab for the given conditions under exterior exposures. Show a longitudinal sketch of your solution, to include bar locations and cutoff points. (It is usually desirable when designing for exterior exposures to add about 1 in. to the estimated slab thickness to account for the increased cover requirements.)

| Prob. No. | Number of Spans | End Support | | Interior Supports | | $f'_c$ psi | Steel Grade | Live Load psf |
|---|---|---|---|---|---|---|---|---|
| | | Type | Clear Span | Type | Clear Span | | | |
| 8.19 | 2 | Beam | 12' 6" | Beam | | 3000 | 40 | 160 |
| 8.20 | 3 | Simple | 13' 6" | Beam | 15' 6" | 3000 | 60 | 140 |
| 8.21 | 4 | Beam | 14' 6" | Beam | 16' 6" | 3000 | 40 | 120 |
| 8.22 | 5 | Simple | 16' 6" | Beam | 18' 6" | 3000 | 60 | 100 |
| 8.23 | 2 | Simple | 12' 6" | Beam | | 4000 | 40 | 160 |
| 8.24 | 3 | Beam | 13' 6" | Beam | 15' 6" | 4000 | 60 | 140 |
| 8.25 | 4 | Simple | 14' 6" | Beam | 16' 6" | 4000 | 40 | 120 |
| 8.26 | 5 | Beam | 16' 6" | Beam | 18' 6" | 4000 | 60 | 100 |

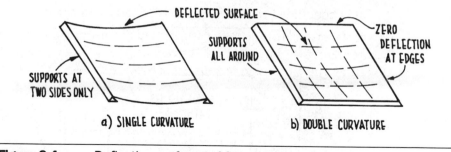

**Figure 8-4**    Deflection surfaces of flat slabs.

## Slab Supports Other Than Simple Supports

To this point, only simply supported slabs have been considered. Such slabs are distinctive in that the deformed surface is curved in only one direction, as indicated in Figure 8-4a. If additional supports are added at the other two sides, as indicated in Figure 8-4b, the surface of single curvature becomes distorted into some undefined surface of double curvature. The problem immediately becomes highly statically indeterminate and correspondingly quite complex.

The support conditions need not be symmetrical, as shown in Figure 8-4. A slab may also be supported only on three sides, as shown schematically in Figure 8-5. Or a slab may be simply supported in one direction and continuous in the other direction; or it may be continuous in all directions, with a corresponding increase in indeterminacy and complexity.

The bending moment diagrams at the supports for a slab continuous in both directions are shown in Figure 8-6. Note that at the corners, the moment becomes negative. Along with negative moment, the reactions at these corners are also negative; the slab is trying to "kick up" at the corners.

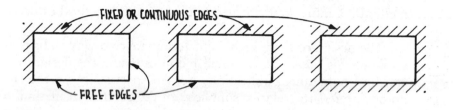

**Figure 8-5**    Support conditions for slabs.

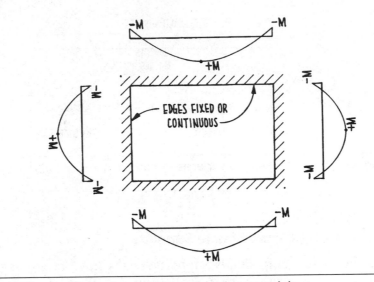

**Figure 8-6**    Moments at supports in a two-way slab.

It should be apparent that the design of slabs can become quite complex whenever the slabs are anything other than simple one-way slabs. Within the industry, however, the use of continuous slabs is a highly desirable feature, both in design and in construction. Long ago, ACI recognized the need for a simplified method to design such slabs and accordingly developed a simplified method for the design of continuous slabs. The use of these slabs is so common that the simplified design method has been made a part of the Code.

Probably the most widely accepted form of this simplified design procedure was the "coefficient method" as it was developed for the 1963 Code. From then until now, the coefficient method of the 1963 Code has been used successfully throughout the world both by highly qualified and minimally qualified practitioners. It is still in use by a large part of the practice even though ACI has, in the 1983 Code, added the "direct design method," a more sophisticated but less general method of designing such slabs.

The design of these highly indeterminate two-way slabs is beyond the scope of this book. Those interested in pursuing the design of such slabs can find numerous publications written specifically on the subject. The author prefers the relative simplicity of the coefficient method for the design of slabs and recommends it over the more recent direct design method.

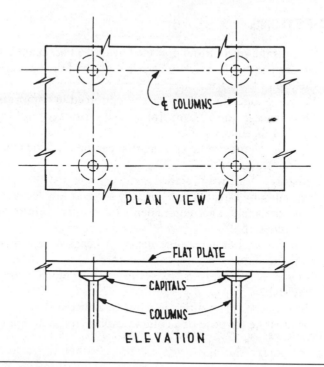

**Figure 8-7**    Flat plate construction.

The success of the coefficient method over some 30 years of use is a strong argument in favor of such a preference.

There is another type of elevated slab called the *flat plate* that is very popular in today's construction. A sketch of a flat plate is shown in Figure 8-7. As indicated, a flat plate is supported directly by columns, without the use of supporting beams or girders. *Column capitals* or *drop panels* are usually required in such designs to distribute the high shears that occur in the vicinity of the columns. The advantage in using a flat plate is that the bottom face of the plate is flat; there is no need for the expensive formwork that would be required for building supporting beams and girders.

The flat plate has another distinct advantage in that it may be used as a part of a rigid frame; that is, the slab can be designed to sustain bending moments due to sidesway. Flat plate structures can become quite limber, however, and will likely have a long period of oscillation when subjected to earthquake excitation. Such technical problems, however interesting, are far beyond the scope of an elementary textbook.

## REVIEW QUESTIONS

1. What is the steel ratio $\rho$ provided by No. 6 bars placed at 6 in. on center in a concrete slab 7 in. thick, where the center of the bars are placed $1\frac{1}{2}$ in. from the top of the slab?
2. Why are concrete slabs so rarely reinforced in compression?
3. How much grade 60 temperature and shrinkage reinforcement is required in a slab 4 in. thick?
4. What is the maximum spacing that can be used for temperature and shrinkage reinforcement?
5. What is a "one-way" slab?
6. Explain why elevated slabs usually serve at low levels of service stress.
7. State the anchorage requirements for positive reinforcement in an elevated concrete slab.
8. State the anchorage requirements for negative reinforcement in an elevated concrete slab.
9. Why is the design of a "two-way" slab so much more complex than a one way slab?
10. How is a flat plate different from an elevated slab?
11. What is the purpose of a column capital in the design of a flat plate?

CHAPTER

# 9

# SPECIAL TOPICS IN FLEXURE

In preceding chapters, the general theory of bending for concrete members was developed. Applications of the theory, however, were limited to a very narrow range of cases. The primary emphasis in those cases was in the design of a member to sustain flexural load.

Other applications can now be considered. One of these special applications involves the design of a section for flexure in combination with a small axial force, either tension or compression. A second topic involves the investigation of a known section to find its capacity in flexure. A third topic involves a combination of the first two topics; that is, the investigation of a section to find its capacity if both a flexural load and an axial load are present. These topics are some of the more common variations in theory that are encountered in routine practice.

## Flexure Plus a Small Axial Load

The foregoing discussions and examples considered rectangular beams loaded only in flexure. In addition to flexure, a beam may be subjected occasionally to an axial load, either in compression or tension. Where the axial load is small, the design of such members can be accomplished by a relatively simple innovation; a separate analytical approach is not necessary.

Although not common, the phenomenon of a beam being subject to a small axial load is not a rare occurrence. A typical example is a basement roof beam being subject to inward axial compression from the soil pressure against the walls. An example of a tension member might be a roof beam over a water tank, where the water pressure exerts an outward load against the walls.

When the term "small axial load" is used, it means that the ultimate axial load $P_n$ is no greater than $0.10f'_c bh/\phi$, where $bh$ is the gross area of the section, neglecting reinforcement. When the axial load exceeds this amount, the member ceases being a beam carrying an axial load and becomes a column carrying a flexural load. The design of column members is quite different from the design of flexural members; columns are presented in Chapter 13.

For small axial loads—that is, for $P_n \leq 0.10f'_c bh/\phi$—Code (9.3.2.2) permits a variable value to be assigned to $\phi$, otherwise $\phi$ must be taken at its lowest value of 0.7 for both moment and axial load. The penalty in using $\phi = 0.7$ for moment (rather than 0.9) is so severe that it is usually worthwhile to use the variable value for $\phi$. As an equation, the variable value for $\phi$ is given by

$$\phi = \frac{0.9}{1 + 0.2\dfrac{P_n}{0.10f'_c bh}} \tag{9-1}$$

with the same value of $\phi$ being applied both to the moment and to the axial load.

The correction for $\phi$ given by Eq. (9-1) applies only when the axial load is in compression. When the axial load is in tension, the value of $\phi$ is not a variable; it is fixed by Code (9.3.2.2) at 0.90 for both moment and axial load.

The design of members subject to a moment $M_n$ and a small axial force $P_n$ can be accomplished using three states of loading, as shown in Figure 9-1.

The design procedure consists of three steps corresponding to the three states of loading:

1. Using regular procedures for flexural design, find the width $b_1$ and the depth $d$ for a rectangular section subject only to the design moment $M_n$ as indicated in Figure 9-1a.
2. Expand the triangular compression block laterally (without reinforcement) by a width $b_2$ such that the design load $P_n$ is exactly opposed, as indicated in Figure 9-1b.

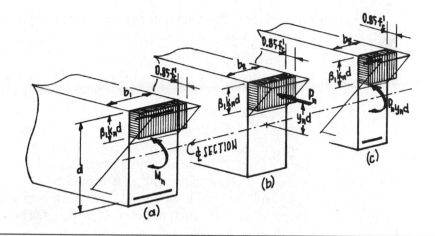

**Figure 9-1**    Flexure plus axial load.

3. Deduct a width $b_3$ to compensate for the moment induced due to the eccentricity of the axial load $P_n$ as indicated in Figure 9-1c.

The final width of the section is then the algebraic sum of the three widths, with the effective depth being the flexural depth computed in the first step.

When the force $P_n$ is in tension rather than compression, the signs of the widths $b_2$ and $b_3$ are simply reversed; steps 2 and 3 become "deduct" and "expand," respectively. For either tension or compression, the line of action for the axial force is assumed to be at the centerline of the section. The angle of rotation of the section is held constant, such that $\beta_1$ and $k_n$ remain constant throughout.

The width $b_1$ shown in Figure 9-1a is found through the ordinary methods of flexural design developed in the preceding sections. In this computation, it is necessary to record the steel ratios $\rho$ and $\rho'$, as appropriate, when the section is chosen. It is premature, however, to select the actual $A_s$ at this time, since the effects of the eccentric axial load $P_n$ have not yet been included.

If a compressive force $P_n$ exists, the triangular compression block shown in Figure 9-1a can be imagined to be expanded laterally (without reinforcement) to oppose this load, as shown in Figure 9-1b. When the compression block is expanded by the width $b_2$, with the rotation of the section held constant, then

$$P_n = 0.85f'_c b_2 \beta_1 k_n d \qquad (9\text{-}2)$$

The width $b_2$ is readily computed from this equation, where $\beta_1 k_n$ is given among the section constants just obtained for $b_1$.

The load $P_n$ is eccentric, however, producing an additional moment on the section. The width $b_3$ shown in Figure 9-1c is the flexural width associated with this additional eccentric moment, $P_n y_n d$. When $P_n$ is compressive, the eccentric moment $P_n y_n d$ acts in the same direction as the applied moment $M_n$, thus "helping out," and allowing a reduction in width. When the axial load is a tensile force $T_n$, the eccentric moment $T_n y_n d$ acts opposite to the applied moment $M_n$, requiring an increase in width.

When both $b_1$ and $b_3$ have been found, the final steel area may be computed, corresponding to the final moment $M_n - P_n y_n d$. The steel ratio $\rho$ is the same for both moments, hence

$$A_s = \rho(b_1 - b_3)d = \rho b_m d \qquad (9\text{-}3)$$

$$\text{where } b_m = b_1 - b_3.$$

The width $b_m$ is called the "least flexural width" and is used in later investigations.

A word of warning is appropriate before proceeding into the examples. Note that the existence of the axial force $P_n$ permits a net reduction in the area of reinforcement $A_s$ due to the eccentric moment. If the force $P_n$ is in fact intermittent, then the section so designed might be inadequate whenever $P_n$ does not exist. In cases where $P_n$ is intermittent, therefore, the section should be designed both with and without $P_n$ and the more conservative section used for the final design.

Some examples will illustrate the design procedure when a section is subject both to flexure and to a small axial force.

## EXAMPLE 9-1.

Design of a rectangular section subject both to flexure and to a small axial compression. No limitation on dimensions.

### Given:

$M_{DL} = 44$ kip-ft, $M_{LL} = 22$ kip-ft

$P_{DL} = 16$ kips, $P_{LL} = 12$ kips

Grade 50 steel, $f'_c = 4000$

**Find:**

Suitable section to sustain the load

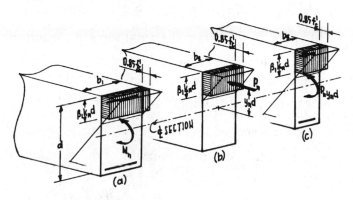

**Solution:**

Estimate $\phi$. The exact value of $\phi$ cannot be determined until $b$ and $h$ are known [Eq. (9-1)]. Since $\phi$ will be somewhere between 0.7 and 0.9, a value of 0.8 is chosen.

Calculate the design values for $M_n$ and $P_n$,

$$M_n = (1.4M_{DL} + 1.7M_{LL})/\phi = (1.4 \times 44 + 1.7 \times 22)/0.8$$
$$= 124 \text{ kip-ft}$$

$$P_n = (1.4P_{DL} + 1.7P_{LL})/\phi = (1.4 \times 16 + 1.7 \times 12)/0.8$$
$$= 53.5 \text{ kips}$$

Determine the required section modulus $Z_c$

$$Z_c = \frac{M_n}{0.85f_c'} = \frac{124,000 \times 12}{0.85 \times 4000} = 438$$

Select the service stress and the section constants to provide a balanced stress condition in flexure,

From Table A-8, with No Compressive Reinforcement,

$$\rho = 0.012, \quad Z_c = 0.161bd^2, \beta_1k_n = 0.177,$$
$$y_n = 0.474, f_{sv} = 2067 \text{ psi}$$

Select sizes for $b_1$ and $d$, assume $b_1 = 0.6d$

$$Z_c = 438 = 0.161bd^2$$
$$d = 16.5 \text{ in.}, \ b_1 = 9.9 \text{ in.}$$

Add width $b_2$ to sustain the axial force $P_n$, using the depth $d$ just selected,

$$P_n = 0.85f'_c b_2 \beta_1 k_n d$$
$$53,500 = 0.85 \times 4000 \times b_2 \times 0.177 \times 16.5$$
$$b_2 = 5.4 \text{ in.}$$

Deduct width $b_3$ due to eccentricity of the axial load $P_n$ which produces a moment $P_n y_n d$, with $d = 16.5$ in.,

$$Z_c = \frac{P_n Y_n d}{0.85 f'_c} = \frac{53,500 \times 0.474 \times 16.5}{0.85 \times 4000} = 0.161 b_3 \times 16.5 \times 16.5$$

$$b_3 = 2.8 \text{ in.}$$

Final width $b = b_1 + b_2 - b_3 = 9.9 + 5.4 - 2.8 = 12.5$ in.

Assume overall height $h = d + 2\frac{1}{2}$ in. $= 19$ in.

Check the assumed value of $\phi$ using Eq. (9-1)

$$\text{Required} \quad \phi = \frac{0.9}{1 + 0.2 \dfrac{P_n}{0.10 f'_c bh}} = \frac{0.9}{1 + 0.2 \dfrac{53,500}{0.10 \times 4000 \times 12.5 \times 19}}$$

$$\phi = 0.81 \text{ (O.K.)}$$

This value of $\phi$ is judged to be close enough to the assumed value of 0.80 to be acceptable.

Determine the required steel area

$$A_s = \rho b_m d = 0.012(9.9 - 2.8)16.5 = 1.42 \text{ in.}^2$$

From Tables A-3 and A-4, use 2 No. 6 and 2 No. 5 bars.

Use $b = 12.5$ in., $d = 16.5$ in., $A_s = $ 2 No. 6 and 2 No. 5 bars.

---

If the computed value of $\phi$ in Example 9-1 had not been close enough to the estimated value to be acceptable, it would have been necessary to revise

$\phi$ to the new value and to repeat all the computations for this new value of $\phi$. However, since the factor $\phi$ affects only loads, and the computed values of $b_1$, $b_2$ and $b_3$ are in all cases directly proportional to loads, the correction can be made simply by multiplying by the ratio $\phi_{OLD}/\phi_{NEW}$. For the values of Example 9-1,

$$\text{Ratio} = \frac{\phi_{OLD}}{\phi_{NEW}} = \frac{0.80}{0.81}$$

Now multiply the loads and widths $b$ by this ratio,

$M_n$ = 122 kip-ft

$P_n$ = 52.8 kip-ft

$b_1$ = 9.78 in., $b_2$ = 5.33 in., $b_3$ = 2.77 in.

$b$ = 12.34 in., $h$ = 19 in.

Now recheck $\phi$,

$$\phi = \frac{0.90}{1 + 0.2 \dfrac{52,800}{0.1 \times 4000 \times 12.34 \times 19}} = 0.81 \text{ (O.K.)}$$

This value of $\phi$ is exactly equal to the assumed value of $\phi$ and is therefore acceptable. The remaining calculations for $A_s$ would then use the revised value for $b$.

If the axial load had been tensile rather than compressive, the signs of $b_2$ and $b_3$ would change. Such a case is shown in the next example. Also, as noted earlier, the value of $\phi$ is not a variable when the axial force is tensile, but remains 0.90 for all values of the axial load.

## EXAMPLE 9-2.

Design of a rectangular section subject to flexure plus a small axial tension. No limitation on dimensions.

**Given:**

$M_{DL}$ = 32 kip-ft    $M_{LL}$ = 35 kip-ft

$T_{DL}$ = 9 kips        $T_{LL}$ = 13 kips

Grade 60 steel, $f'_c$ = 3000 psi

**Find:**

Suitable section to sustain the load

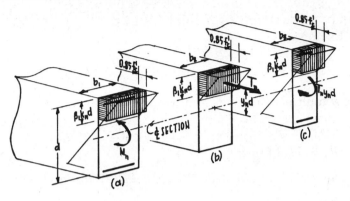

**Solution:**

Calculate the design values for $M_n$ and $P_n$, using $\phi = 0.90$ for both loads

$$M_n = (1.4M_{DL} + 1.7M_{LL})/\phi = (1.4 \times 37 + 1.7 \times 35)/0.90$$
$$= 116 \text{ kip-ft}$$

$$P_n = (1.4P_{DL} + 1.7P_{LL})/\phi = (1.4 \times 9 + 1.7 \times 13)/0.90$$
$$= 39 \text{ kips}$$

Determine the required section modulus $Z_c$

$$Z_c = \frac{M_n}{0.85f'_c} = \frac{116,000 \times 12}{0.85 \times 3000} = 546$$

Select the service stress and the section constants to provide a balanced stress condition in flexure,

Use $f_s/f_c = f_y/0.85f'_c = 60,000/0.85 \times 3000$
$$= 23.53$$

From Table A-8, no compressive reinforcement, select

$$\rho = 0.0060, \quad Z_c = 0.132bd^2, \quad \beta_1 k_n = 0.142$$
$$y_n = 0.492, f_{sv} = 1539 \text{ psi}$$

Select sizes $b_1$ and $d$, assuming $b_1 = 0.8d$

$$Z_c = 546 = 0.132 \times 0.8d \times d^2$$
$$d = 17.3 \text{ in.}, \; b_1 = 13.8 \text{ in.}$$

Deduct the width $b_2$ due to tensile load $t_n$, using the depth $d$ just selected,

$$T_n = 0.85f_c'b_2\beta_1k_nd$$
$$39,000 = 0.85 \times 3000 \times b_2 \times 0.142 \times 17.3$$
$$b_2 = 6.2 \text{ in.}$$

Add the width $b_3$ due to eccentricity of the tensile load $T_n$, which produces a moment $T_ny_nd$, with $d = 17.3$ in.,

$$Z_c = \frac{T_ny_nd}{0.85f_c'} = \frac{39,000 \times 0.492 \times 17.5}{0.85 \times 3000} = 0.132 \times b_3 \times 17.3 \times 17.3$$

$$b_3 = 3.3 \text{ in.}$$

Final width $b = b_1 - b_2 + b_3 = 13.8 - 6.2 + 3.3 = 10.9$ in.

Calculate the required steel area

$$A_s = \rho b_m d = 0.006(13.8 + 3.3)17.3 = 1.77 \text{ in.}^3$$

From Tables A-3 and A-4, use 2 No. 9 bars

Use $b = 11$ in., $d = 18$ in., $A_s = 2$ No. 9 bars

---

Neither Example 9-1 nor 9-2 had any restrictions on the size of the member. Where the depth $d$ is restricted, a common case around air conditioning ductwork, the beams become wide and shallow, as illustrated in the next example.

## EXAMPLE 9-3.

Design of a rectangular section subject to flexure plus a small axial compression. Depth of section limited.

**Given:**

$M_{DL} = 80$ kip-ft; $M_{LL} = 16$ kip-ft

$P_{DL} = 21$ kips; $P_{LL} = 2$ kips

Grade 60 steel; $f_c' = 3000$ psi

Depth $d$ limited to 18 in.

**Find:**

Suitable section to sustain the load

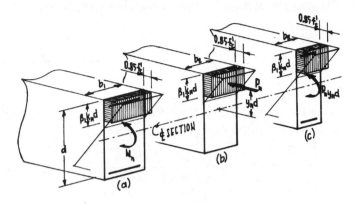

**Solution:**

Estimate $\phi = 0.80$. Calculate the design values for $M_n$ and $P_n$ using $\phi = 0.80$ for both

$$M_n = (1.4M_{DL} + 1.7M_{LL})/\phi = (1.4 \times 80 + 1.7 \times 16)/0.8$$
$$= 174 \text{ kip-ft}$$

$$P_n = (1.4P_{DL} + 1.7P_{LL})/\phi = (1.4 \times 21 + 1.7 \times 2)/0.8$$
$$= 41 \text{ kips}$$

Determine the required section modulus $Z_c$

$$Z_c = \frac{M_n}{0.85f_c'} = \frac{174,000 \times 12}{0.85 \times 3000} = 819$$

Select the service stress and the section constants to provide a balanced stress condition in flexure,

From Table A-8, No Compressive Reinforcement, select

$$\rho = 0.0060, \quad Z_c = 0.132bd^2, \quad \beta_1 k_n = 0.142$$
$$y_n = 0.492, f_{sv} = 1539 \text{ psi}$$

Select size $b_1$, holding $d = 18$ in.,

$$Z_c = 819 = 0.132 \times b_1 \times 18 \times 18$$
$$b_1 = 19.1 \text{ in.}, d = 18 \text{ in.}$$

Add the width $b_2$ due to load $P_n$, using $d = 18$ in.

$$P_n = 0.85f'_c b_2 \beta_1 k_n d$$
$$41,000 = 0.85 \times 3000 \times b_2 \times 0.142 \times 18$$
$$b_2 = 6.3 \text{ in.}$$

Deduct the width $b_3$ due to the eccentricity of the axial load $P_n$, which produces a moment $P_n y_n d$,

$$Z_c = \frac{P_n y_n d}{0.85f'_c} = \frac{41,000 \times 0.492 \times 18}{0.85 \times 3000} = 0.132 b_3 \times 18 \times 18$$
$$b_3 = 3.3 \text{ in.}$$

Final $b = b_1 + b_2 - b_3 = 19.1 + 6.3 - 3.3 = 22.1$ in.

Assume overall height $h = d + 2\frac{1}{2}$ in. $= 20.5$ in.

Check for correctness in assumed value of $\phi$

$$\phi = \frac{0.9}{1 + 0.2\dfrac{P_n}{0.1f'_c bh}} = \frac{0.9}{1 + 0.2\dfrac{41,000}{0.1 \times 3000 \times 22.1 \times 20.5}}$$
$$\phi = 0.85$$

This value of $\phi$ is not close enough to the assumed value of 0.8 to be acceptable. The value of $\phi$ is therefore revised to 0.85. The values of loads and the corresponding values of $b$ are revised by multiplying by the ratio 0.80/0.85, yielding

$$M_n = 164 \text{ kip-ft}, \ P_n = 39 \text{ kips}$$
$$b_1 = 18.0, \ b_2 = 5.9, \ b_3 = 3.1$$
$$b = 20.8 \text{ in.}, \ d = 18 \text{ in.}, \ h = 20.5 \text{ in.}$$

Verify $\phi$,

$$\phi = \frac{0.9}{1 + 0.2\dfrac{39,000}{0.1 \times 3000 \times 20.8 \times 20.5}}$$
$$\phi = 0.85 = \text{ assumed value of 0.85 (O.K.)}$$

Calculate the steel area $A_s$

$$A_s = \rho b_m d = 0.006(18.0 - 3.1)18 = 1.61 \text{ in.}^2$$

From Tables A-3 and A-4 use 4 No. 6 bars.

Use $b$ = 21 in., $d$ = 18 in., $A_s$ = 4 No. 6 bars.

---

It should be noted that the dimension $b$ is directly proportional to both $P_n$ and $M_n$ but that the dimension $d$ is proportional to the square root of the moment. Only the width $b$ may therefore be increased or decreased by the ratio of $\phi_{OLD}/\phi_{NEW}$ as done here; the depth $d$ may not be so changed by simple ratios.

Rather than making the beam wider in Example 9-3, the width $b$ could have been kept narrow by adding compressive steel. Compressive steel is usually avoided where possible, but its use is common in restricted areas. The next example considers a case where both $b$ and $d$ are restricted so much that the use of compressive steel becomes necessary.

## EXAMPLE 9-4.

Design of a rectangular section subject both to flexure and to a small axial compression. Both depth $d$ and width $b$ limited.

**Given:**

$M_{DL}$ = 67 kip-ft;  $M_{LL}$ = 29 kip-ft
$P_{DL}$ = 25 kips;  $P_{LL}$ = 8 kips
Grade 60 steel, $f'_c$ = 4000 psi
Width $b$ limited to 12 in.
Effective depth $d$ limited to 18 in.

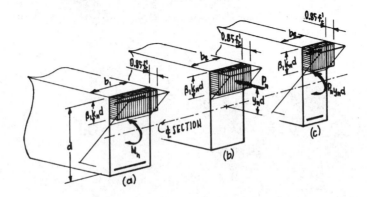

**Find:**

Suitable section to sustain the load

**Solution:**

Estimate $\phi = 0.8$. Calculate the design values for $M_n$ and $P_n$ using $\phi = 0.80$ for both.

$$M_n = (1.4M_{DL} + 1.7M_{LL})/\phi = (1.4 \times 67 + 1.7 \times 29)/0.8$$
$$= 179 \text{ kip-ft}$$

$$P_n = (1.4P_{DL} + 1.7P_{LL})/\phi = (1.4 \times 25 + 1.7 \times 8)0.8$$
$$= 61 \text{ kips}$$

Select the service stress and the section constants to provide the balanced stress condition in flexure.

From this point onward, the solution to this problem becomes a trial-and-error solution to find a width $b = b_1 + b_2 - b_3$ that falls within the allowable 12-in. limitation. As in most trial-and-error solutions, an ordered approach will reduce some of the work.

This problem will be solved by assuming that the width $b$ is not restricted; the procedures of the preceding example will then be used to find a required width $b$. If the solution thus obtained yields a required width $b$ greater than the 12 in. allowed, then compressive steel will be added and a new required width $b$ will be recomputed. If this new solution still yields a required width $b$ greater than the 12 in. allowed, then even more compressive steel will be added. This procedure is repeated until a solution is found that meets the limitations for $b$.

First trial: No compressive reinforcement, $\rho = 0.0089$

Results: $b_1 = 13.5$ in., $b_2 = 6.4$ in., $b_3 = 3.4$ in., $b = 16.5$ in.
Required width $16.5 > 12$ in. allowed (no good)

Second trial: Try $A'_s = 0.60A_s$, $\rho = 0.0128$

Results: $b_1 = 9.5$ in., $b_2 = 6.2$ in., $b_3 = 2.3$ in., $b = 13.4$
Required width $13.4 > 12$ in. allowed (no good)

After further trials, find: $A'_s = 0.87A_s$, $\rho = 0.0159$, $Z_c = 0.252bd^2$;
$b_1k_n = 0.161$, $y_n = 0.482$

Results: $b_1$ = 7.74 in., $b_2$ = 6.19 in., $b_3$ = 1.91 in., $b$ = 12.02 in.
Required width 12.0 = 12 in. allowed (O.K.—use)

For this final solution, $\phi$ = 0.798, almost exactly equal to the assumed value of 0.8. The foregoing values of $b_1$, $b_2$ and $b_3$ will therefore need no further correction for $\phi$.

Use $b$ = 12 in., $d$ = 18 in., $A_s$ = 4 No. 9 bars

$A'_s = 0.85A_s$ = 4 No. 8 bars

## OUTSIDE PROBLEMS

For the following conditions, select a suitable rectangular concrete section to carry the given loads at the balanced stress condition (if possible). Provide a minimum of 1½ in. cover over reinforcement. The c or the t in the tabulation following the axial force denotes compression or tension, respectively.

| Prob. No. | Concrete Strength $f'_c$ psi | Steel Grade | Limits on Width $b$ | Limits on Height $h$ | Dead Load Moment kip-ft | Live Load Moment kip-ft | Axial Dead Load kips | Axial Live Load kips |
|---|---|---|---|---|---|---|---|---|
| 9.1 | 3000 | 40 | None | None | 30 | 21 | 6c | 4c |
| 9.2 | 3000 | 40 | None | None | 59 | 72 | 30c | 39c |
| 9.3 | 3000 | 50 | None | None | 32 | 25 | 8c | 7c |
| 9.4 | 4000 | 50 | None | None | 56 | 65 | 26c | 35c |
| 9.5 | 4000 | 60 | None | 14 in. | 35 | 28 | 10t | 10t |
| 9.6 | 4000 | 60 | None | 14 in. | 52 | 58 | 23t | 32t |
| 9.7 | 5000 | 40 | None | 14 in. | 38 | 32 | 11t | 14t |
| 9.8 | 5000 | 40 | None | 14 in. | 49 | 54 | 22t | 27t |
| 9.9 | 5000 | 50 | 10 in. | 17 in. | 40 | 37 | 13c | 16c |
| 9.10 | 3000 | 50 | 12 in. | 20 in. | 47 | 49 | 19c | 25c |
| 9.11 | 4000 | 60 | 10 in. | 23 in. | 43 | 41 | 15c | 19c |
| 9.12 | 5000 | 60 | 12 in. | 20 in. | 45 | 45 | 17c | 22c |

## Investigation of Known Sections

All the preceding examples were concerned only with the design of a concrete section; that is, the selection of the dimensions or reinforcement that are required if a beam is to carry the prescribed loads. There is another general type of problem that occurs when the size and reinforcement of a section are already known; the problem then becomes that of finding how much load the known section will carry. This "investigation" of a given section is treated here.

There are two cases of reinforcement to be considered when investigating a known section:

1. The area of tensile reinforcement $\rho$ falls within the more common values, such as those given in the tables of the Appendix. For values of $\rho$ less than about $0.75\rho_b$, the compressive reinforcement, if any exists, would serve primarily to reduce service stresses in the concrete and to reduce long-term deflections. For values of $\rho$ greater than $0.75\rho_b$, the compressive reinforcement would serve primarily to increase the moment capacity of the section.

2. The area of tensile reinforcement $\rho$ is much greater than $0.75\rho_b$ and falls outside the range of values given in the tables of the Appendix. For these rarely encountered cases, the beam constants $k_n$ and $Z_c$ must be computed manually.

The following example will illustrate the procedure for investigating a known section when the steel ratios $\rho$ and $\rho'$ are not extreme and they can be found in the tables. All terms and symbols remain those used in the design procedures; the defining sketch for the terms and symbols is Figure 6-1.

## EXAMPLE 9-5.

Investigation of a rectangular section subject only to flexure to determine the nominal ultimate moment $M_n$.

**Given:**

Section as shown

Grade 60 steel

$f'_c = 4000$ psi

**Find:**

Nominal ultimate moment $M_n$

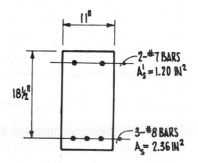

**Solution:**

Calculate the steel ratios $\rho$ and $\rho'$, assuming $\rho' = 0.5\rho$:

$$\rho = \frac{A_s}{bd} \qquad \rho = \frac{2.36}{11 \times 18.5} = 0.012 \qquad \rho' = 0.5\rho \pm$$

Enter Table A-6. It is noted that the computed tensile steel ratio is less than $0.75\rho_b$. For this steel ratio, the plastic section modulus falls within the tabled values;

$$Z_c = 0.192bd^2$$

Determine the nominal ultimate moment from the flexure formula:

$$\frac{M_n}{Z_c} = 0.85f_c' \qquad M_n = 0.85f_c'Z_c = 0.85 \times 4000 \times 0.192 \times 11 \times 18.5 \times 18.5$$

$$M_n = 2460 \text{ kip-in.} = 205 \text{ kip-ft}$$

When the tensile steel ratio of a section falls outside the range of the design tables of the Appendix, the design contains an unusually large amount of tensile reinforcement; that is, $\rho$ is much greater than $0.75\rho_b$. Such extreme designs are rare but they are permitted by code as a means to increase moment capacity when the size of the beam is severely restricted. The investigation of such sections can be somewhat tedious; the investigation requires the computation of the section constants from the original analytical equations, since these constants do not fall within the range of the design tables. Such an investigation follows.

For any section of unknown origin or uncertain design, the first item to be investigated is whether the section conforms to code conditions; that is, whether the tensile reinforcement reaches yield stress before the concrete reaches its ultimate strain of 0.003 in./in. If this condition is not met, then none of the theoretical relationships derived earlier in the strength method are valid. A very quick check can be made simply by verifying that the given steel ratio $\rho$ is within the limits specified by Code; that is,

$$\text{maximum allowable } \rho \leq 0.75\rho_b + \rho'$$

For convenience in making such a check, the values of $0.75\rho_b$ have been included with the design tables of the Appendix.

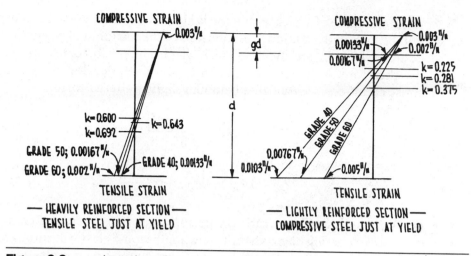

**Figure 9-2**     Location of neutral axis at ultimate load.

Once it is verified that the ACI strength method is valid, the section constants for the given section can be determined. It must be remembered, however, that there are two strain conditions to be considered when the section constants are being computed. One condition occurs when the compressive reinforcement is in yield and the other occurs when the compressive reinforcement is elastic. The next computation must then be to determine which strain condition exists when the given section is at ultimate load.

To determine whether the compressive reinforcement is in yield or whether it is elastic, the strain relationships shown in Figure 9-2 are utilized. In those strain diagrams, the strain in the concrete at ultimate load is taken at 0.003 in./in. as specified by Code (10.2.3); the strain is the same for all values of $f'_c$. For steel, the yield strain is taken at 0.00200 in./in. for grade 60 steel, at 0.00167 for grade 50 steel, and at 0.00133in./in. for grade 40 steel.

As indicated in Figure 9-2, both the tensile and compressive reinforcement will be in yield at ultimate load if the value of $k_n$ falls between the following limits:

$$\text{For grade 60 steel;} \quad 0.375 \le k_n \le 0.600 \qquad (9\text{-}4a)$$

$$\text{For grade 50 steel;} \quad 0.281 \le k_n \le 0.643 \qquad (9\text{-}4b)$$

$$\text{For grade 40 steel;} \quad 0.225 \le k_n < 0.692 \qquad (9\text{-}4c)$$

If the value of $k_n$ lies within the limits given in Eq. (9-4), the neutral axis will be low on the section, the rotations of the section will be small, and the compressive reinforcement will be in yield. The section constants $k_n$ and $Z_c$ are therefore those given by Eq. (6-6) and 6-8), developed in Chapter 6; those equations are repeated below in slightly altered form.

$$k_n = \frac{f_y(\rho - \rho')}{0.85 f'_c \beta_1} \tag{9-5}$$

$$Z_c = \beta_1 k_n \left[ 1 - \frac{\beta_1 k_n}{2} + \rho' \frac{1-g}{\rho - \rho'} \right] bd^2 \tag{9-6}$$

For the lower steel ratios, rotations will be higher; the compressive reinforcement will emerge from yield and become elastic at ultimate loads. For such cases the value of $k_n$ will be less than the lower limits given in Eq. (9-4); that is,

$$\text{For grade 60 steel; } k_n \leq 0.375 \tag{9-7a}$$
$$\text{For grade 50 steel; } k_n \leq 0.281 \tag{9-7b}$$
$$\text{For grade 40 steel; } k_n \leq 0.225 \tag{9-7c}$$

When $k_n$ lies below the limits given in Eq. (9-7), the neutral axis will be high on the section, the rotations of the section will be large, and the compressive reinforcement will be elastic. The section constants for this case are therefore those given by Eq. (6-11) and (6-13), developed in Chapter 6; those equations are repeated below in slightly altered form.

$$k_n = \sqrt{\left[ \frac{f_y \rho - 87,000 \rho'}{2 \times 0.85 f'_c \beta_1} \right]^2 + \frac{87,000 g \rho'}{0.85 f'_c \beta_1}} + \frac{f_y \rho - 87,000 \rho'}{2 \times 0.85 f'_c \beta_1} \tag{9-8}$$

$$Z_c = \left[ \beta_1 k_n \left( 1 - \frac{\beta_1 k_n}{2} \right) + \frac{87,000 \rho'}{0.85 f'_c} (1-g) \frac{k_n - g}{k_n} \right] \tag{9-9}$$

The actual procedure for investigating sections having excess reinforcement is quite direct, though the calculations using the foregoing equations do become somewhat tedious.

1. Determine both possibilities for $k_n$, using Eq. (9-5) and (9-8).
2. Compare the two results obtained for $k_n$ against the limits given by Eq. (9-7). Either both values of $k_n$ will fall above the limiting values

given by Eq. (9-7) or both will fall below the limiting values given by Eq. (9-7). Choose the one that is in its correct domain.

3. Calculate the plastic section modulus $Z_c$ corresponding to the correct value of $k_n$.

4. Determine the nominal ultimate moment $M_n$ for this value of $Z_c$,

$$M_n = 0.85f_c'Z_c$$

It is probably possible to develop a special approach using the tabled values in the Appendix to alleviate some of the calculations in the investigations of these heavily reinforced sections. For the few times in the working life of a designer that such a problem is encountered, however, maintaining familiarity with such specialized approaches is hardly worthwhile. The foregoing procedure is brutal but direct, and is always applicable.

Some examples will illustrate the procedure.

### EXAMPLE 9-6.

Investigation of a section having large levels of tensile reinforcement.

**Given:**

Section as shown
Grade 60 steel
$f_c' = 3000$ psi
$g = 0.125$

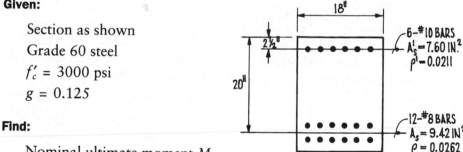

**Find:**

Nominal ultimate moment $M_n$

**Solution:**

Verify that the ACI strength method is valid. The method is valid if:

maximum $\rho \leq 0.75\rho_b + \rho'$.

From Table A-5, $0.75\rho_b = 0.0160$, hence, with $\rho' = 0.0211$,

$0.0262 \leq 0.0160 + 0.0211 \leq 0.0371$ (O.K.)

The ACI strength method is therefore seen to be valid.

Calculate $k_n$ if $0.375 \le k_n \le 0.600$ and compressive reinforcement is in yield.

$$k_n = \frac{f_y(\rho - \rho')}{0.85 f_c' \beta_1} = \frac{60,000(0.0262 - 0.0211)}{0.85 f_c' \beta_1}$$

$= 0.1412$ [This value of $k_n$ is not within the specified domain of Eq. (9-4a); this solution is therefore not valid.]

Calculate $k_n$ if $k_n \le 0.375$ and compressive reinforcement is elastic.

$$k_n = \sqrt{\left[\frac{f_y\rho - 87,000\rho'}{2 \times 0.85 f_c' \beta_1}\right]^2 + \frac{87,000 g \rho'}{0.85 f_c' \beta_1}} + \frac{f_y\rho - 87,000\rho'}{2 \times 0.85 f_c' \beta_1}$$

$= 0.325$ (This value of $k_n$ is within the specified domain of Eq. (9-7a); this equation is therefore valid and the computed value of $k_n$ is the correct one.)

Determine the plastic section modulus for the correct value of $k_n$.

$$Z_c = \left[\beta_1 k_n\left(1 - \frac{\beta_1 k_n}{2}\right) + \frac{87,000\rho'}{0.85 f_c'}(1-g)\frac{k_n - g}{k_n}\right]$$

$= 0.625 b d^2$

Solve for the nominal ultimate moment on the section,

$$M_n = 0.85 f_c' Z_c = 0.85 \times 3000 \times 0.625 \times 18 \times 20 \times 20$$
$$= 11,475 \text{ kip-in.} = 956 \text{ kip-ft}$$

---

## EXAMPLE 9-7.

Investigation of a section having large levels of tensile reinforcement.

**Given:**

Section as shown
Grade 60 steel
$f_c' = 3000$ psi
$g = 0.125$

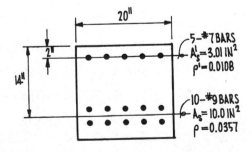

**Find:**

Nominal ultimate moment $M_n$

**Solution:**

Verify that the ACI strength method is valid. The method is valid if:

maximum $\rho \leq 0.75\rho_b + \rho'$

From Table A-5, $0.75\rho_b = 0.0160$, hence, with $\rho' = 0.0108$,

$0.0357 \leq 0.0160 + 0.0108 \leq 0.0268$ (No good)

It is concluded that the ACI strength method is not valid.

[As an alternative method to verify the validity of the ACI strength method, the value of $k_n$ may be computed and compared to the absolute limits given by Eq. (9-4). If $0.375 \leq k_n \leq 0.600$ (compressive steel in yield), then

$$k_n = \frac{f_y(\rho - \rho')}{0.85f'_c\beta_1} = \frac{60,000(0.0357 - 0.0108)}{0.85 \times 3000 \times 0.85}$$

$= 0.689 > 0.375$ [limiting value from Eq. (9-7)]

It is concluded that the compressive reinforcement is in yield and that this solution for $k_n$ is the valid one.

Now compare this value of $k_n$ against the absolute limits given by Eq. (9-4) and note that this value of $k_n$ is greater than the absolute upper limit permitted for $k_n$. It is concluded that the stress in the tensile reinforcement at ultimate load is elastic and that the section does not conform to ACI requirements for ultimate load analysis.]

Since the strength method is not valid, the section can only be investigated for its capacity under elastic conditions. Eq. (5-8) and (5-10) can be used to compute $k_{sv}$ and $S_c$, respectively.

$$k = \sqrt{\left[n\rho + (2n-1)\rho'\right]^2 + 2\left[n\rho + (2n-1)g\rho'\right]} - \left[n\rho + (2n-1)\rho'\right]$$

$k = 0.4698$

For this value of $k$, the elastic section modulus $S_c$ is given by

$$S_c = \left[\frac{k^2}{3} + (2n-1)\rho'\frac{(k-g)^2}{k} + n\rho\frac{(1-k)^2}{k}\right]bd^2$$

$S_c = 0.320bd^2 = 1255 \text{ in}^2$

At a stress in the concrete of $0.60(0.85f_c')$, the service moment is:

$$M_{sv} = 0.60(0.85f_c')S_c = 0.51 \times 3000 \times 1255$$
$$M_{sv} = 1920 \text{ kip-in.} = 160 \text{ kip-ft}$$

## OUTSIDE PROBLEMS

Determine the ultimate moment that the given sections will sustain.

| Prob. No. | Concrete Strength $f_c'$ psi | Steel Grade | Width $b$ in. | Height $h$ in. | Eff. depth $d$ in. | Tensile Reinf. | Compr. Reinf. |
|---|---|---|---|---|---|---|---|
| 9.13 | 3000 | 40 | 12 | 16 | 12 | 6-#9 | 4-#9 |
| 9.14 | 3000 | 50 | 12 | 19 | 16 | 8-#8 | 3-#10 |
| 9.15 | 3000 | 60 | 12 | 23 | 20 | 6-#8 | 2-#5 |
| 9.16 | 3000 | 40 | 16 | 28 | 24 | 10-#10 | 2-#9 |
| 9.17 | 3000 | 50 | 16 | 32 | 28 | 10-#11 | 5-#10 |
| 9.18 | 3000 | 60 | 16 | 36 | 32 | 10-#10 | 5-#8 |
| 9.19 | 4000 | 40 | 14 | 18 | 14 | 8-#10 | 3-#9 |
| 9.20 | 4000 | 50 | 14 | 22 | 18 | 10-#9 | 4-#8 |
| 9.21 | 4000 | 60 | 14 | 26 | 22 | 8-#10 | 3-#9 |
| 9.22 | 4000 | 40 | 18 | 32 | 26 | 18-#10 | 5-#10 |

## Investigation of Beams Subject to Axial Load

The more general case of investigations into known sections will include both a moment $M_n$ and an axial force $P_n$. The following example will illustrate the procedure.

### EXAMPLE 9-8.

Investigation of a given section to determine its capacity both for moment and axial load under balanced stress conditions.

**Given:**

Section as shown
Grade 60 steel
$f_c' = 4000$ psi

**Find:**

Allowable moment $M_n$ and axial load $P_n$ at the balanced stress condition

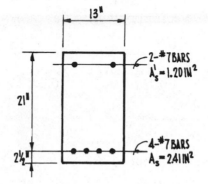

**Solution:**

Enter Table A-8, find section constants at balanced stress conditions.

$$\rho = 0.012, \ Z_c = 0.193bd^2, \ \beta_1 k_n = 0.160,$$
$$y_n = 0.482, f_{sv} = 2043 \text{ psi}, f_s/f_c = 17.65$$

Determine the least flexural width $b_m$, at the balanced stress condition, where $\rho b_m d = A_s$

$$b_m = \frac{A_s}{\rho d} = \frac{2.41}{0.012 \times 21} = 9.56 \text{ in.}$$

Solve for the width $b_2$

$$b_2 = \text{actual } b - b_m = 13.00 - 9.56 = 3.44 \text{ in.}$$

Determine the axial load $P_n$ at the balanced stress condition,

$$P_n = 0.85f'_c b_2 \beta_1 k_n d = 0.85 \times 4000 \times 3.44 \times 0.16 \times 21$$
$$= 39.3 \text{ kips.}$$

Maximum allowable load permitted for a small axial load, $P_n = 0.10f'_c bh/\phi$

Max $P_n = 0.10 \times 4000 \times 13 \times 23.5/0.7 = 174$ kips $> 39.3$ kips (O.K.)

Calculate the total moment capacity of the section

$$M_n = 0.85f'_c Z_c + P_n y_n d = 0.85f'_c \times 0.193 b_m d^2 + P_n y_n d$$
$$= 0.85 \times 4000 \times 0.193 \times 9.56 \times 21 \times 21 + 39300 \times 0.482 \times 21$$
$$= 3164 \text{ kip-in.} = 264 \text{ kip-ft}$$

Determine $\phi$ corresponding to this case of loads,

$$\phi = \frac{0.9}{1 + 0.2 \dfrac{P_n}{0.1 f'_c bh}} = \frac{0.9}{1 + 0.2 \dfrac{39{,}300}{0.1 \times 4000 \times 13 \times 21}}$$

$$\phi = 0.84$$

Final results are then:

Bending moment = 264 kip-ft

Simultaneous axial load = 39.3 kips

Strength reduction factor $\phi$ = 0.84

---

When the stress condition is not known to be balanced, the investigation of a known section to find its capacity will produce an infinite number of combinations of axial load and moment. Such an investigation can sometimes be best accomplished by finding only a representative few of the many possible combinations of load. Such an investigation is presented in the next example.

## EXAMPLE 9-9.

Investigation of a given section not necessarily at balanced stress conditions.

**Given:**

Section as shown
Grade 50 steel, $f'_c$ = 4000 psi

**Find:**

Allowable moment $M_n$ and axial load $P_n$

**Solution:**

For the first of the series of solutions, assume $b_m$ = 10 in., $P_n$ = 0, and $b_2$ = 0

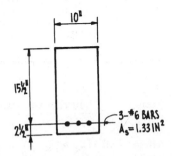

$$\rho = \frac{A_s}{bd} = \frac{1.33}{10 \times 15.5} = 0.0086$$

From Table A-6 of the Appendix,

$\rho = 0.0086$, $Z_c = 0.119 b_1 d^2$, $\beta_1 k_n = 0.127$, $y_n = 0.499$,

$M_n = 0.85 f_c' Z_c = 81$ kip-ft

$P_n = 0$

$\phi = 0.90$

For the next solution in the series, arbitrarily assume the next value of $\rho$ to be the next value that appears in Table A-6 of the Appendix; $\rho = 0.0010$,

$$\rho = 0.0010, \quad Z_c = 0.136 b_1 d^2, \quad \beta_1 k_n = 0.147, \quad y_n = 0.489$$

$$b_m = \frac{A_s}{\rho d} = \frac{1.33}{0.010 \times 15.5} = 8.58 \text{ in.}, \quad b_2 = 10 - 8.58 = 1.42 \text{ in.}$$

$$P_n = 0.85 f_c' \beta_1 k_n b_2 d = 11.0 \text{ kips}$$

$$M_n = 0.85 f_c' Z_c + P_n y_n d = 86.3 \text{ kip-ft}$$

$$\phi = \frac{0.90}{1 + 0.2 \dfrac{P_n}{0.1 \times f_c' bh}} = 0.87$$

Assume next value of $\rho$, $\rho = 0.012$ (the next tabulated value in Table A-6), and repeat the foregoing solution. The procedure is repeated until the maximum allowable value for $P_n$ is exceeded, where maximum $P_n = 0.10 f_c' bh / \phi = 102.9$ kips.

The final results are summarized in the following table:

| Steel Ratio | Section Constants | | | Flex. Width | Load Width | Final Results | | | |
|---|---|---|---|---|---|---|---|---|---|
| $\rho$ | $Z_c$ | $\beta_1 k_n$ | $y_n$ | $b_m$ | $b_2$ | $P_n$ kips | $M_n$ kip-ft | $\phi$ | $f_{sv}$ psi |
| 0.0086 | 0.119 | 0.127 | 0.499 | 10.00 | 0 | 0 | 81.0 | 0.90 | 1703 |
| 0.0100 | 0.136 | 0.147 | 0.489 | 8.58 | 1.42 | 11.0 | 86.3 | 0.87 | 1858 |
| 0.0120 | 0.161 | 0.177 | 0.474 | 7.15 | 2.85 | 26.6 | 94.6 | 0.84 | 2060 |
| 0.0140 | 0.185 | 0.206 | 0.459 | 6.13 | 3.87 | 42.0 | 102.1 | 0.81 | 2246 |

| Steel Ratio | Section Constants | | | Flex. Width | Load Width | Final Results | | | |
|---|---|---|---|---|---|---|---|---|---|
| $\rho$ | $Z_c$ | $\beta_1 k_n$ | $y_n$ | $b_m$ | $b_2$ | $P_n$ kips | $M_n$ kip-ft | $\phi$ | $f_{sv}$ psi |
| 0.0160 | 0.208 | 0.236 | 0.445 | 5.36 | 4.64 | 57.7 | 109.1 | 0.78 | 2417 |
| 0.0180 | 0.230 | 0.265 | 0.430 | 4.77 | 5.23 | 73.0 | 115.3 | 0.75 | 2577 |
| 0.0200 | 0.251 | 0.295 | 0.415 | 4.29 | 5.71 | 88.8 | 120.9 | 0.73 | 2725 |
| 0.0220 | 0.272 | 0.324 | 0.400 | 3.90 | 6.10 | 104.2 | 126.1 | 0.70 | 2863 |

The given section can therefore sustain any of the foregoing combinations of moment $M_n$ and axial load $P_n$ at the indicated levels of service stress. As a matter of interest, the balanced stress condition occurs for $f_s/f_c = 14.71$, for which case $\rho = 0.12$ and $f_{sv} = 2067$ psi.

## OUTSIDE PROBLEMS

Determine the magnitudes of the loads that the given section will sustain.

| Prob. No. | Concrete Strength $f_c'$ psi | Steel Grade | Width $b$ in. | Height $h$ in. | Eff. Depth $d$ in. | Tensile Reinf. | Compr. Reinf. | Type of Load | Balanced Stress Condition |
|---|---|---|---|---|---|---|---|---|---|
| 9.23 | 3000 | 40 | 12 | 14 | 12 | 4-#7 | None | $M_n$ | Unknown |
| 9.24 | 3000 | 50 | 12 | 18 | 16 | 4-#8 | 2-#8 | $M_n$ | Unknown |
| 9.25 | 3000 | 60 | 12 | 22 | 20 | 4-#6 | None | $M_n$ | Unknown |
| 9.26 | 3000 | 40 | 16 | 26 | 24 | 6-#6 | None | $M_n$ | Unknown |
| 9.27 | 3000 | 50 | 16 | 30 | 28 | 3-#8 | None | $M_n$ | Unknown |
| 9.28 | 3000 | 60 | 16 | 34 | 32 | 6-#8 | 3-#8 | $M_n$ | Unknown |
| 9.29 | 4000 | 40 | 14 | 16 | 14 | 4-#5 | None | $M_n$ | Unknown |
| 9.30 | 4000 | 50 | 14 | 20 | 18 | 6-#6 | 4-#6 | $M_n$ | Unknown |
| 9.31 | 4000 | 60 | 14 | 24 | 22 | 3-#7 | None | $M_n,P_n$ | Yes |
| 9.32 | 4000 | 40 | 18 | 28 | 26 | 8-#7 | None | $M_n,P_n$ | Yes |
| 9.33 | 4000 | 50 | 18 | 32 | 30 | 8-#7 | 4-#7 | $M_n,P_n$ | Yes |
| 9.34 | 4000 | 60 | 18 | 36 | 34 | 6-#8 | None | $M_n,P_n$ | Yes |
| 9.35 | 5000 | 40 | 16 | 18 | 16 | 4-#5 | None | $M_n,P_n$ | Yes |
| 9.36 | 5000 | 50 | 16 | 22 | 20 | 4-#8 | 2-#8 | $M_n,P_n$ | Yes |
| 9.37 | 5000 | 60 | 16 | 26 | 24 | 6-#7 | None | $M_n,P_n$ | Unknown |
| 9.38 | 5000 | 40 | 20 | 30 | 28 | 4-#8 | None | $M_n,P_n$ | Unknown |
| 9.39 | 5000 | 50 | 20 | 34 | 32 | 8-#8 | 4-#8 | $M_n,P_n$ | Unknown |
| 9.40 | 5000 | 60 | 20 | 39 | 36 | 6-#11 | None | $M_n,P_n$ | Unknown |

## REVIEW QUESTIONS

1. Give the three cases of loading for a beam that are equivalent to that of a flexural load plus a small axial load.
2. How big can a "small" axial load be and still be classed as "small"?
3. What happens if the "small" axial load exceeds the limit prescribed by Code?
4. What happens to the strength reduction factor $\phi$ when both flexure and a small axial compression are present?
5. What happens to the strength reduction factor $\phi$ when both flexure and a small axial tension are present?
6. How is it that moments, loads, and widths can be reproportioned by simple ratio but that effective depth $d$ cannot be?
7. What is meant by an investigation of a section?
8. How is the investigation of a section different from the design of a section?
9. In the investigation of a beam which has a nominal amount of compressive steel, how does one tell if the compressive steel was used to reduce service stresses and deflections or if it was used to increase moment capacity? What difference does it make?
10. In the investigation of a beam in which the tensile steel ratio exceeds $0.75\rho_b$, why is it necessary to verify that the tensile reinforcement is in yield?

# 10

# SHEAR IN CONCRETE BEAMS

In earlier chapters the subject of flexure in concrete beams was developed, together with methods for designing concrete beams to sustain the flexural loadings. It was found that concrete beams could be designed for flexure using the same concepts that were developed in elementary strength of materials for steel or timber. The lack of tensile strength in the concrete was overcome by adding reinforcement wherever tension was expected to occur. The procedures for designing concrete beams thus became somewhat more detailed than for steel or timber, but the concepts and procedures were familiar ones.

The lack of tensile strength in concrete also has a profound effect on the ability of concrete beams to resist shear. Even the relatively low levels of shear that are encountered in routine buildings can introduce serious tension fields in concrete beams. The design of concrete beams to sustain such shear-induced tension has no counterpart in other common structural materials; no other material is routinely reinforced for the tensions produced by beam shear.

The procedures for designing concrete beams for shear are simple and well developed and their effectiveness has been well proven over the years. Since there are hundreds of combinations of live and dead loads that could vary the shear patterns, the design for shear has evolved into a semiempirical "blanket" method which assures that the member will be capable of carrying the extremes in shear. Although the extremes may not occur in all beams, Code requires that all beams be capable of sustaining them.

A detailed analysis of shear in beams is beyond the scope of this book. Even if it were included, however, it would provide only background information; the actual design is prescribed by Code without requiring a rational analysis. The following sections contain only brief discussions and explanations of shear patterns in beams, intended to identify the sources of the shear problem and the solutions currently being used.

## Shear As a Measure of Diagonal Tension

When a concrete beam is subjected to high shearing forces, tension stresses develop in the beam. It should be noted that using "shear as a measure of diagonal tension" is an accurate summary of the approach used by ACI to predict the magnitude of these tension stresses.

A simply supported rectangular beam subject to applied loadings is shown in Figure 10-1a. A section is removed and shown in Figure 10-1b, demonstrating the shearing force that occurs across the section. Shear on a section is considered to be positive when the left side moves up with respect to the right.

The theory concerning the distribution of shear stress across a beam section is treated in elementary strength of materials texts. The typical distribution of shear stress in a rectangular beam is shown in Figure 10-1c. Shear stress is seen to be maximum at the neutral axis and zero at the extreme fibers, just reverse to the distribution of flexure stress, shown for comparison in Figure 10-1d.

Typical stressed particles are shown in Figure 10-1b. Particle I is in the zone where the concrete is in flexural compression, particle II is at the neutral axis, and particle III is in the zone where the concrete is in flexural tension. In addition to these flexural stresses, the particles are also subject to shear stresses.

The three stressed particles of Figure 10-1b are shown larger in Figure 10-2a, with their flexural stresses $f_c$ and $f_t$ and shear stress $v$ shown on the faces of the particles. Also included as Figure 10-2b, c are two subparticles cut from the original particle, taken at +45° and –45°. At the neutral axis (particle II) it is seen that the shear stress actually becomes a tensile stress at +45° and a compressive stress at –45°. There is no flexural stress at the neutral axis to influence these stresses.

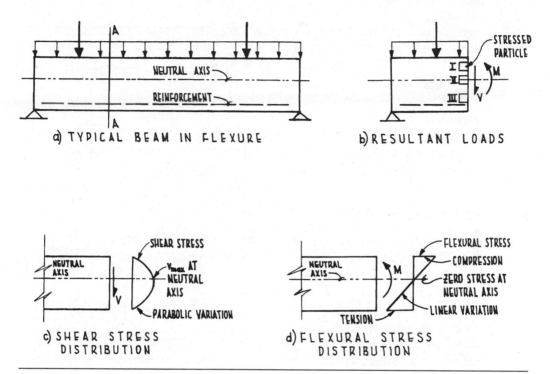

**Figure 10-1**    Beam under load.

When the tensile stress due to shear is combined with the compressive stress due to flexure (particle I, Figure 10-2b), it is seen to reduce the size of the compressive stress. Depending on the magnitudes of these two stresses, the shear stress $v$ could actually cause some tension well into the normally compressive zone, although at some angle away from the flexural compressive stress $f_c$. For the common types of loadings, the magnitude of the shear stress will be significantly greater than the compressive stress only toward the ends of the beam, where shears are high and moments are low.

In particle III, however, the tensile stress due to shear at +45° adds to the tensile stress due to flexure (particle III, Figure 10-2b). Their sum sharply influences the tension cracking in the concrete. It is this region of the cross section where the shear stresses may be investigated and reinforcement must be considered. Note that the worst case is always at +45°; the tensile stress is always less when the angle is –45°.

At the ends of a simply supported beam where there is no flexural stress and the shear stresses are highest, the shears acting alone will produce

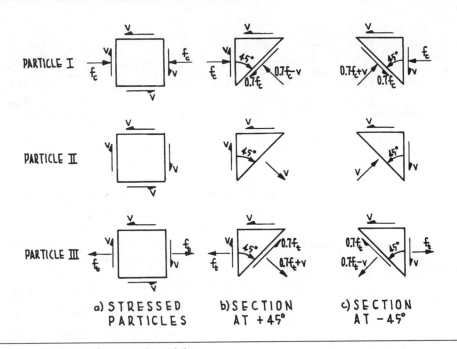

**Figure 10-2**    Stressed particles.

tension stresses acting at 45° from horizontal, as shown in Figure 10-3a. Examining the stress pattern from left to right across the span, it is seen that the angle must slowly change across the span toward the center as the shear decreases to zero and the bending moment increases to maximum. At midspan, where there is no shear on the section, the moment acting alone produces a tension stress acting horizontally.

The tension cracks that can be expected to accompany the shear and moment stresses are also shown in Figure 10-3a. As stated earlier, these cracks are usually quite small, to the point of being invisible to the naked eye. Regardless how small, however, these cracks must exist if the steel reinforcement is to reach any significant levels of stress.

The general direction of tension stresses in a symmetrically loaded beam is shown in Figure 10-3b. Because concrete is so weak in tension, reinforcement must be provided wherever the diagonal tension reaches significant levels. It should be noted that diagonal tension can occur wherever there is shear stress; it is toward the ends of the beam at the neutral axis where the shear stress is highest that the problem is most serious.

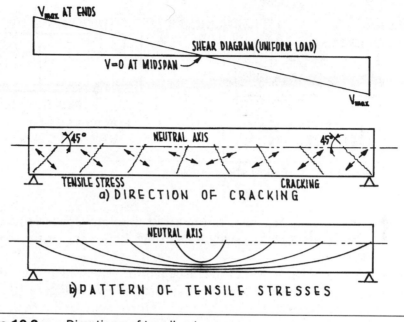

**Figure 10-3**     Directions of tensile stresses.

## Types of Effective Shear Reinforcement

Refer again to Figure 10-3. It is apparent that the directions of the cracks shown in Figure 10-3a will always be perpendicular to the direction of the tensile stresses shown in Figure 10-3b. A potential crack would therefore form first at the bottom of the beam, where tensile stresses due to flexure would be highest (Figure 10-1d). At the lowest point, the shear stresses would be zero (Figure 10-1c), so the crack would start as a vertical crack. The crack would then progress upward, turning gradually toward 45° as it nears the neutral axis. where shear stress is maximum and flexural stress is zero. At all points, the direction of the crack would be perpendicular to the direction of the tensile stress, as shown in Figure 10-3b.

The crack pattern just described is typical of a shear crack in beams, as indicated in Figure 10-4a. In this type of failure, the longitudinal steel is largely ineffective in resisting the separation of the beam into an upper piece and a lower piece. To be effective, any additional reinforcement would have to cross the crack and thereby keep the crack from becoming large enough to be detrimental. As noted before, the crack *must* form

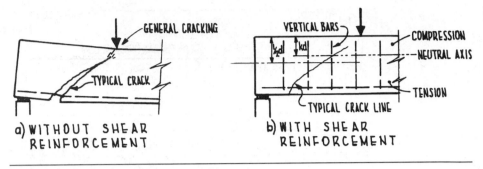

**Figure 10-4**   Typical shear cracking.

before the reinforcement can be stressed to any appreciable amount. The crack need be only the size of a hairline, however, for the steel to become effective.

Ideally, the shape of the tensile reinforcement in a beam should follow the general pattern of the tensile stresses, shown in Figure 10-3b. Spaced at some relatively close spacing, the reinforcement would then cross any potential crack that might form, such as those shown in Figure 10-3a. Such a pattern of steel bars would be difficult to bend and nest into such a shape, but an approximate shape can readily be configured.

The schematic pattern of reinforcement shown in Figure 10-5a follows one such approximate configuration. When compared to the directions of stress shown in Figure 10-3b, it is apparent that this approximation is reasonably close to that desired. The longitudinal flexural reinforcement has simply been bent upward across the web after the moment has decreased to the point that it is no longer needed as flexural reinforcement. (The ends of the bars are bent back into a horizontal plane to produce anchorage.)

Toward the middle of the span, shear stresses become much lower and little, if any, shear reinforcement may be required. The crack pattern in this area is shown in Figure 10-3a and is almost vertical; such tension comes as a result of flexural tension rather than shear. Toward the ends of the beam, the diagonal steel of Figure 10-5a intercepts the cracks almost at right angles and becomes effective reinforcement against diagonal tension.

Less efficient, but still effective, the vertical bars of Figure 10-5b also intercept the potential shear cracks. In combination with the separate bars used for longitudinal reinforcement, the configuration of Figure 10-5b provides only a rude approximation of the directions shown in Figure 10-3b.

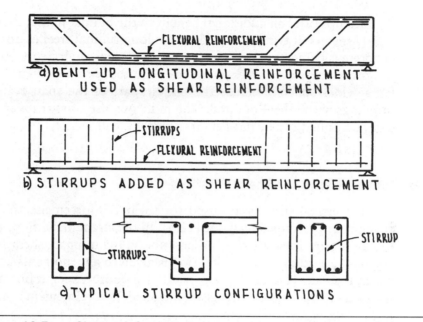

**Figure 10-5**    Shear reinforcement.

However approximate, the crack patterns have nonetheless been intercepted and the desired purpose has been achieved.

The pattern of diagonal shear reinforcement formed by bending the longitudinal reinforcement, as shown in Figure 10-5a, has been in use for many years. Additional diagonal bars may be added to extend the pattern of diagonal bars toward both the center of the span and the ends. In recent years, however, labor costs have increased disproportionately in comparison to materials costs and this method has been replaced by methods requiring less labor.

With few exceptions, the pattern of separate vertical bars shown in Figure 10-5b is used for shear reinforcement in today's practice. The cost of the small additional amount of steel is more than offset by lower labor costs and faster handling times. In this book, only this pattern of vertical shear reinforcement is presented.

Typical configurations of the separate vertical bars are shown in Figure 10-5c. Called *stirrups*, these bars are spaced along the span according to the level of shear stress; where shear stresses are higher, the stirrups are placed closer together. Where shear stresses are nearly zero toward the center of the span, stirrups can theoretically be omitted.

Stirrups offer an additional benefit to the builder in that they provide a sturdy means of tying and holding the longitudinal steel in position while concrete is being cast. For that reason, the builder will frequently use light-weight stirrups even where none are required, or may extend the stirrups (at a wider spacing) across the middle part of the span even when the drawings show them omitted. The judicious use of stirrups in the design can be applied somewhat generously; they will usually be doubly beneficial.

## Code Requirements for Shear

Where vertical stirrups are used for shear reinforcement, their design is stringently prescribed by Code (11.5). In the strength method, a simplified design is permitted which is based only on the ultimate shear force acting across the section; no consideration is given to any moments acting on the section. Code (11.3.2) also prescribes a more exact refinement of the method in which the moments are included; that refinement is not included in the following discussions.

Code (B.7) also permits the design of shear reinforcement using the alternate method, or working stress method, which will be discussed fully in a subsequent section. As usual in the alternate method, Code prescribes allowable stresses rather than allowable loads. Even so, both the strength method and the alternate method use essentially the same approach, but it is emphasized that one is not an exact multiple of the other.

In the strength method, Code (11.1.1) recognizes that the concrete by itself will take a portion of the total shear force even at failure. The concrete takes this transverse shearing force across the cracked section through friction; remember that the cracks are extremely fine and the rough cracked surface will still develop a friction force. The remainder of the shear force is, of course, taken by stirrups. That part taken by concrete alone, $V_c$, is computed by

$$V_c = 2\sqrt{f_c'}b_w d \qquad (10\text{-}1)$$

where $b_w$ is the width of the concrete web (or stem of a tee) and all other symbols are the same as defined earlier. For circular sections, the effective web width $b_w$ is taken as the diameter of the section.

The terms used in Eq. (10-1) are usually interpreted as an "average" ultimate shear stress of $2\sqrt{f_c'}$ acting across the somewhat arbitrary web area (or stem area of a tee) given by $b_w d$. Although this is a convenient interpretation and this "average" ultimate shear stress is a useful parameter, the

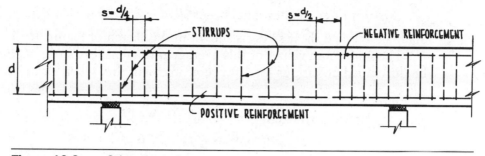

**Figure 10-6**     Stirrup spacing.

Code makes no such claim. The equation is simply a highly generalized but proven means to compute the load taken by the concrete; it applies equally to rectangular beams and slabs as well as tee shapes, circular shapes, and irregular shapes.

The remainder of the nominal ultimate shear force $V_n$ is taken by the vertical steel stirrups. That force, $V_n - V_c$, is designated $V_s$ and is also prescribed by Code (11.5.6), but it is, of course, dependent on the spacing of the stirrups. The horizontal spacing of the stirrups, shown in Figure 10-6, is specified as a portion of the effective depth, such as $\frac{1}{2}d$, $\frac{1}{4}d$, and so on. As the stirrups are spaced closer together, the force $V_s$ is permitted to go higher.

Code (11.5.4) prescribes a maximum spacing of $\frac{1}{2}d$ for stirrups and implies a minimum spacing of $\frac{1}{4}d$, but places no restrictions on the number of intermediate spacings that may be used. Considering the overall level of accuracy of the highly generalized design method, a rigorous calculation using several intermediate stirrup spacings is difficult to justify. In this text, stirrup spacings will be limited to the two Code values, $\frac{1}{2}d$ and $\frac{1}{4}d$.

The force per space $V_s$ that may be taken by a single stirrup is given by Code (11.5.6.2) in terms of the spacing:

$$V_s = V_n - V_c = \frac{A_v f_y d}{s} \qquad (10\text{-}2)$$

where $s$ is the stirrup spacing, shown in Figure 10-6, and $A_v$ is the total cross-sectional area of steel in the stirrup; other symbols are those used previously.

The nominal ultimate shear force $V_n$ that the entire section can develop, to include benefits due to stirrups, can be stated as some overall "average" ultimate shear stress $v_n$ times the cross-sectional area:

$$V_n = v_n b_w d \qquad (10\text{-}3)$$

Such an average shear stress is shown in the sketch of Figure 10-7.

The maximum values of this average ultimate shear stress $V_n$ that can be developed at various levels of reinforcement are, when extracted from the Code:

$$\text{No stirrups:} \quad v_n \leq 2\sqrt{f_c'} \qquad (10\text{-}4a)$$

$$\text{Stirrups at } \tfrac{1}{2}d: \ v_n \leq 6\sqrt{f_c'} \qquad (10\text{-}4b)$$

$$\text{Stirrups at } \tfrac{1}{4}d: \ v_n \leq 10\sqrt{f_c'} \qquad (10\text{-}4c)$$

Code limitations do not permit the average ultimate shear stress to exceed $10\sqrt{f_c'}$.

When the values of Eq. (10-4) are used to compute the shear forces, it is convenient to use the ratio of the stresses rather than the stresses themselves.

$$\text{For concrete alone:} \quad v_c \leq V_n b_w d = 2\sqrt{f_c'}\, b_w d$$

$$\text{Without stirrups:} \quad V_n \leq V_c \qquad (10\text{-}5a)$$

$$\text{With stirrups at } \tfrac{1}{2}d: \ V_n \leq 3V_c \qquad (10\text{-}5b)$$

$$\text{With stirrups at } \tfrac{1}{4}d: \ V_n \leq 5V_c \qquad (10\text{-}5c)$$

The 1:3:5 ratios given in Eq. (10-5) are easy to remember and apply. One first determines the ultimate shear force $V_c$ that the section can sustain without any shear reinforcement, in which case $V_c = 2\sqrt{f_c'}\, b_w d$. Then, if stirrups are added at a spacing of $\tfrac{1}{2}d$, the section can take a total shear force $V_n = 3V_c$. Or, if stirrups are added at a spacing of $\tfrac{1}{4}d$, the section can take a total shear force $V_n = 5V_c$. There is thus no need for complex calculations to find the allowable shear force on a reinforced section.

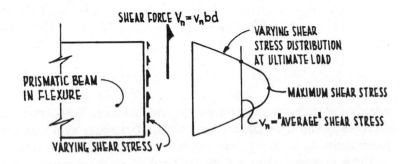

**Figure 10-7**   Average shear stress on a section.

Before proceeding with an example, it is well to list several other requirements and practices:

1.  Sections located closer than a distance $d$ from face of support may be designed for a "critical" shear force $V_{cr}$ computed at a distance $d$ from face of support.
2.  The first stirrup shall be placed within a distance of $\frac{1}{2}d$ from face of support where the required spacing is $\frac{1}{2}d$ and within a distance of $\frac{1}{4}d$ from face of support where the required spacing is $\frac{1}{4}d$.
3.  Minimum shear reinforcement shall be placed wherever the nominal ultimate shear force $V_n$ acting on the section exceeds one-half the shear strength of the concrete $V_c$.
4.  Minimum total cross-sectional area of shear reinforcement in all legs is designated $A_v$ and is computed by

$$A_v = 50\frac{b_w s}{f_y} \tag{10-6}$$

5.  Strength reduction factor $\phi$ for shear is 0.85.

Requirement 3 is something of a curiosity. It states that shear reinforcement is not required at all unless the nominal ultimate shear force exceeds $V_c$, but when the nominal ultimate shear does exceed $V_c$, the shear reinforcement must begin back at $\frac{1}{2}V_c$ rather than at $V_c$.

In addition to the foregoing requirements, there is a practical matter to be considered concerning the placement of stirrups. Heavy stirrups at close spacing are most likely to be required at the ends of heavily loaded girders. At a module point, it can be expected that two such girders with their heavy negative reinforcement and their stirrups will intersect a column with its heavy vertical reinforcement and ties. The congestion of reinforcement in such locations is often formidable. Wherever reasonable, the size of the girders should be kept large enough that the minimum stirrup spacing of $\frac{1}{4}d$ is never required; a spacing no smaller than $\frac{1}{2}d$ should be maintained if possible.

It should be recognized that the strength reduction factor $\phi$ is applied only to external loads, not to any of the computed capacities of the section. For that reason it is recommended that the factor $\phi$ always be included with the shear diagram when making shear computations; it need never be applied thereafter. By this means, there is no confusion later whether the factor $\phi$ should or should not be applied to any other loads that may be under consideration.

## Examples In Reinforcing for Shear

The following examples illustrate the procedure in reinforcing for shear; a relatively simple continuous rectangular beam is chosen as the first example. Shear and moment diagrams are given. A more general coverage is presented in Chapter 11 with tee beams, but it should be recognized that in designing shear reinforcement, all sections are considered to be rectangular; the shear capacity is based on a rectangular area $b_w d$ regardless whether the section is a tee, a rectangle, or a circle.

### EXAMPLE 10-1.

Design of shear reinforcement. Flexural design already completed.

Determine the shear reinforcement for the symmetrically loaded rectangular beam shown below. Reinforcement for flexure has already been

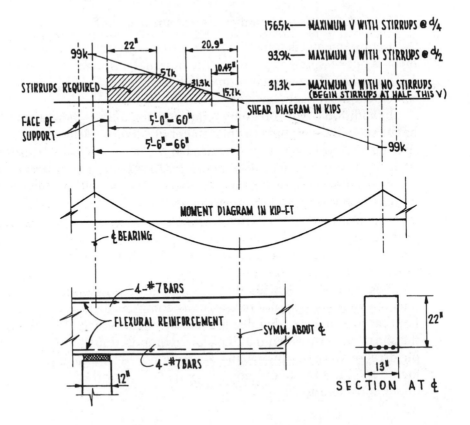

selected as indicated. Shear and moment diagrams include dead load and are drawn for the nominal ultimate values of $V_n$ and $M_n$; they include the strength reduction factor $\phi$. Use grade 60 steel, $f'_c = 3000$ psi.

Calculate (or scale) the critical shear $V_{cr}$ at a distance $d$ from face of support, by ratio;

$$V_{cr} = \frac{60-22}{66}\,99 = 57 \text{ kips (includes } \phi )$$

Determine the allowable shear forces for the three levels of reinforcement:

For concrete alone: $V_c = 2\sqrt{f'_c}b_w d = 2\sqrt{3000} \times 13 \times 22$
$= 31.3$ kips

No stirrups: $\qquad V_n = 1 \times 31.3 = 31.3$ kips
Stirrups at $\tfrac{1}{2}d$: $\qquad V_n = 3 \times 31.3 = 93.9$ kips
Stirrups at $\tfrac{1}{4}d$: $\qquad V_n = 5 \times 31.3 = 156.5$ kips

Determine the horizontal distance to $V_{cr}$ (without stirrups) from the centerline of beam:

$$\text{distance} = \frac{31.3}{99}\,66 = 20.9 \text{ in. from centerline of beam}$$

The foregoing shears and distances have been plotted on the shear diagram above. Note that a stirrup spacing of $\tfrac{1}{2}d$ (11 in.) will be required, beginning within a distance of $\tfrac{1}{2}d$ from face of support and extending to within 10.45 in. of the centerline of the beam. Since the stirrups extend so close to the center, it is chosen in this case to use stirrups across the entire span.

The chosen layout of stirrups is shown in the following sketch.

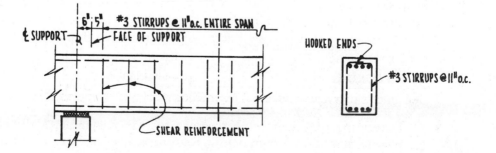

Calculate the required size of the stirrups for the indicated layout, for grade 60 steel. The maximum shear to be carried by any section is 57 kips.

$$V_s = V_{cr} - V_c = \frac{A_v f_y d}{s} \quad A_v = \frac{(V_{cr} - V_c)s}{f_y d}$$

$$A_v = \frac{(57 - 31.3) \times 10^3 \times 11}{60,000 \times 22} = 0.214 \text{ in.}^2 \text{ (in two legs)}$$

Use stirrups, No. 3 bars at 11 in. o.c., as shown in the sketch.

---

The stirrups chosen for Example 10-1 are shown with "hooks" at their upper end. The hooks are an anchorage requirement that is discussed in Chapter 12. It is also an anchorage requirement that the stirrups be anchored within the compression side of the beam. Theoretically, for those parts of the span where the bottom of the beam is in compression, the stirrups would have to be turned upside down, that the hooks are on the bottom. Such a state of stress occurs at the supports, where the moment is negative, but turning the stirrups is rarely done.

Just as it is possible to vary the stirrup spacing across the span to suit the variations in the shearing force, so is it possible to vary the size of the bar used for stirrups where the load is small enough to justify it. Such refinements are rarely made in small buildings. The tonnage of steel that can be saved by such measures would rarely justify the time and effort spent in engineering, drafting, field layout, and in just keeping track of the additional mark numbers of stirrups.

Where a beam is not quite symmetrically loaded but is nearly so, the shear reinforcement is usually laid out symmetrically to avoid the additional labor hours in drafting and field layout. It also obviates any possibility of the stirrups being installed backwards in the beam. On occasion, however, where a beam is distinctly unsymmetrical, the shear reinforcement must be laid out to suit the actual shear diagram. The procedure is the same, just more complex, as illustrated in the next example.

## EXAMPLE 10-2.

Design of shear reinforcement. Flexural design already completed.

Determine the shear reinforcement for the unsymmetrically loaded rectangular beam shown below. Reinforcement for flexure has already been selected as indicated. Shear and moment diagrams include dead load and are drawn for the nominal ultimate values of $V_n$ and $M_n$; the indicated loads include load factors as well as the strength reduction factor $\phi$. Use grade 60 steel and $f'_c = 3000$ psi.

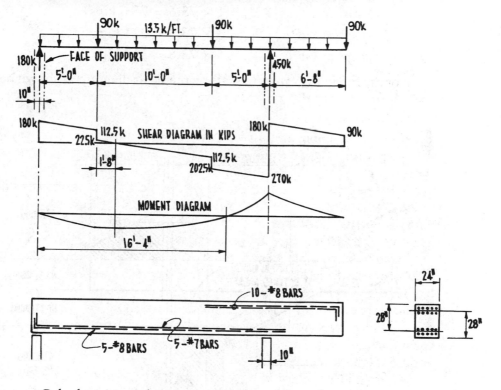

Calculate (or scale) $V_{cr}$ at a distance $d$ from face of support:

At right side of left support, 33 in. from centerline of support:

$$V_{cr} = 180 - (180 - 112.5) \times \frac{33}{60} = 143 \text{ kips}$$

At left side of right support, 33 in. from centerline of support:

$$V_{cr} = 270 - (270 - 202.5) \times \frac{33}{60} = 233 \text{ kips}$$

At right side of right support, 33 in. from centerline of support:

$$V_{cr} = 180 - (180 - 90) \times \frac{33}{80} = 143 \text{ kips}$$

Determine the allowable shear forces for the three levels of reinforcement:

For concrete alone: $V_c = 2\sqrt{f'_c}b_w d = 2\sqrt{3000} \times 24 \times 28$
$= 73.6$ kips

No stirrups: $V_n = 1 \times 73.6 = 73.6$ kips

Stirrups at $\frac{1}{2}d$: $V_n = 3 \times 73.6 = 221$ kips

Stirrups at $\frac{1}{4}d$: $V_n = 5 \times 73.6 = 368$ kips

Plot the foregoing values to scale on the shear diagram as shown below.

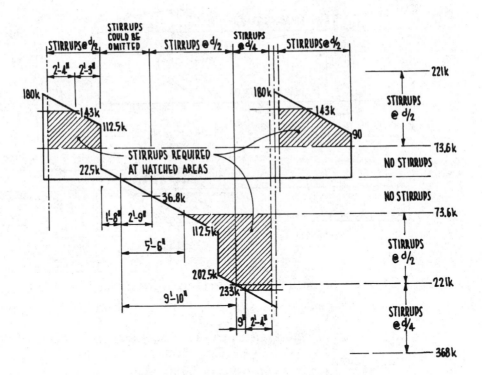

Calculate the required size for the stirrups:

$$V_s = V_n - V_c = \frac{A_v f_y d}{s} \qquad A_v = \frac{(V_n - V_c)s}{f_y d}$$

For spacing at $\frac{1}{4}d$:    $A_v = \dfrac{(233 - 73.6) \times 7}{60,000 \times 28}$

$$= 0.664 \text{ in.}^2$$

For spacing at $\frac{1}{2}d$:    $A_v = \dfrac{(221 - 73.6) \times 14}{60,000 \times 28}$

$$= 1.23 \text{ in.}^2$$

Use four legs, No. 5 bars, $A_v = 1.23$ in.$^2$

Final shear reinforcement is shown in the following sketch.

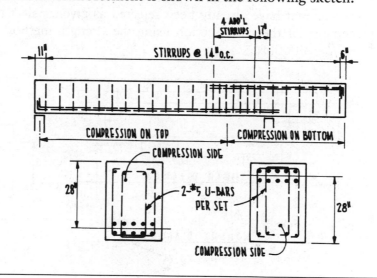

## Shear at Service Levels of Stress

The procedure for the design of shear reinforcement at service levels of stress is identical to that at ultimate levels of load; one is a direct scalar multiple of the other. At service levels of stress,

| | | |
|---|---|---|
| unreinforced, | $v_{sv} = V_n/1.7 = 1.18\sqrt{f_c'}$ | (10-6) |
| no stirrups, | $V_{sv} = 1V_c = 1.18\sqrt{f_c'}b_w d$ | (10-6a) |
| with stirrups @ $\frac{1}{2}d$, | $V_{sv} = 3V_c$ | (10-6b) |
| with stirrups @ $\frac{1}{4}d$, | $V_{sv} = 5V_c$ | (10-6c) |

Since one method is simply a direct numerical multiple (or fraction) of the other, there is no particular reason to prefer one method over the other. As a consequence, only the ultimate strength method is included here. If one wishes, however, the entire design for shear could be performed at service levels, simply by using the reduced values for the allowable stresses in concrete and in steel.

## OUTSIDE PROBLEMS

An interior span in a continuous rectangular beam has the shear and moment diagrams indicated below. The size of the member and its flexural reinforcement have already been selected as given. Select the shear reinforcement for the given section using the strength method. Use grade 60 steel, $f'_c = 3000$ psi.

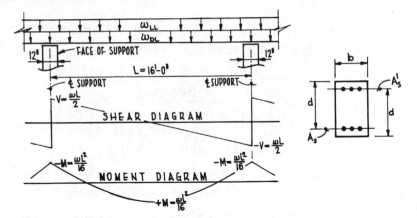

| Prob. No. | Uniform Load DL (kips/ft) | LL (kips/ft) | $d$ (in.) | $b_w$ (in.) | $A_s$ (in.²) | $A'_s$ (in.²) |
|---|---|---|---|---|---|---|
| 10.1 | 3.5 | 15.75 | 36 | 18 | 3.14 | 1.50 |
| 10.2 | 5.25 | 15.75 | 34 | 20 | 4.00 | 2.09 |
| 10.3 | 7.00 | 14.00 | 32 | 22 | 4.68 | 2.53 |
| 10.4 | 8.75 | 14.00 | 30 | 24 | 5.57 | 2.88 |
| 10.5 | 10.50 | 12.25 | 28 | 26 | 6.33 | 3.12 |
| 10.6 | 12.25 | 12.25 | 28 | 28 | 7.33 | 3.76 |
| 10.7 | 14.00 | 10.50 | 26 | 28 | 7.57 | 3.88 |
| 10.8 | 15.75 | 10.50 | 24 | 30 | 8.78 | 4.10 |
| 10.9 | 17.50 | 8.75 | 24 | 30 | 9.37 | 5.07 |
| 10.10 | 19.50 | 8.75 | 24 | 32 | 10.30 | 5.57 |

## REVIEW QUESTIONS

1. When a particle is subjected to a pure shear, how is it that a tension stress occurs?

2. Where in the cross section of a beam is flexure stress highest? Lowest? Where is shear stress highest? Lowest?

3. Where in the span of a simply supported beam is bending moment highest? Where are shearing forces highest?

4. In view of the answers to Questions 2 and 3, explain why the design for shear in a concrete beam may be treated generally independently from the design for flexure.

5. In view of the answers to Questions 2 through 4, explain why the direction of shear cracks will be close to 45° at the ends of a simple span, becoming more nearly vertical toward midspan.

6. How do vertical stirrups provide reinforcement for a diagonal tension stress?

7. On a beam having a circular cross section, what is the shear width $b_w$?

8. In view of the answer to Question 7, determine an effective width for $b_w$ for a hollow circular beam of reinforced concrete.

9. What is the effective width $b_w$ for an I shape, similar to that shown in Figure 3-1? Justify your answer.

10. What is the strength reduction factor $\phi$ for shear?

11. Where is the strength reduction factor $\phi$ applied in the ultimate strength design method?

12. Where is the strength reduction factor $\phi$ applied in the working stress design method?

# CHAPTER
# 11
# TEES AND JOISTS

$T$he topics treated in this chapter utilize the theoretical concepts introduced in earlier chapters. Since there is no development of any new theory, this chapter might be viewed as a "how-to" chapter on the design of tee beam floor and roof systems. Joists are also included, but it will be seen that they are simply tee beams that meet special conditions, thereby gaining some design advantages under the Code.

Tee beams are one of the more important features that may be used in concrete construction, important enough that an entire chapter devoted to their design is justified. The advantage of having a concrete deck working monolithically with its concrete beams offers a significant increase in efficiency of materials. For comparison, designing a steel deck to be stressed monolithically with its steel supporting member would rarely be feasible outside aircraft or ship design.

A typical concrete tee beam is shown in Figure 11-1. A certain width of the concrete floor slab becomes the compression flange whenever the beam is in positive moment (tension on bottom). This compressive width, designated $b'$ in Figure 11-1, is the effective width for flexure only; the width of the web, $b_w$, is still the effective width for shear.

A system of repetitive tee beams might well be considered a thick slab that has had its reinforcement lumped together at regular intervals and in addition has had the concrete in the unreinforced areas removed to save weight. Since all concrete in the tension zone ceases to exist, insofar as the flexural analysis is concerned, it does not really matter whether these

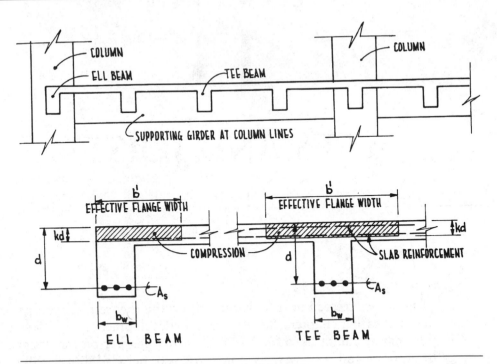

**Figure 11-1**     Typical tee beams and ell beams.

"dead" areas are solid or void. The tee beam is thus equivalent to a thick slab with its dead weight considerably reduced.

Although the removal of the recessed areas makes little difference in the capacity to resist flexure, it causes a serious reduction in the capacity to resist shear. Tee beams are quite weak in shear compared to slabs and usually require shear reinforcement. The spacing of repetitive tee beams should therefore be set with due regard to the shear capacity of each tee beam.

## Tee Beams Subject to Negative Moment

Design for flexure and design for shear were developed independently of each other in earlier chapters. Both may now be combined in the design of a typical tee-beam stem subject to a large shear and a small negative moment. Such cases commonly occur where spans are short and loads are heavy.

Under negative moment, the compressive area of a tee beam is at the

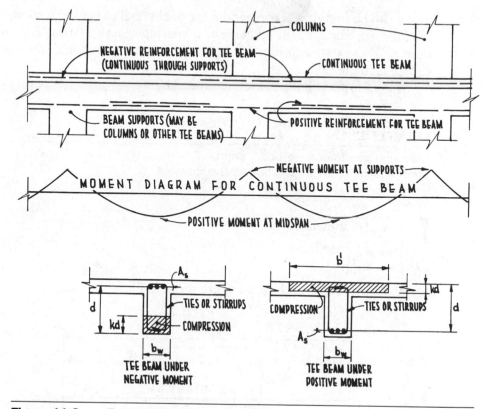

**Figure 11-2**     Tee beam subject to moments.

bottom of the stem, as shown in Figure 11-2. The top of the section is in tension and therefore disappears as far as the flexural analysis is concerned. The section responds to load exactly like an upside-down rectangular section.

For continuous rectangular beams, the negative moment and the high shears at the supports can be expected to be larger than the positive moment and low shears at midspan; conditions at the supports will therefore usually govern the design. For continuous tee beams, where the compressive area is so much smaller at the support than at midspan, the conditions at the supports will almost always govern the size of the tee-beam stem.

The size of the section is usually determined first for flexural considerations; in this solution the section modulus is found and solved for $b_w$ and $d$. Then, from the shear criteria, a separate and independent solution for $b_w$

and $d$ can be found, where $b_w$ and $d$ are deliberately kept large enough that stirrups are not needed, or at best, a spacing closer than $d/2$ is not required. These two solutions for $b_w$ and $d$ are then compared and the final choice is made.

An example will illustrate the procedure.

### EXAMPLE 11-1.

Selection of size of a tee-beam stem.
Stem subject to shear and to negative moment.

**Given:**

Beam as shown
Grade 60 steel
$f'_c = 3000$ psi
Values of shear and moment are those at ultimate load

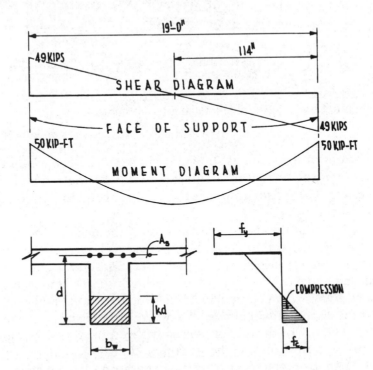

**Find:**

Required size of stem

**Solution:**

Sizes are first determined for flexural load, $M_n$.

At the face of the support, $M_n$ = 50 kip-ft

From Table A-8 for the balanced stress condition,

$$\rho = 0.0060, \quad Z_c = 0.132bd^2$$

$$\frac{M_n}{0.85f_c'} = Z_c, \quad \frac{50,000 \times 12}{0.85 \times 3000} = 0.132bd^2$$

Use $b = 0.5d$, solve for $d = 15.3$ in.

For flexure, use $b = 8$ in., $d = 15.5$ in.

Second, determine sizes for the end shear, $V_n$ = 49 kips.

At a distance $d$ from face of support, with $d = 15.5$ in. as a first approximation for $d$,

$$V_{cr} = \frac{114 - 15.5}{114} \times 49 = 42.3 \text{ kips}$$

Determine the required size for shear, with stirrups at $d/2$ and $b_w = 0.5d$

$$V_{cr} = v_n b_w d; \quad 42,300 = 3 \times 2\sqrt{3000} \times 0.5d \times d$$

$$d = 16.0$$

For shear, use $b = 8$ in., $d = 16.0$ in.

The solution for shear is compared to the solution for flexure. The larger size is chosen for the final design.

From shear requirements

Use $b = 8$ in., $d = 16$ in., stirrups at $\frac{1}{2}d$.

Using these dimensions, solve for the final value of $\rho$. In this particular case, the flexural solution does not change, so,

Use $b = 8$ in., $d = 16$ in., $\rho = 0.0060$, stirrups @ $\frac{1}{2}d$.

It is again emphasized that the first calculation for the required size of the stem is made for flexure. The second calculation is made for shear. It was noted earlier that for continuous beams, such as this example, the negative moment can be expected to be larger than the positive moment and will therefore govern the size of the stem. The shears are then reviewed to see that shear reinforcement is acceptable.

The foregoing solution would be identical if the beam had been specified as a rectangular section rather than a tee section. Such rectangular beams occur in structural systems where a cast-in-place rectangular concrete frame is used to support flooring made of precast prestressed "planks." In these systems, which are in relatively common use, the supporting member is obviously a rectangular section throughout its length.

## Code Requirements for Tee and Ell Shapes

When the moment on a tee beam becomes positive, the tee section suddenly gains a great deal more compressive material, as shown in Figure 11-3. When the neutral axis falls within the top slab, or flange, as indicated in Figure 11-3b, the beam still responds to load as does any other rectangular section, but as one having a compressive width of $b'$. All the design procedures discussed previously are still valid; the section simply happens to be wide and shallow.

A complication arises when the neutral axis falls below the soffit of the slab, as shown in Figure 11-3c. For this case the compressive area is no

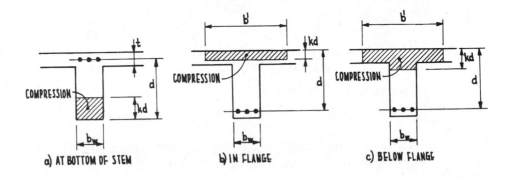

**Figure 11-3**   Compression zones in tee sections.

longer rectangular. The amount of error between this case and the truly rectangular case will be seen later to be quite small at service levels and in some cases can be ignored.

Before beginning the design of a tee section subject to positive moment, it is first necessary to define the effective dimensions of a tee beam. Typical sections of the two most common types of tee beams are shown in Figure 11-4. One of these types is commonly called an ell beam but it functions like a tee beam. At midspan the moment is positive and the floor slab is in compression; the floor slab is seen to be the compression flange of the tee beams.

As shown in Figure 11-4, not all of the slab will be effective as a compression flange. Code (8.10.2) prescribes the width of the slab that may be considered to be effective in compression; these limits and requirements are based on extensive performance tests. Test results indicate that the effective flange will not extend halfway to the next stem unless the stems are spaced fairly close together.

Code (8.10) limitations concerning the effective compression area are given in the following requirements.

1.  The total width of slab effective as a tee beam flange or as an ell beam flange (dimension $b'$ in Figure 11-4) shall not exceed one-fourth the span length of the beam.
2.  The effective overhanging flange width on each side of the web (or stem) shall not exceed eight times the slab thickness or one-half the clear distance to the next web.

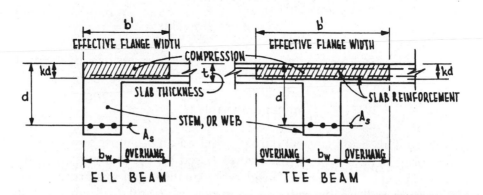

**Figure 11-4**    Effective flange widths.

3. For beams having a slab on one side only (ell beams), the effective overhanging flange width shall not exceed one-twelfth the span length of the beam, six times the slab thickness, or one-half the clear distance to the next web.

4. Isolated beams in which the tee shape is used to provide a flange for additional compressive area shall have a flange thickness no less than one-half the width of the web and an effective flange width no more than four times the width of the web.

Requirement 4 is applicable to the mixed precast "planks" and cast-in-place rectangular frames mentioned in the preceding section. Where an isolated framing member is required, a tee section offers a great deal more lateral stability than does a rectangular section.

The foregoing requirements are obviously somewhat arbitrary, since there can be no sharp break between the stress in the effective flange and the stress in the immediately adjacent parts of the slab. The rather rough approximation does not matter, however, since the total area of concrete is so high that concrete stresses at service levels are invariably quite low. The service stress in the concrete is rarely even checked in practice, but will be checked here as a matter of academic interest.

Occasionally, compressive reinforcement is used in a tee beam. Compressive reinforcement can be effective in reducing deflections. Tee beams, which are often comparatively shallow, invite problems with deflections.

One additional point of the Code deserves repeating. The minimum steel ratio $\rho$ is given by Code (10.5.1):

$$\text{minimum allowable } \rho \geq \frac{200}{f_y} \qquad (6\text{-}15)$$

In checking for the minimum steel ratio in tee beams, the width of the stem is used (see Figure 11-4):

$$\text{minimum usable } \rho_{\min} = \frac{A_s}{b_w d} \geq \frac{200}{f_y}$$

This artificial steel ratio $\rho_{\min}$, based on $b_w$, must be applied to tee sections. The real steel ratio, when based on the compressive width $b'$ (see Figure 11-4), may be considerably smaller than this artificial value. This point will be illustrated in Examples 11-2 and 11-3 when the steel ratio based on $b'$ must be used when selecting a tee section.

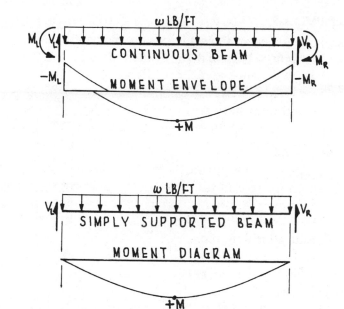

**Figure 11-5**    Common tee beam load systems.

## Tee Beams Subject to Positive Moment

Two common cases of loading on tee beams will be considered for tee-beam design. These cases are shown schematically in Figure 11-5.

1.  *Continuous tee beam with negative moment at supports.* The tee beam is subject to negative moment at its ends and to positive moment at midspan.
2.  *Simple span.* The tee beam is never subject to negative moment.

For case 1, the size of the stem is invariably set by shear and moment considerations at the support; such a calculation was presented in Example 11-1. At midspan, since the size of the stem is fixed, the only remaining calculation is for the area of reinforcement to take the positive moment. This calculation was first demonstrated in Example 6-5 where both $b_w$ and d were fixed, but will be repeated in the next example to show the effects of the wider compression area. Tee beams in positive moment can rarely be designed for the balanced stress condition.

## EXAMPLE 11-2

Design of a tee beam subject to positive moment.
Size of stem fixed by conditions at supports.

**Given:**

Tee section as shown

$M_{DL}$ = 57 kip-ft

$M_{LL}$ = 61 kip-ft

Grade 60 steel

$f'_c$ = 3000 psi

Span = 23 ft 0 in.

Tees at 10 ft 0 in. o.c.

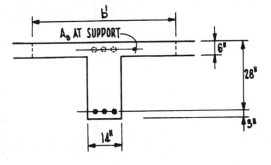

**Find:**

Required reinforcement

**Solution:**

Determine the effective width of the flange (in inches):

$$b' < \frac{\text{span}}{4};$$ $\qquad$ $$b' < \frac{23 \times 12}{4} \text{ or } 69 \text{ in.}$$

$$b' < 8 \times \text{flange} \times 2 + b_w;$$ $\qquad$ $$b' < 8 \times 6 \times 2 + 14 \text{ or } 110 \text{ in.}$$

$$b' < \text{tee spacing};$$ $\qquad$ $$b' < 10 \times 12 \text{ or } 120 \text{ in.}$$

Use $b'$ = 69 in.

Calculate the nominal ultimate moment:

$$M_n = \frac{1.4M_{DL} + 1.7M_{LL}}{\phi} = \frac{(1.4 \times 57 + 1.7 \times 61)}{0.9}$$

$$= 204 \text{ kip-ft}$$

Since both $b'$ and $d$ are known, Solve for the coefficient of $Z_c$:

$$\frac{M_n}{0.85f'_c} = Z_c \qquad \frac{204,000 \times 12}{0.85 \times 3000} = \text{coeff.} \times b'd^2 = \text{coeff.} \times 69 \times 28 \times 28$$

coeff. = 0.018 (note that this is not a balanced stress condition)

From Table A-5, select $Z_c = 0.018b'd^2$ at $\rho = 0.00075$ (interpolated)

Calculate the required steel area:

$$A_s = \rho b'd = 0.00075 \times 69 \times 28 = 1.45 \text{ in.}^2$$

From Table A-3, select 2 No. 8 bars, $A_s$ furnished $= 1.57$ in.$^2$

check: $\rho_{min} = 200/f_y = 0.00333$;

$\rho$ furnished $= 1.57/(14 \times 28) = 0.0040 > 0.0033$ (O.K.)

Use $d = 28$ in., $b' = 69$ in., $b_w = 14$ in., 2 No. 8 bars

---

Example 11-2 is quite direct. It demonstrates that the minimum steel requirement given by Code, $\rho_{min} = 200/f_y \leq A_s/b_wd$, must always be computed using the stem width $b_w$ rather than the actual compressive width $b'$. The actual design, however, is still based on the actual compressive width $b'$.

The second case to be considered where tee beams are subject to positive moment is the case where the beam is simply supported and is therefore subject to positive moment throughout the span. For this case the size of the stem has not been set by moment conditions at the support as it was in the first case. However, the shear at the support must still be considered in selecting the size of the stem.

Other considerations must also enter the selection. Depending on the magnitude of loads and length of span, the shear conditions might not be serious and may not therefore require a very large section. Consequently, a small shallow section might suffice if the end shear were to be the only consideration.

Unusually shallow tee sections, however, can introduce problems in deflections and may require excessive amounts of flexural reinforcement; they should be used with caution. To satisfy most of the routine requirements, the overall depth of a tee section should be about four to five times the thickness of the slab it supports. For shear requirements, the depth should be roughly equal to one-tenth to one-twelfth the span length. For deflection control, the minimum depths given in Table 3-1 will apply. Obviously, such rules must vary considerably with span length, intensity of load, and spacing of tee stems.

Such considerations are not usually a formal part of the calculations. All such limitations are usually juggled mentally as the designer makes the final choice. The following example, however, includes these items as formal considerations for the sake of demonstration.

## EXAMPLE 11-3

Design of a tee beam subject to positive moment.
Simply supported span, slab thickness fixed.

**Given:**

> Grade 60 steel
> $f'_c$ = 4000 psi
> Concrete weight 145 pcf
> $\phi$ = 0.90 for flexure and 0.85 for shear

**Find:**

> Suitable section to sustain the load

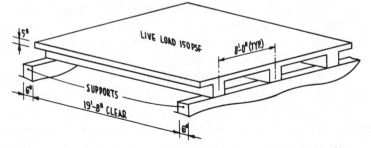

**Solution:**

Estimate the stem size (for dead-load calculations):

$$\text{approximate} \quad h = \frac{\text{span}}{10} \text{ to } \frac{\text{span}}{12} \qquad h = 20 \text{ to } 24 \text{ in.}$$

$$\text{approximate} \quad h = 5 \times \text{slab thickness} \qquad h = 20 \text{ to } 25 \text{ in.}$$

$$\text{minimum} \qquad h = \frac{\text{span}}{16} \text{ (Table 3-1)} \qquad h = 15 \text{ in.}$$

Try $h$ = 22 in., $d$ = 19 in., $b$ = 10 in.

$$\text{weight per foot} = b(h-t) \times \frac{145}{144} = 171 \text{ plf}$$

Determine the dead load and live load for the shear and moment diagrams:

uniform live load    = 150 psf × 8        = 1200 plf
uniform dead load  = ($5/12$) × 145 × 8 = 483 plf
uniform stem load                        = 171 plf
Total $w_{DL}$ = 654 plf

For the shear and moment diagrams, compute the ultimate shear and moment:

$$M_{LL} = \frac{w_{LL} \times L^2}{8} = \frac{1200 \times 19.67^2}{8} = 58 \text{ kip-ft}$$

$$M_{DL} = \frac{w_{DL} \times L^2}{8} = \frac{654 \times 19.67^2}{8} = 32 \text{ kip-ft}$$

$$M_n = \frac{1.4 \times 32 + 1.7 \times 58}{0.9} = 159 \text{ kip-ft}$$

$$V_{LL} = \frac{w_{LL} \times L}{2} = \frac{1200 \times 19.67}{2} = 159 \text{ kips}$$

$$V_{DL} = \frac{w_{DL} \times L}{2} = \frac{654 \times 19.67}{2} = 6.5 \text{ kips}$$

$$V_n = \frac{1.4 \times 6.5 + 1.7 \times 12}{0.85} = 35 \text{ kips}$$

Draw the shear and moment diagrams from the calculated values:

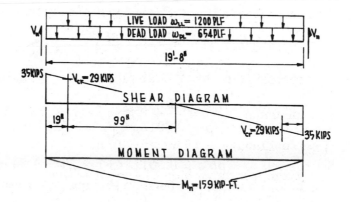

With no negative moment, the stem size is not controlled by flexure; stem size is therefore set to suit shear requirements. The dimensions $b$ and $d$ are computed first for the case where no stirrups are needed and second for the case where stirrups are placed at $\frac{1}{2}d$; the two results are then compared.

With no stirrups:  $v_n = 2\sqrt{f_c'} = 126$ psi
$V_n = v_n b_w d\ 29{,}000 = 126 \times 0.5d \times d$
Solve for $d$: $d = 21$ in., $b = 10\frac{1}{2}$ in.

With stirrups at $d/2$:  $v_n = 6\sqrt{f_c'} = 380$ psi
$V_n = v_n b_w d\ 29{,}000 = 380 \times 0.5d \times d$
Solve for $d$: $d = 12.4$ in., $b = 6.2$ in.

For the final choice between the two cases, it is elected to use the section that does not require stirrups. Critical shear is now corrected for $d = 21$ in. and the increase in dead load is added. Hence use $h = 24$ in., $d = 21$ in., $b = 12$ in. (The increase in the dead load above the initial estimate was found to be negligible.) Determine the effective width of the compression flange.

$$b' < \frac{\text{span}}{4};\qquad b' < \frac{242}{4} = 61 \text{ in.}$$

$b' < 8 \times t \times 2 + \text{stem}\quad b' < 8 \times 5 \times 2 + 12 = 92 \text{ in.}$

$b' < \text{tee spacing}\qquad b' < 96 \text{ in.}$

$\quad$ Use $b' = 61$ in.

Select reinforcement by finding the required section modulus:

$$Z_c = \frac{M_n}{0.85 f_c'}\quad \text{coeff.} \times b'd^2 = \frac{M_n}{0.85 f_c'}$$

$$\text{coeff.} \times 61 \times 21 \times 21 = \frac{159{,}000 \times 12}{0.85 \times 4000}$$

Solve for coeff.: coeff. $= 0.021$

From Table A-6, $\rho = 0.0012$, $k_n \beta_1 = 0.021$ (not at the balanced stress condition)

Select steel; $A_s = \rho b' d = 0.0012 \times 61 \times 21 = 1.54$ in.$^2$

Use 2 No. 7 and 2 No. 4 bars, no stirrups, $A_s$ furnished $= 1.60$ in.$^2$

Check: $\rho_{min} = 200/f_y = 0.0033$;

Actual $\rho = 1.6/(12 \times 24) = 0.0056 > 0.0033$ (O.K.)

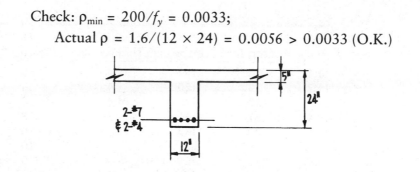

In Example 11-3 there were no limitations imposed on the overall height of the beam. Had such a limitation existed, a shallower section might have been required. A shallower section would, in turn, require shear reinforcement, probably stirrups at a spacing of $d/2$.

## OUTSIDE PROBLEMS

An interior span of a continuous tee beam has the shear and moment diagrams indicated below. The live load and the beam spacings are listed in the tabulation. The dead load of the stem must be estimated, then confirmed later when design is final. Using the strength method, select a suitable tee section for the given conditions, to include both positive and negative reinforcement and any shear reinforcement required. $f'_c = 3000$ psi, grade 60 steel.

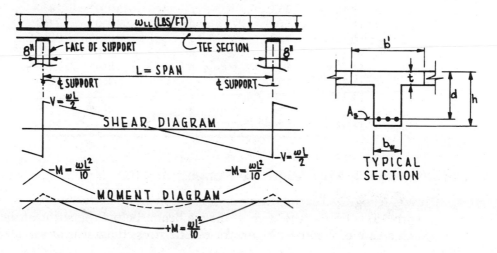

| Prob. No. | Span ft | LL psf | Slab Thickness in. | Tee Spacing ft |
|---|---|---|---|---|
| 11.1 | 24 | 100 | 4 | 8 |
| 11.2 | 23 | 120 | 4 | 8 |
| 11.3 | 22 | 140 | 4 | 8 |
| 11.4 | 21 | 160 | 4 | 8 |
| 11.5 | 20 | 180 | 5 | 10 |
| 11.6 | 19 | 200 | 5 | 10 |
| 11.7 | 18 | 220 | 5 | 10 |
| 11.8 | 17 | 240 | 5 | 10 |
| 11.9 | 16 | 260 | 6 | 12 |
| 11.10 | 15 | 280 | 6 | 12 |
| 11.11 | 14 | 300 | 6 | 12 |
| 11.12 | 13 | 320 | 6 | 12 |

Design a tee-beam floor system for a simple span with the conditions and live loads shown below. The dead load of the system must also be included. Select the flexural reinforcement and any shear reinforcement required. $f'_c$ = 4000 psi, grade 60 steel.

| Prob. No. | Live Load (psf) | Slab Thickness (in.) | Tee Spacing | Clear Span |
|---|---|---|---|---|
| 11.13 | 250 | 6 | 12′ 0″ | 18′ 0″ |
| 11.14 | 250 | 6 | 12′ 0″ | 16′ 0″ |
| 11.15 | 200 | 5 | 10′ 0″ | 22′ 0″ |
| 11.16 | 200 | 5 | 10′ 0″ | 18′ 0″ |
| 11.17 | 150 | 5 | 10′ 0″ | 24′ 0″ |
| 11.18 | 150 | 5 | 10′ 0″ | 20′ 0″ |
| 11.19 | 125 | 4 | 8′ 0″ | 28′ 0″ |
| 11.20 | 125 | 4 | 8′ 0″ | 24′ 0″ |
| 11.21 | 100 | 4 | 8′ 0″ | 30′ 0″ |
| 11.22 | 100 | 4 | 8′ 0″ | 26′ 0″ |
| 11.23 | 75 | 4 | 8′ 0″ | 32′ 0″ |
| 11.24 | 75 | 4 | 8′ 0″ | 28′ 0″ |

## Tees with Neutral Axis Below the Flange

It was recognized earlier that a complication arises in the analysis when the neutral axis falls in the stem of the tee beam, below the soffit of the slab. Such a case is shown in Figure 11-6. The stress diagrams at service levels

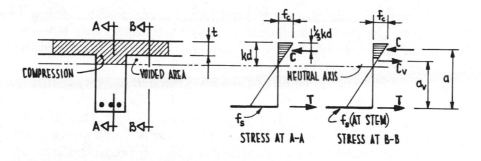

**Figure 11-6**     Neutral axis of tee shape.

are also shown; section A-A is taken within the stem and section B-B is taken through the overhang at some distance from the stem.

It is evident from Figure 11-6 that the compressive area is now irregular; it was assumed to be rectangular in the flexure analysis of Chapter 6. The effect of the new void areas, however, is quite small where the penetrations of the neutral axis into the stem is small.

Now consider the loss in compressive force, $C_v$, shown in the stress diagram of Figure 11-6. This loss in compressive force is the product of a relatively low stress acting on a relatively thin strip of area. The loss in moment is then this small force $C_v$ acting over a reduced moment arm $a_v$. In the elastic range, the total reduction in moment, $C_v a_v$, is obviously small when compared to the total elastic moment.

As load increases, the rotation of the section continues and the ultimate moment is reached; the stress diagrams shown in Figure 11-7 then become the final state of stress. Note here that the loss in moment is the full ulti-

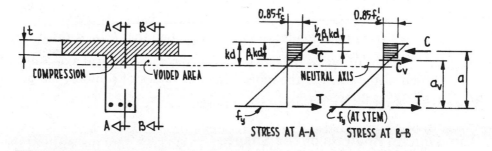

**Figure 11-7**     Tee beam at ultimate state.

mate stress acting over the void area, but again with a reduced moment arm. The loss in moment can be judged to be higher than at working levels and that it would be a small loss only as long as the penetration of the neutral axis into the stem is small.

A close examination of Tables A-5 through A-7 reveals that for a given stress ratio $\rho$, the depth of the compressive area of ultimate load, $k_n\beta_1 d$, is significantly less than the depth at elastic levels, $k_{sv}d$. The difference is more marked at lower steel ratios, where tee beams would normally fall. It may be concluded that as rotation of the section progresses from the elastic range into the ultimate state, the neutral axis shifts significantly upward and the area in compression becomes thinner.

For tee beams, this shift means that many beams having the neutral axis in the stem at elastic levels will have the neutral axis shift back into the slab at ultimate load. Consequently, the strength method of design is likely to have its compression block completely outside the stem.

The incidence of the neutral axis falling below the slab into the stem of the tee beam is somewhat unusual. It occurs where circumstances permit a very thin slab to be used, but it also requires that the stems be deep, closely spaced, and heavily reinforced. Its treatment here would require additional design tables, which, since this case is not common, are not included in this limited work. Since this case is not included in this book, the following limits provide a means to recognize when the neutral axis is in the stem of the tee and some means to avoid such a design.

Refer again to the elastic stress diagrams of Figure 11-6. It is seen that the neutral axis stays within the slab where the depth of the compression block is deliberately kept less than the slab thickness $t$. Similarly, for the ultimate stress diagrams of Figure 11-7, the compression block stays rectangular where $k_n\beta_1 d$ is kept less than the slab thickness $t$. These observations produce the following rules,

- At elastic levels, select $\rho$ low enough that $k_{sv} < t/d$.
- At ultimate loads, select $\rho$ low enough that $k_n\beta_1 < t/d$.

If $k_n\beta_1$ or $k_{sv}$ exceeds these values of $t/d$, the neutral axis will be in the stem of the tee. For the more common values of $t$ and $d$, the problem is rarely encountered.

## Approximate Analysis of Tee Shapes

The conclusions of the preceding section suggest that an approximate analysis can be made, based on the stress in the tensile steel. Refer to Figure

11-8 for the state of stress due to loads at elastic levels. The sum of moments about the compressive force C yields, at elastic levels,

$$M_{sv} = f_s A_s \left[ d - \frac{k_{sv}d}{3} \right] \qquad (11\text{-}1a)$$

Similarly, for the ultimate load condition shown in Figure 11-8,

$$M_n = f_y A_s \left[ d - \frac{k_n \beta_1 d}{2} \right] \qquad (11\text{-}1b)$$

It should be recognized that an approximate value of $k_n\beta_1 d$ or $k_{sv}$ in Eq. (11-1a) and (11-1b) could afford some useful analytical equations.

In establishing approximate values of $k_n\beta_1$ and $k_{sv}$, the values of $k_n\beta_1$ and $k_{sv}$ given in Tables A-5 through A-7 are first examined. It is recognized that the lower range of values for steel ratios are the values that would normally be applicable for tee beams. In this range it is observed that the value of $k_{sv}$ is no higher than about 0.30 and the value of $k_n\beta_1$ is no higher than about 0.10. These values are consistent with the limits assumed for $t/d$ by the general rules of the preceding section, where $d$ is taken at about $4t$ to $5t$. Substitute these values into Eq. (11-1a) and Eq. (11-1b) to find:

At service levels: $M_{sv} = f_{sv}A_s(0.90d)$      (11-2a)

At ultimate loads: $M_n = f_y A_s(0.95d)$      (11-2b)

Relative positions of the couples associated with these moments are shown in the sketches of Figure 11-9.

Eq. (11-2a) and (11-2b) provide a means to make a fast manual check on steel stresses or steel areas. The check does not require a reference to the design tables.

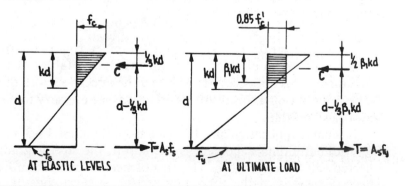

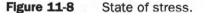

**Figure 11-8**    State of stress.

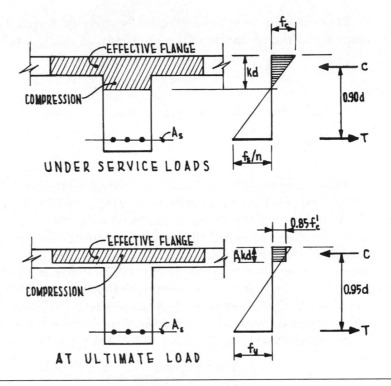

**Figure 11-9**   Approximate couples on a tee beam.

For a typical application, refer to Example 11-2. The nominal ultimate moment $M_n$ is shown as 204 kip-ft and depth is fixed at 28 in. The area of grade 60 steel can be calculated directly from Eq. (11-2b):

$$M_n = f_y A_s (0.95d) \qquad 204,000 \times 12 = 60,000 \times A_s \times 0.95 \times 28$$

Hence $A_s$ = 1.53 in². This result compares to an area of 1.45 in² obtained in the example, derived from the design tables of the Appendix. Note that it was not necessary to know the effective width of compression, $b'$, before obtaining this result. Neither was it necessary to have access to a set of design tables.

A second application may be taken from Example 11-3. In that example, the nominal ultimate moment is 159 kip-ft, the steel is grade 60, and the selected value for $d$ is 21 in. The area of steel is found from Eq. (11-2b):

$$M_n = f_y A_s (0.95d) \quad 159{,}000 \times 12 = 60{,}000 \times A_s \times 0.95 \times 21$$

Hence $A_s = 1.59$ in.$^2$ This area compares to an area of 1.54 in.$^2$ in the exact solution.

The approximate Eq. (11-2a) and (11-2b) are seen to be useful in making rough calculations and in making a quick independent check on existing calculations. They may be readily committed to memory by remembering that the moment arm for steel in the elastic range is about 90% of $d$, shifting to 95% of $d$ at ultimate load. The equations are also quite useful when there are no design tables immediately available.

When compression steel is used, its center of gravity will necessarily be close to the top face of the tee. The resultant force on this compressive reinforcement will therefore be quite close to C, the resultant force on the concrete (Fig. 11-9). As a consequence, the approximate Eq. (11-2a) and (11-2b) lose but little accuracy in beams reinforced in compression.

Many responsible designers use yet another approximate method for selecting the steel area for positive moment in all tee beams. They simply recognize that the stress in the concrete in such tee beams is always quite low, and that the approximate solution is adequate for all cases. For such a practice, the center of the compressive force C is taken at mid-height of the slab as shown in Figure 11-10.

From the geometry of Figure 11-10,

$$M_n = T(d - \tfrac{1}{2}t) = f_y A_s (d - \tfrac{1}{2}t) \qquad \text{11-3)}$$

Solve for required $A_s$ to find

$$A_s = \frac{M_n}{f_y \left( d - \tfrac{1}{2}t \right)} \qquad (11\text{-}4)$$

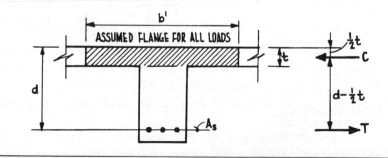

**Figure 11-10**    Assumed couple on a tee-beam slab.

Eq. 11-4 is then used for all cases, without regard for the location of the neutral axis. The result is almost always conservative, however, and will produce slightly higher values for the required steel area $A_s$ than would be obtained using more exact methods.

An example will illustrate the method of designing tee beams, using both the exact and the approximate methods of design.

### EXAMPLE 11-4.

Selection of tee beam reinforcement.
Simply supported span.
Size of stem fixed by conditions at supports.

**Given:**

Tee section as shown
$M_{DL} = 88$ kip-ft; $M_{LL} = 68$ kip-ft
Grade 60 steel, $f'_c = 4000$ psi
Span = 24' 0", tees @ 12' 0" o.c.

**Find:**

Tensile reinforcement for positive moment

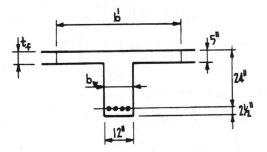

**Exact Solution:**

Determine the effective flange width

$$b' < \frac{\text{span}}{4}, \quad b' < \frac{24 \times 12}{4} \text{ or } 72 \text{ in.}$$
$$b' < 8t_F \times 2 + b_w, \ b' < 8 \times 5 \times 2 + 12 \text{ or } 92 \text{ in.}$$

$b'$ < tee spacing, $b'$ < 12' 0" or 144 in.

Use $b' = 72$ in.

Calculate the nominal ultimate moment

$$M_n = \frac{1.4M_{DL} + 1.7M_{LL}}{\phi} = \frac{1.4 \times 88 + 1.7 \times 68}{0.9}$$

$$= 265 \text{ kip-ft}$$

Determine the required magnitude of the plastic section modulus $Z_c$,

$$Z_c = \frac{M_n}{0.85f'_c} = \frac{265,000 \times 12}{0.85 \times 4000}$$

$$= 923 \text{ in.}^3$$

With both $b'$ and $d$ known, calculate the required coefficient of the plastic section modulus $Z_c$,

$Z_c = \text{coeff.} \times b'd^2$; $935 = \text{coeff.} \times 72 \times 24 \times 24$

coeff. = 0.023

Enter Table A-6, select a section

$\rho = 0.0013$, $Z_c = 0.023$

Determine the required steel area

$A_s = \rho b'd = 0.0013 \times 72 \times 24 = 2.25 \text{ in.}^2$

From Tables A-3 and A-4,

Use 4 No. 7 bars, $A_s$ furnished = 2.41 in.²

Approximate Solution No. 1 [Eq. (11-2)]

Calculate the nominal ultimate moment as before,

$M_n = 265$ kip-ft

Using Eq. (6-2b), solve for $A_s$

$M_n = f_y A_s \times 0.95d$,

$$A_s = \frac{265,000 \times 12}{60,000 \times 0.95 \times 24} = 2.32 \text{ in.}^2$$

From Tables A-3 and A-4, select reinforcement,

Use 4 No. 7 bars, $A_s$ furnished = 2.41 in.²

Approximate Solution No. 2 [Eq. (11-4)]

Calculate the nominal ultimate moment as before,

$M_n = 265$ kip-ft

Using Eq. (6-4) solve for $A_s$,

$$A_s = \frac{M_n}{f_y\left(d - \frac{1}{2}t\right)}; \quad A_s = \frac{265,000 \times 12}{60,000(24 - 2.5)}$$

$$A_s = 2.47 \text{ in.}^2$$

From Tables A-3 and A-4, select reinforcement,

Use 2 No. 9 + 2 No. 6 bars, $A_s$ furnished $= 2.88$ in.$^2$

For the three solutions, it is seen that the required area of steel varies but little; any one of the three solutions is acceptable.

## OUTSIDE PROBLEMS

Determine the required area of positive flexural reinforcement $A_s$ for the following tee beams using the exact methods developed earlier. Then recalculate the area of reinforcement using the approximate equations (11-2) and (11-4). Compare the three sets of results.

| Prob. No. | Concrete Strength $f'_c$ psi | Steel Grade | Dead Load Moment kip-ft | Live Load Moment kip-ft | Stem Width $b_w$ in. | Effective Depth $d$ in. | Slab Thick. $t$ ft | Clear Span ft | Tee Spacing ft |
|---|---|---|---|---|---|---|---|---|---|
| 11.25 | 3000 | 40 | 58 | 39 | 14" | 28" | 5" | 18' 0" | 6' 0" |
| 11.26 | 3000 | 50 | 82 | 73 | 16" | 32" | 6" | 18' 0" | 6' 0" |
| 11.27 | 3000 | 60 | 99 | 106 | 18" | 36" | 7" | 20' 0" | 6' 0" |
| 11.28 | 3000 | 60 | 118 | 141 | 20" | 40" | 8" | 20' 0" | 8' 0" |
| 11.29 | 4000 | 40 | 58 | 39 | 12" | 26" | 5" | 22' 0" | 6' 0" |
| 11.30 | 4000 | 50 | 82 | 73 | 14" | 28" | 5" | 22' 0" | 6' 0" |
| 11.31 | 4000 | 60 | 99 | 106 | 14" | 30" | 6" | 24' 0" | 8' 0" |
| 11.32 | 4000 | 60 | 118 | 141 | 16" | 32" | 8" | 24' 0" | 12' 0" |
| 11.33 | 5000 | 40 | 58 | 39 | 12" | 24" | 5" | 26' 0" | 8' 0" |
| 11.34 | 5000 | 50 | 82 | 73 | 14" | 27" | 5" | 26' 0" | 8' 0" |
| 11.35 | 5000 | 60 | 99 | 106 | 16" | 30" | 6" | 28' 0" | 8' 0" |
| 11.36 | 5000 | 60 | 118 | 141 | 18" | 33" | 8" | 28' 0" | 12' 0" |

## One-way Joist Construction

When tee beams are spaced at close intervals, the stems can be set just wide enough to develop the necessary strength in shear without requiring stirrups. At such close spacing, the effective flange width $b'$ will always be taken halfway to the two adjacent stems; the other requirements never govern. In addition, the floor slab can be made quite thin, although practical considerations usually will not permit a slab less than 2 in. thick to be used. Such a floor system, called a one-way joist system, is shown in Figure 11-11a.

In building such a configuration, the filled spaces between stems can be formed by any available block or building tile that can be held in position while the concrete is being cast around them, such as the tiles shown in Figure 11-11b. Or special steel forms for forming void space can be rented that can be handled easily, placed manually, and subjected to multiple uses. These joists are cast in place at the same time and in the same plane as their supporting girders (without having to sit on top of their supports), thereby reducing the story height by a considerable amount (Figure 11-11c).

A floor or roof system of this type is also called a *pan-joist* system or a *tile-filler* system of floor construction. It is one of the more popular floor systems in the industry and represents a major advantage in the use of concrete construction. Other materials currently being used to form the joists are such things as fiberglass pans, corrugated paper blocks, foamed concrete blocks, Styrofoam blocks, and kraft paper tubes. When fillers are used, the bottom surface is usually just sprayed to form the finished ceiling.

There are many innovations in common use. The ends of the joists can be widened or tapered as shown in Figure 11-11d to accept unusually heavy shears (or negative moments). Or the tile filler blocks can simply be omitted over short distances where shears and moments are too high for the thin stem. Shallower filler tiles or pans can be placed where a plumbing fitting requires extra slab thickness or where special structural attachments are to be placed.

Distribution members, or bridging, are required by Code at the interior of longer spans, as shown in Figure 11-11a. Such members are easily formed simply by leaving a space between the tiles or pans when they are located and fixed in the forms. An individual cross-rib for attaching special ceiling-mounted equipment or balcony hangers can be formed the same way.

The design of joist systems is the same as that for other tee-beam sys-

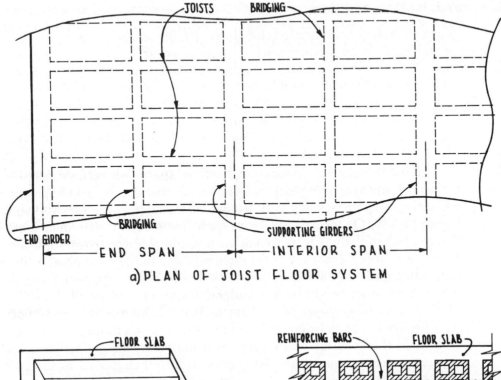

a) PLAN OF JOIST FLOOR SYSTEM

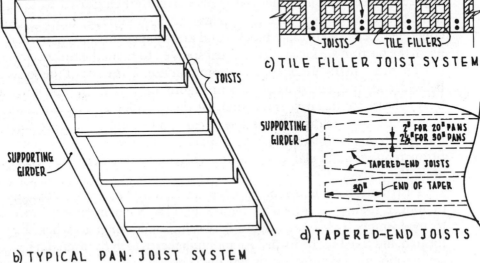

b) TYPICAL PAN·JOIST SYSTEM

c) TILE FILLER JOIST SYSTEM

d) TAPERED-END JOISTS

**Figure 11-11**   Joist systems.

tems, but with some additional Code (8.11) requirements. There are, however, some benefits to be had, such as a higher allowable shear on the section. The Code requirements that are of immediate interest are:

1.  Ribs shall be not less than 4 in. in width.
2.  Ribs shall have a depth no less than 3.5 times the least width.
3.  Clear spacing between ribs shall not exceed 30 in.
4.  When removable pans are used, slab thickness shall not be less than one-twelfth the clear distance between ribs but no less than 2 in.; slab reinforcement normal to the ribs shall be provided as required for flexure but shall be no less than the minimum amount required for shrinkage and temperature.
5.  When permanent clay tile fillers are used that have a compressive strength at least as great as that of the concrete, the vertical shells of fillers in contact with ribs may be included in the flexural calculations; slab thickness over permanent fillers may not be less than one-twelfth the clear distance between ribs nor less than 1½ in.
6.  Shear stress in the working stress method may not exceed $1.2\sqrt{f'_c}$.
7.  Joists not meeting the requirements shall be designed as ordinary slabs and tee beams.

In the United States, the "pans" used to form the void space have become reasonably well standardized (Concrete Reinforcing Steel Institute, 1989). The standardization has been made in Imperial units and does not seem likely to change in the near future. The dimensions shown in Figure 11-12 are generally accepted standard dimensions in the industry.

In countries using metric units, there seems to be little standardization. American-manufactured pans are usually available in most countries; design calculations simply adjust these dimensions to the nearest few millimeters. Slab thicknesses are commonly 60, 80, or 100 mm and rib widths are commonly 120, 140, 160, 180, and 200 mm.

Depths of joist systems can be considerably less (about one-third less) than that of widely spaced tee-and-slab systems. Deflection problems with joist systems are common, however, and deflection limitations should always be checked.

Code permits shear reinforcement to be used with joist systems as shown in Figure 11-13, but the deep narrow stems make placement of such reinforcement difficult. It is far simpler to slightly increase the stem width where shears are excessive than to place numerous stirrups in each of the many tall, thin stems. Too, it was noted earlier that one of the attractive

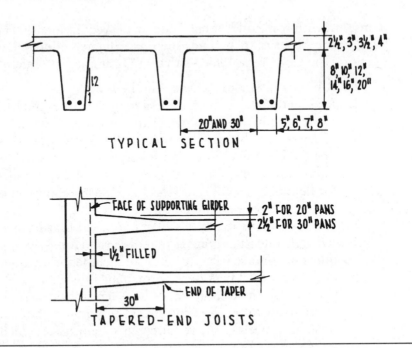

**Figure 11-12**    Standard joist dimensions.

features of joist systems was that it becomes possible to eliminate shear reinforcement; to use shear reinforcement with a joist system effectively discards one of the attractive features of the system.

Where shears are light, it may be desirable to use the minimum widths of joist. In such cases, stacking the reinforcement vertically, as shown in Figure 11-13c, is a reasonable alternative. The minimum widths of a section for various bar sizes placed in horizontal layers are given in Table A-4; the column headed "with no stirrups" is intended to apply particularly to joist systems.

Solid blocking, or bridging, perpendicular to the joists is always placed at support lines and where partitions are located in the span. Lines of bridging to stabilize the slender joists are also placed at the interior of the span where spans are long or loads are heavy. Interior bridging is not usually used with roof joists. For floor joists:

- No interior lines of bridging are required for spans less than 20 ft.
- One interior line of bridging is required for spans 20 to 30 ft (at midspan).

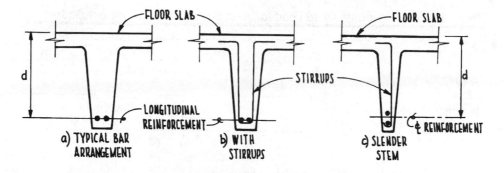

**Figure 11-13**    Shear reinforcement in tee shapes.

- Two interior lines of bridging are required for spans longer than 30 ft (at third points).

Where bridging is required, its dead load may be estimated at about 2 psf.

Due to the large number of combinations of slab thicknesses, stem widths, stem heights, and joist spacings, the dimensions of joist systems are almost always selected from handbooks or, in unusual configurations, designed using one of the many computer programs available. Deflection criteria are always included in these standard design aids.

For the sake of demonstration, the design of a joist system is presented here. Before beginning the example, it will be necessary to have a table of the average uniform dead weights of a joist system for various configurations. Such a tabulation is presented in Table 11-1 for common dimensions.

It should be noted that in Table 11-1 joist systems are not entered where the total height of stem is more than five times the slab thickness. Also, joist systems are not entered where the pan width (clear span of the slab) is more than 12 times the thickness of the slab nor where the pan height is greater than 3.5 times the least width of the stems. The systems listed in Table 11-1 are therefore permissible under all the limitations imposed earlier in this chapter.

The following example uses a typical system. The span is long, 32 ft, and the live load is relatively heavy at 160 psf. It should be noted that the entire system must be selected before the calculations can be started. The calculations are thereafter limited to selection of reinforcement and a review for shear.

**Table 11-1**   Average Dead Load for Joist Systems in Pounds per Square Foot

| Pan Height (in.) | Slab Thickness (in.) | Total Height (in.) | Pan Widths 20 — Stem 4 / Joist Sp. 24 | Stem 5 / 25 | Stem 6 / 26 | Stem 7 / 27 | Stem 8 / 28 | Pan Widths 30 — Stem 4 / Joist Sp. 34 | Stem 5 / 35 | Stem 6 / 36 | Stem 7 / 37 | Stem 8 / 38 |
|---|---|---|---|---|---|---|---|---|---|---|---|---|
| 8 | 2½ | 10½ | 51 | 54 | 57 | 60 | 62 | | | | | |
|   | 3 | 11 | 57 | 60 | 63 | 66 | 68 | 51 | 54 | 56 | 58 | 60 |
|   | 3½ | 11½ | 63 | 66 | 69 | 72 | 75 | 57 | 60 | 62 | 64 | 67 |
|   | 4 | 12 | 69 | 73 | 76 | 78 | 81 | 64 | 66 | 69 | 71 | 73 |
| 10 | 2½ | 12½ | 56 | 60 | 64 | 68 | 71 | | | | | |
|   | 3 | 13 | 63 | 67 | 70 | 74 | 77 | 55 | 58 | 61 | 64 | 67 |
|   | 3½ | 13½ | 69 | 73 | 77 | 80 | 83 | 62 | 65 | 67 | 70 | 73 |
|   | 4 | 14 | 75 | 79 | 83 | 86 | 89 | 68 | 71 | 74 | 76 | 79 |
| 12 | 2½ | 14½ | 63 | 67 | 72 | 76 | 79 | | | | | |
|   | 3 | 15 | 69 | 74 | 78 | 82 | 86 | 60 | 63 | 67 | 70 | 73 |
|   | 3½ | 15½ | 75 | 80 | 84 | 88 | 92 | 669 | 73 | 76 | 79 | |
|   | 4 | 16 | 81 | 86 | 90 | 94 | 98 | 72 | 76 | 79 | 82 | 86 |
| 14 | 3 | 17 | 75 | 81 | 86 | 90 | 95 | 64 | 68 | 72 | 76 | 80 |
|   | 3½ | 17½ | 81 | 87 | 92 | 97 | 101 | 70 | 75 | 79 | 82 | 86 |
|   | 4 | 18 | 88 | 93 | 98 | 103 | 107 | 77 | 81 | 85 | 89 | 92 |
| 16 | 3½ | 19½ | 88 | 94 | 100 | 105 | 110 | 75 | 80 | 84 | 89 | 93 |
|   | 4 | 20 | 94 | 101 | 106 | 112 | 117 | 81 | 86 | 91 | 95 | 99 |
| 20 | 4 | 24 | | | 124 | 130 | 136 | | | 103 | 109 | 114 |

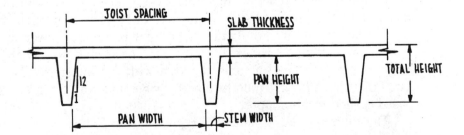

## EXAMPLE 11-5.

Design of a pan joist system.
Simple span, no limitations on dimensions.

### Given:

Simple span, 31 ft clear

Seats on 12 in. bearing wall

Live load 160 psf

Grade 60 steel

$f'_c$ = 4000 psi

Floor tile 4 psf

Exterior exposure

### Find:

Suitable Pan Joist System

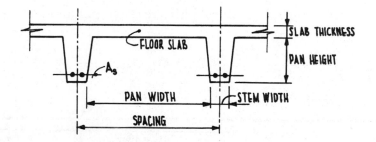

Assume that two lines of bridging will be required for a span more than 30 ft; assume that the dead load of bridging is 2 psf. For deflection control, minimum $h$ = L/16 or 24 in. (Table 3-1). Try depth about 66% of the usual depth; approximate depth 0.66 × L/12 or 21 in.; use 24 in. Try pans 30 in. wide, 20 in. deep; slab 4 in. thick; stem 7 in. wide; joist spacing 37 in.; dead load = 109 psf.

### Solution:

Determine the necessary values for drawing the shear and moment diagrams:

Dead load (psf)

| | |
|---|---|
| Joist system | 109 |
| Floor tile | 4 |
| Bridging | 2 |
| Total | 115  psf |

For joist spacing of 37 in.:

$$w_{DL} = 115 \times 3.1 = 357 \text{ plf}; \; w_{LL} = 160 \times 3.1 = 496 \text{ plf}$$

Determine the working shear and moment:

$$V_{DL} = \frac{w_{DL}L}{2} = \frac{357 \times 31}{2} = 5.5 \text{ kips}; \quad V_{LL} = \frac{w_{LL}L}{2} = \frac{496 \times 31}{2} = 7.7 \text{ kips}$$

$$M_{DL} = \frac{w_{DL}L^2}{8} = \frac{357 \times 31 \times 31}{8} = 42.9 \text{ kip-ft}$$

$$M_{LL} = \frac{w_{LL}L^2}{8} = \frac{496 \times 31 \times 31}{8} = 59.6 \text{ kip-ft}$$

Determine the nominal ultimate shear and moment:

$$V_n = \frac{1.4V_{DL} + 1.7V_L}{0.85} = 24.5 \text{ kips/ft}; \quad M_n = \frac{1.4M_{DL} + 1.7M_{LL}}{0.90} = 179 \text{ kip-ft}$$

Sketch the shear and moment diagrams:

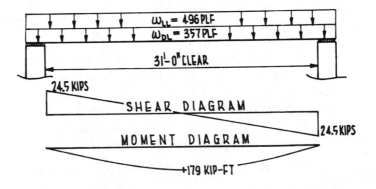

Determine $V_{cr}$ at a distance $d$ from face of support. Assume that $d = h - 3 = 21$ in.; find $V_{cr}$ at 21 in. from face of support.

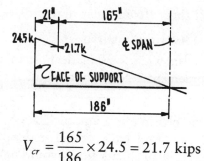

$$V_{cr} = \frac{165}{186} \times 24.5 = 21.7 \text{ kips}$$

Calculate the average stem width and flange width:

$$b_w = 7 + \frac{20}{12} = 8.7 \text{ in. (avg.)}$$

$b' = \text{joist spacing} = 37 \text{ in.}$

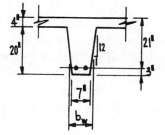

Solve for the coefficient of the section modulus:

$$Z_c = \frac{M_n}{0.85f_c'} \quad \frac{179,000 \times 12}{0.85 \times 4000} = \text{coeff.} \times b'd^2 = \text{coeff.} \times 37 \times 21 \times 21$$

$$\text{coeff.} = 0.039$$

From Table A-6, select $\rho = 0.0023$, $Z_c = 0.039b'd^2$ (not at balanced stress condition)

Solve for the steel area:

$$A_s = b'd = 0.0023 \times 37 \times 21 = 1.79 \text{ in.}^2$$

From Table A-3, select 2 No. 9 bars, $A_s = 2.00$ in.$^2$

Review for shear:

With no stirrups = $v_c = 2\sqrt{f_c'} = 126$ psi

Concrete shear $V_c = v_c b_w d = 126 \times 8.7 \times 21 = 23$ kips

Actual shear (from shear diagram) = 21.7 kips < 23 kips

(Section O.K. without stirrups)

Use pans 36 in. wide × 20 in. deep; stem width 7 in. at base, spacing 37 in.; reinforcement 2 No. 9 bars, 2-in. cover (minimum).

---

It would be possible in Example 11-5 to reduce the bar size by using two layers of bars, four bars total. The depth $d$ would then be reduced to about 19½ in., and all calculations would have to be repeated for this new value of $d$.

Even with the long span and relatively heavy live load in Example 11-5, it can be seen that a moderately sized stem of 7 in. is adequate for the shear loads. It could easily be increased if necessary just by opening up the spacing of the pans. It may be concluded that relatively high shears can be sustained by joist systems with little penalty in increased stem size.

It should be apparent that one-way joist systems are one of the more adaptable floor systems in the industry. Where a supporting structure occurs at all four sides of a floor system in a reasonably square pattern, joists may be placed in both directions. Such two-way systems, called "waffle" systems, permit comparatively heavy loads to be carried on comparatively long spans with maximum economy of materials. Two-way joist systems are beyond the scope of this book, not due to unusual complexity but to lack of space.

## OUTSIDE PROBLEMS

Select a one-way pan-joist system for the following sets of conditions. Assume simple supports and exterior exposure. Surface load includes floor finishes, piping, ducts, etc., but does not include bridging.

| Prob. No. | Concrete Strength $f'_c$ psi | Steel Grade | Live Load psf | Span Feet | Surface Load psf |
|---|---|---|---|---|---|
| 11.37 | 3000 | 40 | 75 | 24 | 6 |
| 11.38 | 3000 | 60 | 100 | 28 | 6 |
| 11.39 | 4000 | 40 | 125 | 24 | 8 |
| 11.40 | 4000 | 60 | 150 | 30 | 8 |
| 11.41 | 5000 | 40 | 175 | 24 | 10 |
| 11.42 | 5000 | 60 | 200 | 30 | 10 |

## REVIEW QUESTIONS

1. A floor slab is 4 in. thick. If it is used as part of a tee-beam system, what maximum depth should be observed in selecting the depth of the stem? Why?

2. How can one accurately determine whether the neutral axis of a tee beam is actually in the slab or whether it falls in the stem?

3. Under what circumstances would one suspect that the neutral axis of a particular tee beam might fall in the stem of the tee?

4. For simple spans, how is the size of the tee-beam stem usually determined?

5. For continuous beams, how is the size of the tee beam usually determined?

6. Tee beams are commonly spaced at about one-fourth of the span length. Why?

7. What width of beam is used to compute the minimum allowable steel ratio $\rho$? How is it used in regions of positive moment?

8. Visualizing the "internal couple" concept for moment in beams as shown in Figure 11-6, what is the approximate length of the moment arm $jd$ under service levels of load? At ultimate load?

9. What are the most common widths of steel pans used to form one-way pan-joist systems?

10. What is the usual means to increase the shear capacity of a pan-joist system without resorting to the use of shear reinforcement?

# CHAPTER

# 12

# ANCHORAGE OF REINFORCEMENT

$\mathbf{T}$he term *development length*, denoted $\ell_d$, is used frequently in subsequent discussions. The development length of a reinforcing bar is the length of bar that must be bonded to the concrete to develop the full strength of the bar. Ideally, the end of a bar could be embedded in concrete to a depth equal to its development length and when the protruding end is loaded to failure, the bar itself will fail just as it starts to pull out of the concrete block. A typical state of this equilibrium loading is shown on the reinforcing bar of Figure 12-1.

The adhesion of concrete to its reinforcement is called *bond*. It is pointed out in Chapter 1 that bond strength is one of the more uncertain properties of concrete that must be used in design. To reduce the dependence on theoretical values when using this uncertain property, the development lengths of reinforcement under a wide variety of circumstances have been determined by extensive tests; requirements for embedment have been established from these test results and are prescribed by Code.

There are several special circumstances defined by the Code where the development lengths may be reduced somewhat. Those special circumstances that are encountered frequently enough to justify the time and effort to learn them are included in the subsequent sections. Other cases, infrequently applicable, are omitted to avoid a confusion of special cases. The results herein may therefore be slightly conservative at times.

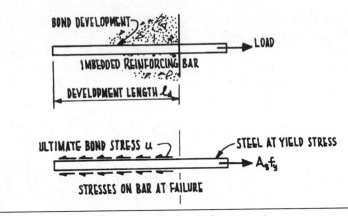

**Figure 12-1**    Development length of reinforcement.

Code does not have a separate means for determination of development lengths at elastic levels. For the working stress method, for example, the Code (B.4) simply says to use the strength method for development lengths and splices. There is therefore only one set of requirements for bond and anchorage; that set is applied at both ultimate load and at elastic levels.

In earlier years, Code requirements were based on an embedment length given in terms of a certain number of diameters of the bar. For example, under certain circumstances, a development length may have been required to be 48 bar diameters. The computation of this embedment length, say 48 times ¾ or 36 in. for a No. 6 bar, was then repeated each time the information was needed.

Present Code requirements are prescribed by empirical formulae, to be used for computing the required embedment length under various circumstances. These lengths have been computed for various bar sizes and are tabulated in Tables A-12 and A-13 of the Appendix. The lengths were determined from the criteria given in the following discussions. They are equally applicable whether the design is being done according to the strength method or at elastic stress levels.

The following discussions are limited to bar sizes No. 11 or smaller for flexural reinforcement and to bar sizes No. 5 or smaller for stirrups and ties. Bar sizes larger than these are subject to additional Code requirements requiring additional design considerations and conditional checks. Fur-

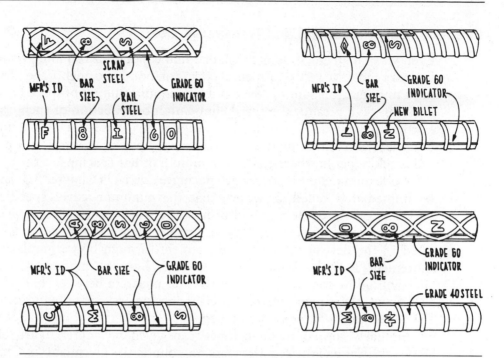

**Figure 12-2**    Deformation patterns on reinforcement (adapted from Ref. 11).

ther, these larger sizes would apply primarily to heavy construction, which is beyond the scope of this book.

In addition, consideration here is limited only to deformed bars; smooth wire and mesh are not included. Code requirements concerning the patterns and sizes of the deformations are quite stringent. Several approved patterns of deformations are shown in Figure 12-2.

Though not included here, ACI 318-89 prescribes several additional special formulas and checks that apply to the use of smooth bars and mesh. As a matter of perspective, as late as 1979 the British Code of Practice CP 110 handled the matter simply by limiting the anchorage load on plain bars to 70% of that for deformed bars of the same size.

The strength reduction factor $\phi$ is rarely required in determination of development lengths. The empirical formulas adopted by ACI inherently include the effects of placement tolerances and other factors that might affect ultimate strength.

## Effects of Cover, Spacing, and Transverse Ties

Even very small cracks in the concrete immediately surrounding reinforcing bars can have very detrimental effects on bond and anchorage. Providing a minimum amount of cover and a minimum amount of space between bars can do a great deal toward elimination or reduction of such cracks. Further, placing transverse ties or stirrups at very close spacing around the longitudinal reinforcement will provide a very positive control over potential cracking, even where cover is minimum or bar spacing is close.

Code minimums for cover and spacing are given in Chapter 3. It should be immediately noted, however, that the minimum cover specified in Chapter 3 is the cover that will protect the reinforcement from weather, oxidation, and intrusion of salts; that requirement for clear cover is unrelated to the development of reinforcement through bond. Similarly, requirements for bar spacing given in Chapter 3 are the minimum distances that will allow free movement of the wet concrete between bars during placement; those requirements for clear spacing are generally unrelated to the development of the reinforcement.

Similarly, the requirements for stirrups given in Chapter 10 are related to beam shear and the requirements for column ties given later in Chapter 13 are related to buckling of reinforcement. Neither set of requirements is concerned with development of reinforcement through bond.

The cracking or "splitting" of concrete around its reinforcement is shown in Figure 12-3. Where cover over the bars is inadequate, a splitting crack outward to the surface will occur as indicated in Figure 12-3a. Where

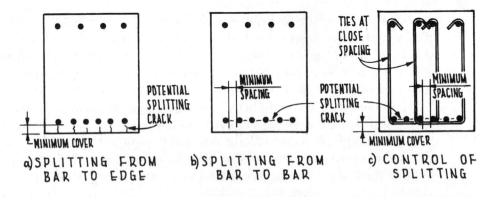

**Figure 12-3**    Cover, spacing, and transverse ties.

bar spacing is too close, a splitting crack will propagate between bars as shown in Figure 12-3b. Such splitting may be controlled, however, by providing closely spaced ties at alternate bars as shown in Figure 12-3c.

The topics of cover, spacing, and transverse ties are considered in the following sections. The three items are intimately interrelated; inadequacies in one item can sometimes be corrected by providing excesses in another. The approach, however, is a "cookbook" approach, using certain configurations that have been tried and tested and are therefore known to work.

## Development Length of Straight Bars in Tension

There are many variables that can have an effect on the required development lengths for reinforcement. The more common variables recognized by Code are:

- Clear concrete cover to the nearest exposed face (for beams, columns, slabs, walls)
- Clear spacing between bars (for beams, columns, slabs, walls)
- Transverse reinforcement (existence of stirrups or ties)
- Distance from bottom of bar to bottom face of member (whether top bars or bottom bars)
- Type of aggregate (whether regular or lightweight)
- Coatings on bars (whether epoxy coated or clean)
- Excess reinforcement at a section (whether $A_s$ provided is greater than $A_s$ required)
- Bundling of reinforcement (whether bars are grouped in bundles of 2, 3, or 4 bars)

Code (12.2.2) establishes a "basic" development length $\ell_{db}$ based on standard sets of assumed conditions. If any one of the assumed conditions is not met, a prescribed multiplier is used to extend the development length, thereby supplying additional length to correct for the inadequacy. In the event that the actual conditions are better than the assumed conditions, a reduction of the development length may be permitted; the reduction is similarly obtained by multiplying the basic development length by a prescribed modifier. After all extensions and reductions have been applied, the resulting development length $\ell_d$ is then checked against the absolute minimum development length required by Code. The end result of such a computation is a development length for deformed bars in tension that is closely conformed to the actual design conditions.

The individual Code requirements are presented in somewhat tedious detail in the discussions immediately following. A summary of these requirements is then presented in tabular form following the discussions.

The basic development length $\ell_{db}$ for deformed bars in tension is prescribed by Code (12.2.2) for bars no larger than No. 11,

$$\ell_{db} = \frac{0.04 A_b f_y}{\sqrt{f'_c}} \qquad (12\text{-}1)$$

where $A_b$ is the cross-sectional area of the bar, and all other symbols are those used previously.

Code (12.2.3.1) establishes four sets of standard conditions for which the basic development length of Eq. (12-1) applies:

1. For beams and columns, the basic development length $\ell_{db}$ prescribed by Eq. (12-1) applies to bars for which:
   a) Clear cover over bars conforms to minimum requirements (as listed in Chapter 3).
   b) Clear spacing between bars is no less than $3d_b$, where $d_b$ is the diameter of the bar.
   c) Ties and stirrups conform to minimum requirements for shear (as given in Chapter 10) or for column ties (given later in Chapter 13).

2. Alternatively for beams and columns, the basic development length $\ell_{db}$ prescribed by Eq(12-1) applies to bars for which:
   a) Clear cover over bars conforms to minimum requirements (as listed in Chapter 3).
   b) Clear spacing between bars conforms to minimum requirements (as listed in Chapter 3).
   c) The area of the transverse reinforcement $A_{tr}$; $A_{tr}$ along the length $\ell_{db}$ is at least as much as the following minimum,

$$A_{tr} \geq \frac{d_b s N}{40} \qquad (12\text{-}2)$$

where   $d_b$ is the bar diameter in inches;

              $s$ is the spacing in inches of the stirrups or ties; and

              $N$ is the number of bars in a layer that are being spliced or developed.

3. For slabs or walls, the basic development length $\ell_{db}$ prescribed by Eq(12-1) applies to bars in the inner layer of slab or wall reinforcement for which:
   a) Clear cover over bars conforms to minimum requirements (as listed in Chapter 3).
   b) Clear spacing between bars is not less than $3d_b$, where $d_b$ is the bar diameter.
4. For all types of members, the basic development length $\ell_{db}$ prescribed by Eq. (12-1) applies to bars for which:
   a) Clear cover over bars is not less than $2d_b$.
   b) Clear spacing between bars is not less than $3d_b$.

Code (12.2.3.2) requires an extended development length when the requirements for minimum cover or minimum spacing are not met.

For bars with cover of $d_b$ or less, or with clear spacing of $2d_b$ or less, the basic development length $\ell_{db}$ shall be multiplied by a factor of 2.

With detailed study, one can find a few gaps in the foregoing criteria. For example, there is no provision for bars having clear cover greater than $d_b$ but less than $2d_b$. For such gaps, Code (12.2.3.3) provides a catchall:

For bars not included in the foregoing criteria, the basic development length $\ell_{db}$ shall be multiplied by a factor of 1.4.

Code (12.2.3.4) recognizes that benefits may be derived where the minimum requirements for cover and spacing are not only met but are significantly exceeded. For such cases, a reduction in the basic development length $\ell_{db}$ is permitted.

For bars having a clear spacing not less than $5d_b$ and with cover not less than $2.5d_b$ from face of member to edge bar (measured in the plane of the bars), the basic development length $\ell_{db}$ may be modified by a factor of 0.80

Note that in the terminology of the Code, extensions to the basic development length are computed through the use of a "multiplier," while reductions are computed through the use of a "modifier."

Code (12.2.3.5) also recognizes that benefits may be derived where lateral or transverse ties exceed the minimum requirements for stirrups (Chapter 10) or for column ties (Chapter 13). For such cases, a reduction in the basic development length $\ell_{db}$ is permitted.

For bars enclosed within ties or stirrups, where such ties or stirrups are made of No. 4 bars or larger and spaced no more than 4 in. on center and arranged such that alternate bars shall have support provided by the corner of a tie or hoop having an included angle no more than 135°, the basic development length $\ell_{db}$ may be modified by a factor of 0.75.

Code (12.2.3.6) also recognizes that a build up in reductions could, in some circumstances, produce development lengths that would be too small. After all the permissible corrections have been applied, the basic development length must therefore be checked against a minimum length prescribed by Code.

For all bars, after all corrections for cover, spacing, and transverse ties have been applied, the basic development length $\ell_{db}$ shall be taken not less than the minimum allowable value,

$$\text{minimum } \ell_{db} \leq \frac{0.03 d_b f_y}{\sqrt{f_c'}} \qquad (12\text{-}3)$$

At this point, the basic development length $\ell_{db}$ has now been established to include any special conditions involving cover, spacing, and transverse ties. There remain, however, several physical or design conditions which must also be accounted for including the position of the reinforcement within the beam (top or bottom), the type of concrete (regular or lightweight), the use of protective coatings (epoxy-coated bars), and the existence of excess steel area ($A_s$ furnished being greater than $A_s$ required). In this text, lightweight concrete and epoxy-coated bars are considered to be specialty items and are not included in further discussions.

Horizontal bars having 12 in. or more of fresh concrete cast below them can lose a considerable amount of bond strength due to collections of air, water, and laitance at their lower side. These impurities rise and collect under the bars during compaction of the concrete. Additional development length must be provided for such bars, called "top" bars in the Code (12.2.4.1).

For horizontal bars so placed that more than 12 inches of fresh concrete is cast in the member below the development length or splice, the basic development length $\ell_{db}$ shall be multiplied by a factor of 1.3.

In detailing concrete, it commonly occurs that the area of flexural reinforcement will be selected at a point where loads are high, but the bars will then be extended into areas where the load is significantly less. For this and other such cases, Code (12.2.5) permits the development length to be reduced to reflect a reduction in stress levels.

> Where reinforcement in a flexural member is in excess of that required by analysis, except where anchorage or development for $f_y$ is specifically required or except under earthquake loading, the basic development length $\ell_{db}$ may be modified by a factor of ($A_s$ required)/($A_s$ provided).

After all multipliers and modifiers have been applied, the final development length $\ell_d$ so computed must be checked against the absolute minimum development length permitted by Code (12.2.1).

$$\text{In tension, minimum } \ell_d = 12 \text{ inches} \tag{12-4}$$

The foregoing requirements for computing development lengths of deformed bars in tension are summarized in Table 12-1. A detailed examination of Table 12-1 indicates that as a very general guide, splitting of concrete cover can be controlled by maintaining a minimum cover of $2d_b$ over all reinforcement. Similarly, splitting from bar-to-bar along a line of reinforcement can be controlled by maintaining a minimum spacing of $3d_b$ between bars. If the cover and spacing do not meet these minimums, then additional development length will be required to offset the possibility of splitting.

The criteria for development length have been incorporated into the development lengths listed in Table A-13 of the Appendix, insofar as possible. Table A-13 was prepared assuming that at least one of the four basic sets of standard conditions has been met. The only multiplier appearing in the table is that for top bars; no other multiplier or modifier has been included.

Other multipliers may be applied as appropriate to extend the development lengths given in Table A-13. For reductions, however, the only modifier that may be applied to the values given in Table A-13 is the modifier to account for excess reinforcement. Development lengths that include excess cover or excess spacing must be recomputed manually in their entirety. Such computations must include appropriate checks against the Code minimums, $\ell_{db} \leq 0.03\, d_b f_y / \sqrt{f_c'}$ and $\ell_d \leq 12$ inches, where appropriate.

Some examples will demonstrate the use of Table A-13 in finding development lengths.

**Table 12-1**    Development Lengths in Tension

- Compute basic development length $\ell_{db}$ for deformed bars in tension under standard conditions of service,

$$\ell_{db} \geq \frac{0.04 A_b f_y}{\sqrt{f_c'}}$$

- The foregoing value of $\ell_{db}$ is valid if any one of the four following sets of standard conditions is met:

    1. Beams and columns
       Clear cover meeting Code minimums
       Clear spacing $\geq 3d_b$
       Ties or stirrups meeting Code minimums
    2. Beams and columns
       Clear cover meeting Code minimums
       Clear spacing meeting Code minimums
       Ties or stirrups having $A_{tr} \geq d_b sN/40$
    3. Slabs or walls
       Clear cover meeting Code minimums
       Clear spacing $\geq 3d_b$
    4. Any type of member
       Clear cover $\geq 2d_b$
       Clear spacing $\geq 3d_b$

- Should an inadequacy exist in the applicable set of standard conditions,
    multiply $\ell_{db}$ by a factor of 2 if cover $\leq d_b$
    multiply $\ell_{db}$ by a factor of 2 if spacing $\leq 2d_b$
    multiply $\ell_{db}$ by a factor of 1.4 for any other inadequacy

- Should a significant excess exist in the applicable set of standard conditions,
    modify $\ell_{db}$ by a factor of 0.8 if spacing $\geq 5d_b$ and cover $\geq 2.5\ d_b$
    modify $\ell_{db}$ by a factor of 0.75 if No. 4 ties at 4 in. o.c. exists

- Check $\ell_{db}$ against Code minimum: $\ell_{db} \geq 0.03 d_b f_y / \sqrt{f_c'}$

- Multiply $\ell_{db}$ by a factor of 1.3 for top bars

- Modify $\ell_{db}$ by a factor of $(A_s$ required$)/(A_s$ provided$)$ for excess $A_s$

- Check final $\ell_d$ against absolute Code minimum: $\ell_d \geq 12$ inches

## EXAMPLE 12-1.

Determination of development lengths.

**Given:**

Cantilever slab as shown
Grade 60 steel, $f_c' = 4000$ psi
Structure not exposed to weather

**Find:**

Required imbedment length of longitudinal bars

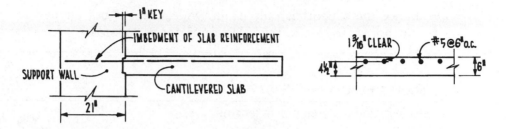

**Solution:**

Clear cover = $1\frac{3}{16}"$ clear or $1.9d_b$

Clear spacing = $6" - \frac{5}{8}" = 5\frac{3}{8}"$ or $8.6d_b$

Conclude that the set of standard conditions given in Table 12-1 as set No. 3 have been met.

With only $4\frac{3}{16}"$ of concrete below the bars, these bars are not top bars. From Table A-13, find $\ell_d = 18$ in.

Use imbedment 18 in. into support for 100% development.

---

### EXAMPLE 12-2.

Determination of development lengths.

**Given:**

Cantilever slab of Example 12-1 but with No. 7 bars @ 12 in., providing roughly the same $A_s$ per foot of width

**Find:**

Whether this selection of reinforcement is permissible with straight bars as shown

**Solution:**

Clear cover = $1\frac{1}{16}"$ clear or $1.2d_b$

Clear spacing = $6" - \frac{7}{8}" = 5\frac{1}{8}"$ clear or $5.86d_b$

Conclude that the set of standard conditions given in Table 12-1 as set No. 3 have been met.

From Table A-13, find $\ell_d = 25$ in.

Since the support is only 21 in. wide, there is not enough room to anchor these bars so this selection is not permissible.

---

### EXAMPLE 12-3.

Determination of development lengths.

### Given:

Rectangular section under negative moment, 12 in. × 20 in.

Required negative reinforcement is 2.31 in.²

Assume clear cover of 2 in. to top face of section

Exterior exposure, Grade 60 steel, $f'_c = 3000$ psi

### Find:

Required development length at support

### Solution:

The bar arrangement is shown in the sketch

Assume 4 No. 7 bars for reinforcement, $A_s = 2.4$ in.²

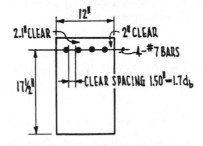

Since spacing of the bars is less than $3d_b$, the set of standard conditions given in Table 12-1 for set No. 1 or set No. 2 has not been met by this steel arrangement. A multiplier of 2 must therefore be used for the values given in Table A-13.

From Table A-13, find $\ell_d$ = 37 in. (for top bars)

With multiplier of 2, $\ell_d$ = 74 in.

Note that the development length required in Example 12-3 is more than 6 ft, half of which is due to the fact that the clear distance is too small. Such a severe penalty often makes it worthwhile to juggle the bar sizes to reduce the development lengths. In Example 12-3, for example, the use of 3 No. 8 bars will meet the required steel area of 2.31 in.$^2$, but will require a smaller development length of only 63 in. Such a reduction can sometimes be a worthwhile alternative. Verification of this reduction is left to the reader.

## EXAMPLE 12-4.

Determination of development lengths in tension.

### Given:

Rectangular beam, 16 in. wide, $f'_c$ = 4000 psi, grade 50 steel

4 No. 6 bars in tension, classed as "other than top" bars,

Required $A_s$ = 1.01 in.$^2$, exterior exposure.

Clear cover 2.5 in. both vertical and horizontal

### Find:

Required development length of bars in tension

### Solution:

Compare cover requirements to cover actually provided

Code minimum cover for No. 6 bars, exterior exposure, must be $\geq$ 2.0 in.

Required cover if modifier is allowed must be $\geq 2d_b$

Cover actually provided = 2.50 in. $\geq 2d_b$

Conclude:  cover meets minimum requirements

cover OK for modifier if spacing $\geq 3d_b$

Compare spacing requirements to actual spacing

Code minimum spacing $\geq 1d_b$ but not less than 1 in.

Required spacing if modifier is allowed must be $\geq 3d_b$ or 2.13 in.

Clear spacing actually provided:

$$\text{Spacing} = \frac{b - 2 \times \text{clear cover} - \text{No. bars} \times d_b}{\text{No. bars} - 1}$$

$$= \frac{16 - 2 \times 2.50 - 4 \times 0.75}{4 - 1} = 2.67 \text{ in.} \geq 3d_b$$

Conclude:  Spacing meets code minimums

No multipliers are required

Spacing and cover permit use of modifier

Compare required steel area to area actually provided

Required steel area = 1.01 in.$^2$

Area of steel provided = 1.77 in.$^2$

Conclude:  Development lengths may be reduced

due to excess steel area.

Calculation of required length of imbedment

From Table A-13, $\ell_d$ = 18 in.

No multipliers required due to insufficient spacing

Apply modifier of 0.8 for excess cover and spacing

$\ell_d = 18 \times 0.8 = 14.4$ in.

Apply modifier for excess steel area

$\ell_d = 14.4 \times 1.01/1.77 = 8.22$ in.

Absolute minimum for No. 6 bars in tension, $\ell_d \geq 12$ in.

Use $\ell_d$ = 12 in.

## OUTSIDE PROBLEMS

Determine the required development length for straight deformed bars used as flexural tensile reinforcement in the given rectangular beams, used in an exterior exposure. Assume that the given clear cover is maintained both horizontally and vertically.

| Prob. No. | Concrete Strength $f'_c$ psi | Steel Grade | Beam Width in. | Top or Other Bars | Bars | Required $A_s$ in.² | Clear Cover in. |
|---|---|---|---|---|---|---|---|
| 12.1 | 3000 | 40 | 12″ | Other | 4 No. 5 | 1.19 | 1.5 |
| 12.2 | 3000 | 50 | 12″ | Top | 4 No. 6 | 1.66 | 1.5 |
| 12.3 | 3000 | 60 | 18″ | Other | 4 No. 9 | 3.91 | 2.5 |
| 12.4 | 3000 | 60 | 16″ | Top | 3 No. 6 | 0.64 | 2.5 |
| 12.5 | 4000 | 40 | 18″ | Other | 4 No. 8 | 3.01 | 2.5 |
| 12.6 | 4000 | 50 | 15″ | Top | 3 No. 10 | 3.06 | 2.5 |
| 12.7 | 4000 | 60 | 18″ | Other | 3 No. 6 | 0.61 | 3.0 |
| 12.8 | 4000 | 60 | 18″ | Top | 4 No. 8 | 2.09 | 3.0 |
| 12.9 | 5000 | 40 | 21″ | Other | 3 No. 11 | 4.60 | 3.0 |
| 12.10 | 5000 | 50 | 21″ | Top | 5 No. 9 | 4.88 | 2.5 |
| 12.11 | 5000 | 60 | 16″ | Other | 3 No. 8 | 1.09 | 2.5 |
| 12.12 | 5000 | 60 | 18″ | Top | 5 No. 10 | 6.09 | 2.5 |

## Development Length of Straight Bars in Compression

Development lengths in compression are established much like development lengths in tension: a basic development length is determined, then multipliers or modifiers are applied to correct for any inadequacies; the end result is then checked against an absolute minimum prescribed by Code. The computations for development lengths in compression, however, are much simpler than for those in tension. In compression, there are no circumstances where the development length must be lengthened, and there are only two circumstances where the development length may be reduced.

As before, the individual Code requirements are presented in the following discussions. Following these discussions, the requirements are summarized in tabular form.

The basic development length $\ell_{db}$ for deformed bars in compression is prescribed by Code (12.3.2) for all bar sizes,

$$\ell_{db} = \frac{0.02d_b f_y}{\sqrt{f'_c}} \tag{12-5}$$

but $\ell_{db}$ cannot be less than $0.0003d_b f_y$.

The basic development length $\ell_{db}$ computed from Eq.(12-3) may be reduced to reflect the reduced stress levels where excess reinforcement occurs.

Where compressive reinforcement is in excess of that required by analysis, the basic development length $\ell_{db}$ may be modified by a factor of $(A_s \text{ required})/(A_s \text{ provided})$.

The basic development length $\ell_{db}$ for compressive reinforcement may also be reduced where ties are provided in accordance with the minimum requirements for column ties (Chapter 13).

For bars enclosed within No. 4 ties that meet the requirements for column ties, where ties are spaced at not more than 4 inches on center, the basic development length $\ell_{db}$ may be modified by a factor of 0.75.

After all modifiers have been applied, the final development length $\ell_d$ so computed must be checked against the absolute minimum development length permitted by Code.

$$\text{In compression, minimum } \ell_d = 8 \text{ inches} \qquad (13\text{-}6)$$

All of the foregoing requirements for computing development lengths of deformed bars in compression are summarized in Table 12-2 for ready reference.

The criteria for deformed bars in compression are included in the development lengths given in Table A-13 of the Appendix. It should be noted in Table A-13 that there is no Code provision for top bars in compression reinforcement.

Some examples will demonstrate the use of Table A-13 in finding the required development length of deformed bars in compression.

---

**Table 12-2**   Development Lengths in Compression.

- Compute basic development length $\ell_{db}$ for deformed bars in compression, where minimum cover, spacing, and ties as required by code have been provided:

$$\ell_{db} \geq \frac{0.02 d_b f_y}{\sqrt{f_c'}}$$

- Check $\ell_{db}$ against Code minimum: $\ell_{db} \geq 0.0003 d_b f_y$
- Modify $\ell_{db}$ by a factor of $(A_s \text{ required})/(A_s \text{ provided})$ for excess $A_s$
- Modify $\ell_{db}$ by a factor of 0.75 if No. 4 ties @ 4 in. o.c. exist
- Check final $\ell_d$ against absolute Code minimum: $\ell_d \geq 8$ inches

## EXAMPLE 12-5.

Determination of development lengths.

### Given:

No. 8 reinforcing bar imbedded 12 in. vertically in a concrete foundation, protruding ½ in. above the concrete, to be loaded by a concentric axial force, Grade 60 steel, $f'_c$ = 4000 psi

### Find:

The ultimate compressive load that can be distributed into the concrete by the single reinforcing bar.

### Solution:

The development length for a No. 8 bar is found from Table A-13 to be 19 in. At 19 in. of imbedment, the bar will develop its yield stress of 60,000 psi; the load is then $f_y A_s$,

$$P_{19} = f_y A_s = 60,000 \times \pi \times 0.5^2 = 47,000 \text{ lb}$$

The bar is not imbedded to the required 19 in., however, but is only imbedded 12 in. The load at 12 in. imbedment is taken to be proportional to the imbedment length,

$$\frac{P_{19}}{19} = \frac{P_{12}}{12} \qquad P_{12} = P_{19} \times 12/19 = 29,800 \text{ lb}$$

It is assumed that this bar will, at ultimate load, distribute 29,800 lbs of force into the concrete over its embedment length of 12 in.

It is pointed out that the assumed ratio of force to imbedment length used in Example 12-5 is not proposed by Code. It is based on the assumption that at full development, the stress in the bar will be entering yield just as the bond between concrete and steel begins to fail. In compression, there is undoubtedly a certain amount of end bearing at the bottom of the bar, suggesting that load is not exactly proportional to length and that the assumption used here is therefore somewhat conservative. (If the load were tensile, however, there would be no end effects and the assumed linear ratio of force to imbedment length would probably be more accurate). In the absence of a Code provision, the designer's judgement becomes the final authority in such matters; some rationale such as that used here to solve Example 12-5 then becomes necessary.

### EXAMPLE 12-6.

Determination of development length.

#### Given:

Square pedestal 20 in. × 20 in. × 24 in. high, reinforced with 4 No. 9 bar at each corner, Grade 60 steel, $f'_c$ = 5000 psi; the pedestal bears on its foundation pad built of concrete having $f'_c$ = 3000 psi. The pedestal is subject to compressive loads only.

#### Find:

Required length of imbedment for the No. 9 bars.

#### Solution:

From Table A-13 for $f'_c$ = 5000 psi, $\ell_d$ = 20 in.

From Table A-13 for $f'_c$ = 3000 psi, $\ell_d$ = 25 in.

Extend No. 9 bars 20 in. into pedestal, 25 in. into foundation.

---

In finding the development lengths of Example 12-6, it should be noted that potential loads on the bars were never considered. The bars were simply imbedded far enough to develop their full strength, regardless of whether their full strength would ever be needed. Such is the approach commonly used in the industry for anchorage of reinforcement; imbedded reinforcement is detailed to provide full development regardless of loading.

### EXAMPLE 12-7.

Determination of development lengths in compression.

#### Given:

Rectangular beam, 14 in. wide, $f'_c$ = 4000 psi, grade 60 steel

3 No. 10 bars in compression, required $A_s$ = 2.20 in.$^2$, exterior exposure.

Clear cover 2.5 in. both vertical and horizontal

**Find:**

Required development length of bars in compression

**Solution:**

Compare cover requirements to cover actually provided

Code minimum cover for No. 10 bars, exterior exposure, must be $\geq 2.0$ in.

Cover actually provided $= 2.50$ in. $\geq 2.0$ in.

Conclude: cover meets minimum requirements

Compare spacing requirements to actual spacing

Code minimum spacing $\geq 1d_b$ but no less than 1 in., or 1.27 in. for No. 10 bars

Clear spacing actually provided:

$$\text{Spacing} = \frac{b - 2 \times \text{clear cover} - \text{No. bars} \times d_b}{\text{No. bars} - 1}$$

$$= \frac{14 - 2 \times 2.50 - 3 \times 1.27}{3 - 1} = 2.60 \text{ in.} \geq 1.27$$

Conclude: Spacing meets code minimums

Compare required steel area to area actually provided

Required steel area $= 2.20$ in.$^2$

Area of steel provided $= 3.80$ in.$^2$

Conclude: Development lengths may be reduced due to excess steel area.

Calculation of required length of imbedment

From Table A-13, $\ell_d = 24$ in.

No multipliers required due to insufficient spacing

Apply modifier for excess steel area

$$\ell_d = 24 \times 2.20/3.80 = 13.9 \text{ in.}$$

Absolute minimum for No. 10 bars in compression, $\ell_d \geq 8$ in.

Use $\ell_d = 14$ in.

## OUTSIDE PROBLEMS

Determine the required development length for the given straight deformed bars used as compression reinforcement in beams in an exterior exposure. Assume the clear cover is maintained both horizontally and vertically.

| Prob. No. | Concrete Strength $f'_c$ psi | Steel Grade | Member Width in. | Bars | Required $A_s$ in.$^2$ | Clear Cover in. |
|---|---|---|---|---|---|---|
| 12.13 | 3000 | 40 | 10 | 4 No. 5 | 1.19 | 1.5 |
| 12.14 | 3000 | 50 | 10 | 4 No. 6 | 1.66 | 1.5 |
| 12.15 | 3000 | 60 | 14 | 4 No. 9 | 3.91 | 2.5 |
| 12.16 | 3000 | 60 | 9 | 3 No. 6 | 0.64 | 2.5 |
| 12.17 | 4000 | 40 | 12 | 4 No. 8 | 3.01 | 2.5 |
| 12.18 | 4000 | 50 | 12 | 3 No.10 | 3.06 | 2.5 |
| 12.19 | 4000 | 60 | 10 | 3 No. 6 | 0.61 | 3.0 |
| 12.20 | 4000 | 60 | 14 | 4 No. 8 | 2.09 | 3.0 |
| 12.21 | 5000 | 40 | 13 | 3 No.11 | 4.60 | 3.0 |
| 12.22 | 5000 | 50 | 16 | 5 No. 9 | 4.88 | 2.5 |
| 12.23 | 5000 | 60 | 12 | 3 No. 8 | 1.09 | 2.5 |
| 12.24 | 5000 | 60 | 18 | 5 No.10 | 6.09 | 2.5 |

## Development Length of Bundled Bars

Bundling of longitudinal reinforcement is discussed in Chapter 3. One of the penalties incurred when bars are bundled together is the loss of surface area for bonding the steel to the concrete. The end result of such a loss is that development lengths must then be extended to compensate for the reduction in surface area.

Code (7.6.6) requirements for the extended lengths are quite simple. The basic development lengths are those computed earlier for single bars in compression or tension, multiplied by a factor to account for the effects of bundling.

Development length of individual bars in a bundle, whether in tension or in compression, shall be that for the individual bar, increased 20 percent for a three-bar bundle and 33 percent for a four-bar bundle.

For determining the appropriate factors to be applied when computing $\ell_{db}$, a unit of bundled bars shall be treated as a single bar having a diameter that produces the same total area.

**EXAMPLE 12-8.**

Determination of cover and spacing.

**Given:**

Bundle of 4 No. 6 bars

**Find:**

Cover requirements for $2_{db}$ minimum cover

Spacing requirements for $3_{db}$ minimum spacing

**Solution:**

From Table A-3 for 4 No. 6 bars, $A_s = 1.77$ in.$^2 = \pi D^2/4$

Equivalent diameter $D = \sqrt{4A_s/\pi} = 1.50$ in.

For calculating cover and spacing, use $d_b = D = 1.50$ in.

Provide clear cover $2 \times 1.50 = 3.00$ in. outside of bundle.

Provide clear spacing $3 \times 1.50 = 4.50$ in. between bundles.

## Development Length of Standard Hooks

Where high-strength steels are used with low-strength concrete, development lengths of straight bars can become prohibitively long. The amount of steel that is duplicated in these straight development lengths can add considerably to the total tonnage of reinforcement. A more efficient but not always less expensive method of providing development of reinforcement consists of bending the end of the bar into a "hook"—a hook is simply a bend configuration that is known to improve the anchorage of the bar in the concrete.

The two types of hooks prescribed by Code (7.1) are shown in Figure 12-4. Both types of hooks produce full development of the bar and both are used extensively throughout the industry, but the 90° hook is easier to tie in the forms and is probably the more frequently used. As a general rule, the cost of bending the reinforcement into hooks is at least partially offset by the savings in materials. In some cases, however, there simply is not enough room to provide for straight development lengths and the use of a hook becomes necessary regardless of cost.

A hook transfers a rather large concentration of load from the steel into

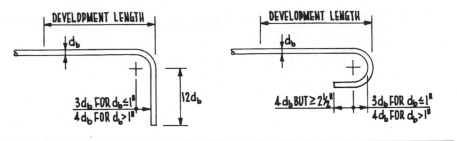

**Figure 12-4** Standard tension hooks.

the concrete in a relatively small distance. As a consequence, a rather confused stress pattern develops locally around hooks in which there will almost certainly be high tensile stresses. Splitting of the concrete in the vicinity of hooks is likely to occur when cover is small, particularly where several hooks occur together. In all cases, particular care is warranted when detailing hooked bars to assure that the required cover is maintained.

Hooks are not effective in compression.

Development lengths $\ell_{db}$ for hooked bars in tension are established much like other development lengths: a basic development length $\ell_{hb}$ is determined, then multipliers or modifiers are applied to correct for any inadequate conditions. Development lengths for hooks, however, are quite simple. As indicated in subsequent discussions, there are no cases where multipliers must be used to extend the required development length (except for lightweight concrete which is not included here), and there are only two cases where modifiers may be used to reduce the development length. After the modifiers are applied, however, the result must be checked against Code minimums.

The basic development length $\ell_{hb}$ for hooked deformed bars in tension is prescribed by Code (12.5.2, 12.5.3.1) for all bar sizes,

$$\ell_{hb} = \frac{0.02d_b f_y}{\sqrt{f_c'}} \qquad (12\text{-}7)$$

The basic development length $\ell_{hb}$ computed from Eq. (12-4) may be reduced when improved conditions exist regarding clear cover around the bar.

For hooked bars No. 11 and smaller, where side cover normal to the plane of the hook is not less than 2½ in., and for 90° hook, cover over the bar extension beyond the hook is not less than 2 in., the basic development length $\ell_{hb}$ may be modified by a factor of 0.7.

A reduction in the basic development length $\ell_{hb}$ is also permitted when transverse ties meet certain conditions.

For hooked bars No. 11 and smaller, where the hook is enclosed vertically or horizontally within ties or stirrups placed along the full development length $\ell_{dh}$, such ties or stirrups being spaced not more than $3d_b$ apart where $d_b$ is the diameter of the hooked bar, the basic development length $\ell_{hb}$ may be modified by a factor of 0.8.

The foregoing reduction due to transverse ties is further limited, however, by a subsequent Code provision.

For hooks located at discontinuous ends of members where both side and top (or bottom) cover over the hook is less than 2½ in., the hooked bar shall be enclosed within ties or stirrups placed along the full development, length $\ell_{dh}$ at a spacing not greater than $3d_b$, where $d_b$ is the diameter of the hooked bar. For this case, the foregoing modification factor of 0.8 for transverse ties shall not apply.

As with straight bars, the basic development length $\ell_{hb}$ for hooked bars in tension may be reduced where excess reinforcement occurs, regardless of whether the excess is accidental or deliberate.

Where reinforcement is in excess of that required by analysis, except where anchorage or development of $f_y$ is specifically required, the basic development length $\ell_{hb}$ for hooked bars in tension may be modified by a factor of ($A_s$ required)/($A_s$ provided).

After all modifications have been applied, the final computed development length $\ell_{dh}$ must be checked against Code (12.5.1) minimums,

$$\text{Minimum } \ell_{dh} \geq 8d_b \qquad (12\text{-}8)$$

but $\ell_{dh}$ shall not be less than 6 in.

The foregoing requirements for computing the development lengths of hooked bars in tension are summarized in Table 12-3.

The basic development lengths given by Eq. (12-7) and limited by Eq. (12-8) have been computed and tabulated in Table A-12 of the Appendix. As before, if modification factors are to be used, the entire computation for development length must be repeated to include appropriate checks against the Code minimums of Eq. (12-5).

**Table 12-3**  Development Lengths for Hooked Bars.

■ Compute basic development length $\ell_{hb}$ for hooked deformed bars in tension, where minimum cover, spacing, and ties as required by code have been provided:

$$\ell_{hb} \geq \frac{0.02 d_b f_y}{\sqrt{f'_c}}$$

■ Modify $\ell_{hb}$ by a factor of 0.7 for side cover $\geq 2\frac{1}{2}$ inches and, for 90° hook only, cover $\geq 2$ in. over the extension beyond the hook.
■ Modify $\ell_{hb}$ by a factor of 0.8 for ties spaced $\leq 3 d_b$, but not to include hooks located at discontinuous ends of members having cover $\leq 2\frac{1}{2}$ inches.
■ Modify $\ell_{hb}$ by a factor of $(A_s$ required$)/(A_s$ provided$)$ for excess $A_s$.
■ Check final $\ell_{dh}$ against Code minimums:  $\ell_{dh} \geq 8 d_b$
$\ell_{dh} \geq 6$ inches.

Some examples will demonstrate the use of Table A-12 in finding development lengths of hooked bars in tension.

**EXAMPLE 12-9.**

Selection of anchorage for reinforcement.

**Given:**

Rectangular beam as shown
Grade 40 deformed steel
$f'_c = 3000$ psi

**Find:**

Required anchorage for
the reinforcement

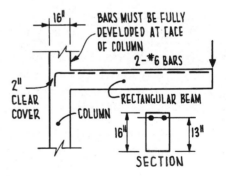

**Solution:**

The required development length for a straight No. 6 bar is given in Table A-13. Since the bars shown in the sketch have more than 12 in. of fresh concrete below them, they are classed as top bars, and $\ell_d = 21$ in. in tension. There is not enough room at the support to provide this development length. If a hook is used, the development length from Table A-12 is 11 in., which can be provided. Use hooked bars with at least 2 in. of cover, as shown; the actual embedment length then becomes 14 in.

**EXAMPLE 12-10.**

Selection of embedment for a lifting eye.

**Given:**

Precast tee as shown

Grade 40 steel

$f'_c = 3000$ psi

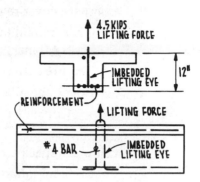

**Find:**

Required anchorage for the lifting eye

**Solution:**

To develop full strength of the embedded No. 4 bar, a development length of 12 in. is required for a straight bar or 7 in. for a hooked bar. Since there is again not enough room for the straight bar, it must be hooked as shown. The capacity is then $2f_yA_s = 2 \times 40,000 \times 0.2 = 16$ kips.

In Example 12-10 it should be recognized that the solution did not include the actual load on the lifting eye. The bar is simply imbedded to develop its full capacity. Nor do the tables include any provision for loads other than full capacity. A further discussion of this omission is presented toward the end of this chapter.

**EXAMPLE 12-11.**

Determination of development lengths of hooked bars in tension.

**Given:**

Rectangular beam, 16 in. wide, $f'_c = 4000$ psi, grade 50 steel

4 No. 6 bars in tension, classed as "other than top" bars

Required $A_s = 1.01$ in.$^2$, bars are to be hooked

Clear cover 2.5 in. both vertical and horizontal, exterior exposure

**Find:**

Required development length of hooked bars in tension

**Solution:**

Compare cover requirements to cover actually provided

Code minimum cover for No. 6 bars, exterior exposure, must be $\geq 2.0$ in.

Required cover if modifier is allowed must be $\geq 2.50$ in.

Cover actually provided = 2.50 in. $\geq 2.50$ in.

Conclude:  cover meets minimum requirements
cover OK for modifier

Compare spacing requirements to actual spacing

Code minimum spacing $\geq 1d_b$ but no less than 1 in.

Clear spacing actually provided:

$$\text{Spacing} = \frac{b - 2 \times \text{clear cover} - \text{No. bars} \times d_b}{\text{No. bars} - 1}$$

$$= \frac{16 - 2 \times 2.50 - 4 \times 0.75}{4 - 1} = 2.67 \text{ in.} \geq 3d_b$$

Conclude:  Spacing meets code minimums
No multipliers are required

Compare required steel area to area actually provided

Required steel area = 1.01 in.$^2$

Area of steel provided = 1.77 in.$^2$

Conclude:  Development lengths may be reduced due to excess steel area.

Calculation of required length of imbedment

From Table A-13, $\ell_d$ = 12 in.

No multipliers required due to insufficient spacing

Apply modifier of 0.7 for excess cover

$\ell_d = 12 \times 0.7 = 8.4$ in.

Apply modifier for excess steel area

$\ell_d = 8.4 \times 1.01 / 1.77 = 5.0$ in.

Absolute minimum for No. 6 hooked bars, $\ell_d \geq 8d_b$ but not less than 6 in.

Use $\ell_d$ = 6in.

## OUTSIDE PROBLEMS

**12.25** Flexural reinforcement in a beam consists of 12 No. 6 bars having at least $2d_b$ cover and at least 3 $d_b$ spacing. The bars are grade 60 top bars in tension, $f'_c = 3000$ psi. What would be the development length, cover requirements, and spacing requirements if the bars are collected into 6 bundles? 4 bundles? 3 bundles?

**12.26** Compressive reinforcement placed uniformly around a circular pedestal consists of 15 No. 8 bars placed in a circle of 12 in. diameter. What would be the required cover if the bars were collected into 5 bundles uniformly spaced around the pedestal?

**12.27–12.38.** For Problems 12.1 through 12.12, respectively, determine the required development length of the bars if they were to be hooked. Assume that the given cover is maintained everywhere, both vertically and horizontally.

## Cutoff Points

When determination of tensile reinforcement was made (Chapters 5 and 6), the amount of reinforcement was determined on the basis of maximum moment on the section. At some distance away from this maximum moment, the moment may become reduced or may even become zero. At such points, it is possible to cut off a part of the tensile reinforcement, retaining only that amount of reinforcement required for the reduced moment in that particular area. Code places strict limits, however, on the minimum amount of reinforcement that must be retained.

A typical moment envelope for a flexural member is shown in Figure 12-5b. The moment envelope is defined as the outermost moment diagram at any point resulting from any one of the possible load cases; any other load case will produce a lesser value of moment at that point. (Moment envelopes are discussed more fully in Chapter 9.) Note that the inflection point can shift a significant distance laterally under different loadings.

The point at which the bar must attain its full yield strength is called the *critical section*. The critical section is defined by Code (12.10.2) as the point of maximum stress, or the point within the span where adjacent reinforcement terminates or is bent. The development length must of course lie entirely outside this point. For example, the critical section for bars mk c in Figure 12-3 is at midspan, and for bars mk d is at face of support.

Typical cutoff points for flexural reinforcement are shown in Figure 12-5b. In most cases, these cutoff lengths can be determined graphically

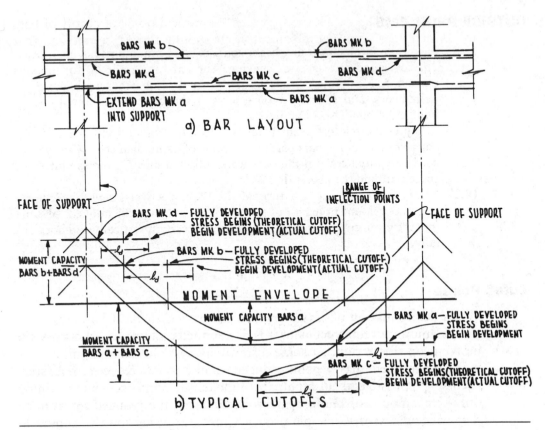

**Figure 12-5**    Typical flexural reinforcement.

simply by scaling the moment diagram. A higher degree of accuracy is difficult to defend in view of the empirical nature of the design.

It is required by Code that the reinforcement that is to be cut must be extended past its theoretical cutoff point (the point where it is theoretically no longer needed) by a distance not less than d nor less than 12 $d_b$. By this means, most of the strength of the bar is available where it is first needed and the remaining strength will be available further along the bar. At these theoretical cutoff points, however, the bars that have not been cut must carry the full moment and must be fully developed at such points.

It should be recognized that all cutoff points are measured from the theoretical point where the bar is needed, not from the actual cutoff points of adjacent bars. Such a case is shown for bars mk a in Figure 12-5b, where the cutoff length is measured from the theoretical point of cutoff of bars mk c, not their actual point of cutoff. It should also be recognized that more than one criterion may apply; the cutoff point for bars mk b must satisfy the requirement of extending at least a distance d or $12d_b$ beyond the point of inflection as well as extending at least their full development length beyond the theoretical point of cutoff of bars mk d. The development lengths are of course those of Table A-13 of the Appendix, as discussed earlier.

## Abbreviated Criteria for Cutoff Points

In addition to requirements for development of strength introduced in the preceding section, Code also has requirements for embedment of reinforcement (and cutoff lengths) which must be met regardless of strength, stress levels, or load capacities. A few of these general requirements have been discussed elsewhere but are repeated below for ready reference. Other practices that are permitted but are not required are also included in the following list.

1. Tension reinforcement may be anchored by bending it across the web to provide a suitable anchorage length.
2. Tension reinforcement may be made continuous with reinforcement on the opposite face of the member.
3. Flexural reinforcement shall extend past the point where it is no longer needed by a distance no less than $d$ or less than $12d_b$, whichever is greater, except at supports of simple spans and at free ends of cantilevers.
4. Continuing reinforcement shall have an embedment length no less than the development length $\ell_d$ beyond the theoretical cutoff of adjacent noncontinuing reinforcement.
5. Flexural reinforcement shall not be terminated in a tension zone unless one of the following conditions is satisfied:
   (a) Shear at the cutoff point does not exceed two-thirds of that permitted, to include effects of any shear reinforcement.
   (b) Continuing reinforcement provides double the area required for flexure at the theoretical cutoff point and shear does not exceed three-fourths of that permitted.
6. At least one-third of the positive moment reinforcement in simple members and one-fourth of the positive moment reinforcement in continuous members shall extend along the same face of the mem-

ber into the support; in beams, such reinforcement shall extend into the support no less than 6 in.

7. At simple supports and at point of inflection, positive moment tension reinforcement shall be limited to a diameter such that the development length $\ell_d$ satisfies the following equation:

$$\phi\,\ell_d \leq \frac{M_n}{V_n} + \phi\,\ell_a \tag{12-9}$$

where $M_n$ = nominal ultimate moment capacity of the section at the support with its reduced $A_s$

$V_n$ = nominal ultimate shear force on the section

$\phi$ = strength reduction factor for shear

$\ell_a$ = embedment beyond center of support, or, at a point of inflection, depth $d$ or $12d_b$, whichever is greater

8. The value of $M_n/V_n$ in Eq. (12-9) may be increased 30% when the ends of the reinforcement are confined by a compressive reaction. Eq. (12-9) need not be satisfied for reinforcement terminating beyond the centerline of simple supports by a standard hook, or by mechanical anchorage at least equivalent to a standard hook.

9. Negative moment reinforcement in a continuous, restrained, or cantilever member shall be anchored in or through the supporting member by embedment length, hooks, or mechanical anchorage.

10. At least one-third of the total tension reinforcement provided for negative moment at a support of a continuous beam shall have an embedment length beyond the point of inflection no less than the depth $d$ or less than $12d_b$ or less than one-sixteenth the clear span, whichever is greater.

The foregoing list of requirements may be worth learning if one is deeply involved in detailing concrete every working day. Otherwise, a simplified approach is needed that can quickly be relearned whenever needed. One such approach is to make steel cutoffs of up to one-half of the positive area of steel at one time, or all of the negative area of steel at one time. The foregoing list then reduces to the following rules.

1. In regions of negative moment:
   (a) Continue the entire area of flexural steel past the point of inflection by a distance $d$, $12d_b$, or span/16, whichever is farthest.
   (b) At integral columns, provide full development for all flexural steel at face of support by providing full development length in

or through the column, hooking as necessary, keeping reinforcement diameters small enough that full development is achieved.

2.  In regions of positive moment:
    (a) At interior supports, extend at least one-fourth of the area of flexural steel past the center of supports by a distance no less than 6 in., or past the face of support by a hook; extend the remaining flexural steel past the point of inflection by a distance no less than d nor less than $12d_b$, whichever is greater.
    (b) At end columns or simple supports, extend all tension reinforcement past the center of the support by a distance not less than $\ell_d$ or by a standard hook.

The foregoing abbreviated rules are summarized in Figure 12-6. Such an abbreviated set of rules is obviously conservative and will require more steel than if the full set of criteria were applied. In small projects, however, the savings in steel is generally offset by the additional labor hours required

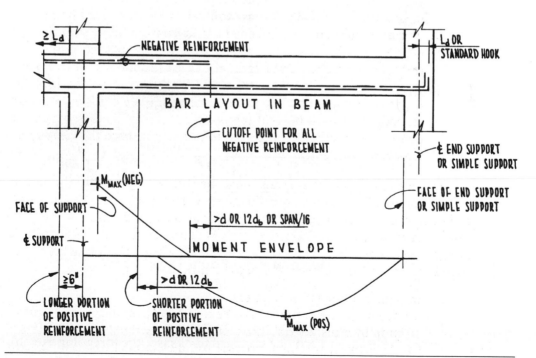

**Figure 12-6**    Potential cutoff points.

to fabricate and place the extra mark numbers. In larger projects, the savings would be significant and the more refined criteria could become worthwhile.

For an example in the use of the foregoing criteria, the beam of Example 10-2 is used. The final reinforcement sizes and stirrup sizes have already been determined, but no attempt has been made to check the development of the reinforcement. For the example, the various distances on the moment diagram have been computed; in practice, they would probably be scaled.

### EXAMPLE 12-12.

Reinforcement cutoff points.

Determine the cutoff points of the flexural reinforcement of Example 10-2. The beam and the moment diagram are shown below. Steel is grade 60, $f'_c$ is 3000 psi.

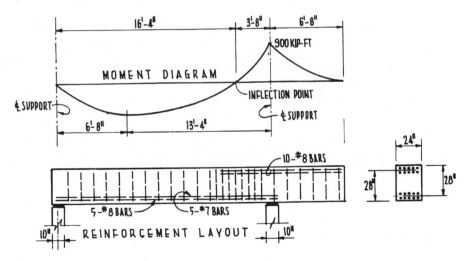

Determine the development lengths from Table A-13:

For No. 8 bars: $\ell_d$ = 34 in. or 48 in. for top bars

For No. 7 bars: $\ell_d$ = 26 in.

For negative reinforcement at the right support, extend the bars to the end of the beam and provide standard hooks (see the foregoing sketch).

For negative reinforcement to the left of the right support,

Cutoff point:  $\geq d$ (or 28 in.) past inflection point

$\geq 12d_b$ (or 12 in.) past inflection point

$\geq \dfrac{\text{span}}{16}$ (or 15 in.) past inflection

Use a cutoff point 28 in. past the inflection point, or 6 ft 0 in. from the centerline of support (see the following sketch).

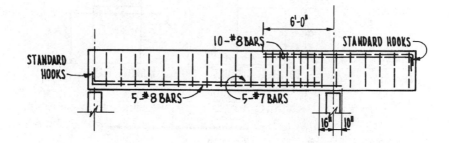

For positive reinforcement at the left support, extend all bars past centerline of support and provide standard hooks (see the foregoing sketch). At the right support, extend the bottom layer of bars 6 in. past the centerline of the support. For the remainder of positive bars at the right support,

cutoff point:  $\geq d$ (or 28 in.) past the inflection point

$\geq 12d_b$ (or 10½ in.) past inflection point

Use a cutoff point 28 in. past the inflection point or 16 in. from centerline of support.

## OUTSIDE PROBLEMS

For the following rectangular beams, determine the required areas of positive and negative reinforcement at the balanced stress condition. Find also the cutoff points (or anchorage requirements) for both positive and negative reinforcement. Use the abbreviated criteria for cutoffs and anchorages. Assume that the given dead load includes an allowance for the weight of the rectangular beam. It is not necessary to include a design for any required shear reinforcement. $f'_c$ = 4000 psi, grade 60 steel.

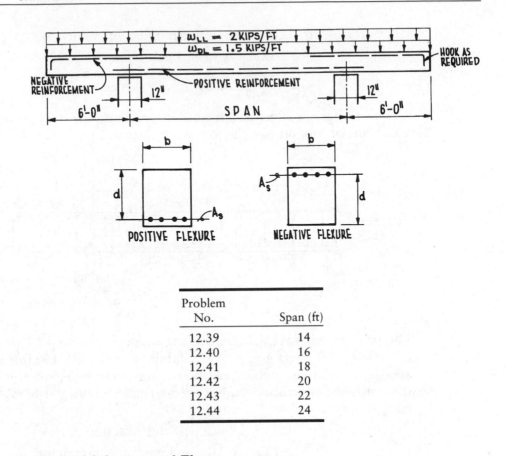

| Problem No. | Span (ft) |
|---|---|
| 12.39 | 14 |
| 12.40 | 16 |
| 12.41 | 18 |
| 12.42 | 20 |
| 12.43 | 22 |
| 12.44 | 24 |

## Development of Stirrups and Ties

Development lengths for stirrups and ties are of course comparable to those of flexural reinforcement. The problem is complicated, however, by the fact that the space available to provide anchorage for stirrups is severely limited.

A typical stirrup arrangement is shown in Figure 12-7, with a general case of loading. The load in the stirrup occurs as a result of diagonal cracking starting at the tension side of the beam. The actual size of the crack need be only a hairline, but once such a crack forms, regardless how thin, the stirrups bridging the crack are fully loaded.

The close spacing used in the design of stirrups inherently limits any large buildup of forces in a single stirrup, which in turn serves to keep the size of stirrups within the smaller bar sizes (No. 5 bars and smaller). For these smaller bar sizes, anchorage is much simplified. Code (7.2.1, 7.2.2)

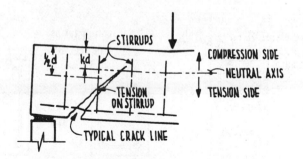

**Figure 12-7**    Loading on stirrups.

establishes separate anchorage criteria for these smaller stirrup bars, to include both hook criteria and bend radii.

Requirements for standard hooks in stirrups and ties are shown in Figure 12-8 for bar sizes No. 5 and smaller. For the smaller bar sizes used for stirrups, Code (7.2.2) also permits a sharper bend radius, also indicated in Figure 12-8.

Code (12.13.2.1) requires that the ends of single leg, simple U, or multiple U stirrups be anchored by providing a standard hook around longitudinal bars. Such anchorage is shown in the sections of Figure 12-8. Where there is no longitudinal reinforcement to hook the stirrups over, it is necessary to provide additional No. 4 bars longitudinally on which to hang the stirrups. Called *hangers*, these additional bars are in fact a part of the shear reinforcement rather than the flexural reinforcement. Or, if conditions are suitable, two of the flexural reinforcing bars are sometimes ex-

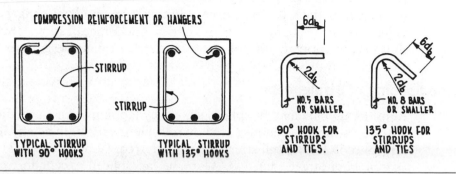

**Figure 12-8**    Typical stirrup and tie bends.

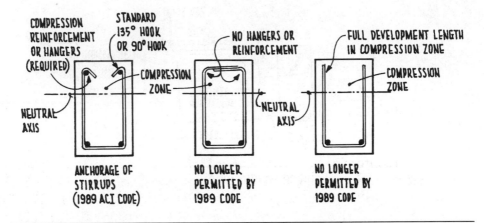

**Figure 12-9** Stirrup anchorage in the compression zone.

tended across the span to provide hangers; such an alternative is quite commonly used.

At ultimate load, the diagonal cracking shown in Figure 12-7 can penetrate quite far into the compression side of the beam. To assure that the anchorage at the top of the stirrups remains always in the compression zone, Code (12.13.1) requires that the anchorage for the stirrup be held as close as possible (with proper cover) to the compression face of the member. Some common anchorage arrangements permitted in the 1983 Code were dropped in the 1989 Code; two such arrangements that are no longer permitted are pointed out in Figure 12-9.

From the foregoing requirements, it is seen that there are no choices to be made when anchoring web reinforcement. One simply hooks the ends of the stirrups or ties over the longitudinal steel. If there is no longitudinal reinforcement to hook the ends over, additional longitudinal hangers are provided on which to hang the stirrups or ties. Anchorage of stirrups and ties has thus been reduced to a detailing practice rather than a design problem.

## Lap Splicing of Straight Bars

Reinforcing bars may be spliced simply by lapping them a specified length within a concrete member, as shown in Figure 12-10. The efficiency of a lap splice is 100%; full strength is transferred. Code (12.14.2.1) does not permit bars larger than No. 11 to be lap spliced except in exceptional circumstances.

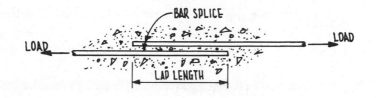

**Figure 12-10**    Lap splice.

Reinforcing bars need not contact each other in a lap splice. For example, in Figure 4-3 it may be noted that the No. 4 bars at the back of the cantilevered retaining wall are not at the same spacing as the No. 5 dowels projecting from the footing, nor are the two sets of bars the same size, yet a full transfer of load occurs. In this particular splice, 100% of the bars are spliced at the construction joint, although in other arrangements a splice may only involve 50% of the bars or even less.

For noncontact lap splices such as those of Figure 4-3, Code (12.14.2.3) specifies only that:

> The transverse spacing of flexural reinforcing bars spliced by non-contact lap splices shall be not farther apart than one-fifth the required lap splice length nor farther apart than 6 inches.

Individual bars within a bundled set of bars may also be lap spliced. In such splices, only one bar may be spliced at a time, and no part of successive splices may overlap. Further, the entire bundle may not be spliced at a single location. Also, the computed splice length for bundled bars must be increased 20% for a three-bar bundle and 33% for a four-bar bundle, as discussed earlier.

Code (12.14.3) permits welded splices in reinforcement. Code (12.14.3) also permits mechanical (patented) splice connections to be used. In both cases, the splice is required to develop 125% of the yield strength of the bar rather than the full ultimate strength of the bar.

## Development of Lap Splices in Tension

Code separates lap splices in tension into two classes, Class A and Class B. Class A splices of deformed straight bars in tension are those in which

a)   $(A_s$ provided)/$(A_s$ required) $\geq 2$ over the required lap length.
b)   Less than one-half the $A_s$ is spliced within the required lap length.

Splices not meeting the requirements for Class A splices are classified by default as Class B splices.

Required length of lap for the two classes of splices are prescribed by Code (12.15.1),

$$\text{Class A: lap length} = 1.0\ell_d \qquad (12\text{-}10a)$$

$$\text{Class B: lap length} = 1.3\ell_d \qquad (12\text{-}10b)$$

$$\text{Class A and B: minimum lap} = 12 \text{ in.} \qquad (12\text{-}10c)$$

where $\ell_d$ is the development length for deformed bars in tension (see Table 12-1). When computing $\ell_d$, however, the modification factor for excess $A_s$ may not be used but multipliers for top bars, epoxy coating, and lightweight concrete must be applied. The absolute minimum lap length after all multipliers have been applied is 12 in.

The required length of lap splices in tension is included in Table A-13 of the Appendix, to include extensions for top bars. It is well to observe in Table A-13 that splice lengths of 8 ft or more can occur in larger bar sizes; the cost of making a lap splice 8 feet long can sometimes make the use of welded splices or mechanical connectors an economical alternative.

## Development of Lap Splices in Compression

The criteria of this section applies to bars that are always in compression. If during a load reversal the bars should go into tension, no matter how slightly, the criteria of this section no longer apply. For those cases where tension may sometimes occur, the criteria of the next section apply.

There are no Class A or Class B categories for lap splices that are always in compression. For steels grade 60 and below, with concrete strengths 3000 psi and above, the lap length is prescribed by Code

$$\text{Lap length} = 0.0005 f_y d_b \qquad (12\text{-}11a)$$

$$\text{Minimum lap} = 12 \text{ in.} \qquad (12\text{-}11b)$$

When bars of different sizes are lap spliced in compression, the lap length shall be the larger of the development length $\ell_d$ for the larger bar size or the splice length given by Eq. (12-11) for the smaller bar size.

The required length of lap splices for deformed bars in compression are included in Table A-13 of the Appendix. Note that there is no classification of "top bars" for bars in compression.

Welded splices and mechanical connectors are permitted by Code for lap splices in compression. The criteria given earlier for such welds and connections remain applicable.

## Development of Lap Splices in Columns

Code (12.17) makes a special case of lap splices in columns since the steel at one side of a column can undergo a stress reversal and go into tension. Such cases occur in rigid frames where wind or earthquake loads produce high bending moments on the columns. For structures braced against sidesway (which are the primary structures treated in this book), these high bending loads on columns are not likely to occur, but even in these braced structures it is possible on occasion for reinforcement at one side of a column to go slightly into tension. A discussion of the topic is therefore warranted.

For an ultimate load $P_n$ placed concentrically on a column section ($e_n = 0$), the entire section is in yield; the corresponding strains are shown by Line OA in Figure 12-11. For this case of strain, there is no moment. All of the strength of the column section is used just to sustain the concentric axial load $P_n$.

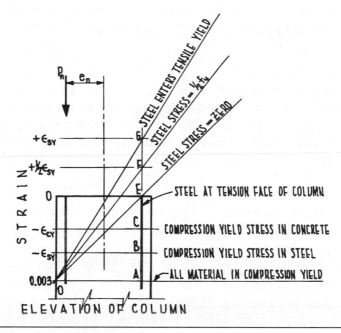

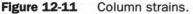

**Figure 12-11**    Column strains.

Code also prescribes a second case of loading in which the axial load $P_n$ is reduced and placed slightly eccentric on the section, producing a moment $M_n = P_n e_n$. The load $P_n$ and eccentricity $e_n$ are set such that the section rotates up to the line OE. For these values of load $P_n$ and eccentricity $e_n$, the stress in the steel on the "tension" side is now seen to be elastic, but having a magnitude of zero.

Code further prescribes a third case of loading in which the axial load $P_n$ is reduced even further and placed even more eccentric on the section, increasing the moment $M_n$ even more. The load $P_n$ and eccentricity $e_n$ are set such that the section rotates up to the line OF. For these values of load $P_n$ and eccentricity $e_n$, the stress in the steel on the tension side is again seen to be elastic, but now having a magnitude of $+\frac{1}{2}f_y$,

And finally, the axial load $P_n$ is again reduced and the eccentricity $e_n$ again increased until the section rotates up to the Line OG. For these values of load $P_n$ and eccentricity $e_n$, the stress in the steel on the tension side is seen to be barely elastic, having just reached the yield stress $f_y$. (This condition is the same as the "balanced strain" condition discussed in Chapter 6.)

The strain conditions shown in Figure 12-11 by lines OE and OF are those selected by Code for controlling the splice lengths of reinforcement. Rather than refer to these conditions of strain, however, Code refers to the corresponding stress levels in the steel on the tension side; that is, Code (12.17.2) uses $f_s = 0$ and $f_s = +\frac{1}{2}f_y$ on the tension side to describe its limiting cases.

For load cases where the stress in the steel is always compressive ($f_s \leq 0$), compression lap splices shall be used.

For load cases where the stress in the steel has entered tension but the stress is still less than $+\frac{1}{2}f_y$, that is, $0 \leq f_s \leq \frac{1}{2}f_y$, lap splices shall be Class B tension lap splices if more than half the bars are spliced. If fewer than half the bars are spliced and alternate lap splices are staggered a distance $\ell_d$ along the length of the bars, lap splices may be Class A tension lap splices.

For load cases where the stress in the steel exceeds $+\frac{1}{2}f_y$, lap splices shall be Class B tension lap splices.

The end result of the rather complex special requirement for column splicing is to permit Class A lap splices to be used for a rather narrow band of loads. And even then, the lap splices must be staggered by a length $\ell_d$ if the Class A lap splice is to be permitted. The benefit, of course, is a 30%

**Table 12-4**    Values of $e_n/h$ for Columns When $f_s = 0$ and $f_s = \frac{1}{2}f_y$ in Steel at Tension Face.

|  |  | $A_s/bb$ | Steel at Flexure Faces Only | | | | | | | | Steel Uniformly Distributed | | | | | | | |
|---|---|---|---|---|---|---|---|---|---|---|---|---|---|---|---|---|---|---|
|  |  | Steel ratio $A_s/bh$ | Values of $e_n/h$ for $f_s = 0$ and $f_s = +\frac{1}{2}f_y$ | | | | | | | | Values of $e_n/h$ for $f_s = 0$ and $f_s = +\frac{1}{2}f_y$ | | | | | | | |
|  |  | $e_n/h$ | 0.01 | 0.02 | 0.03 | 0.04 | 0.05 | 0.06 | 0.07 | 0.08 | 0.01 | 0.02 | 0.03 | 0.04 | 0.05 | 0.06 | 0.07 | 0.08 |
| 3000 | 40 | 0 | 0.14 | 0.17 | 0.18 | 0.20 | 0.21 | 0.23 | 0.24 | 0.25 | 0.13 | 0.14 | 0.15 | 0.16 | 0.17 | 0.18 | 0.18 | 0.19 |
|  |  | $+\frac{1}{2}f_y$ | 0.25 | 0.30 | 0.35 | 0.40 | 0.44 | 0.48 | 0.52 | 0.55 | 0.23 | 0.26 | 0.29 | 0.32 | 0.35 | 0.37 | 0.39 | 0.40 |
|  | 50 | 0 | 0.15 | 0.17 | 0.20 | 0.21 | 0.23 | 0.24 | 0.25 | 0.26 | 0.14 | 0.16 | 0.17 | 0.18 | 0.19 | 0.20 | 0.20 | 0.21 |
|  |  | $+\frac{1}{2}f_y$ | 0.28 | 0.34 | 0.40 | 0.46 | 0.50 | 0.55 | 0.59 | 0.62 | 0.26 | 0.31 | 0.35 | 0.39 | 0.42 | 0.45 | 0.47 | 0.50 |
|  | 60 | 0 | 0.15 | 0.18 | 0.21 | 0.22 | 0.24 | 0.25 | 0.26 | 0.27 | 0.15 | 0.17 | 0.19 | 0.20 | 0.21 | 0.22 | 0.23 | 0.24 |
|  |  | $+\frac{1}{2}f_y$ | 0.30 | 0.38 | 0.45 | 0.51 | 0.56 | 0.60 | 0.64 | 0.68 | 0.29 | 0.36 | 0.42 | 0.47 | 0.51 | 0.55 | 0.58 | 0.61 |
| 4000 | 40 | 0 | 0.15 | 0.16 | 0.18 | 0.19 | 0.21 | 0.22 | 0.23 | 0.24 | 0.14 | 0.15 | 0.16 | 0.16 | 0.17 | 0.17 | 0.18 | 0.18 |
|  |  | $+\frac{1}{2}f_y$ | 0.24 | 0.29 | 0.33 | 0.37 | 0.40 | 0.44 | 0.47 | 0.50 | 0.23 | 0.26 | 0.28 | 0.31 | 0.33 | 0.35 | 0.37 | 0.38 |
|  | 50 | 0 | 0.15 | 0.17 | 0.19 | 0.20 | 0.22 | 0.23 | 0.24 | 0.25 | 0.14 | 0.16 | 0.17 | 0.18 | 0.19 | 0.19 | 0.20 | 0.21 |
|  |  | $+\frac{1}{2}f_y$ | 0.27 | 0.32 | 0.37 | 0.42 | 0.46 | 0.50 | 0.54 | 0.57 | 0.25 | 0.30 | 0.33 | 0.37 | 0.39 | 0.42 | 0.45 | 0.47 |
|  | 60 | 0 | 0.15 | 0.18 | 0.20 | 0.21 | 0.23 | 0.24 | 0.25 | 0.26 | 0.15 | 0.17 | 0.18 | 0.20 | 0.21 | 0.22 | 0.22 | 0.23 |
|  |  | $+\frac{1}{2}f_y$ | 0.29 | 0.36 | 0.41 | 0.47 | 0.51 | 0.55 | 0.59 | 0.63 | 0.28 | 0.34 | 0.39 | 0.43 | 0.47 | 0.51 | 0.54 | 0.57 |
| 5000 | 40 | 0 | 0.15 | 0.16 | 0.18 | 0.19 | 0.20 | 0.21 | 0.22 | 0.23 | .14 | 0.15 | 0.16 | 0.16 | 0.17 | 0.18 | 0.18 | 0.18 |
|  |  | $+\frac{1}{2}f_y$ | 0.24 | 0.28 | 0.31 | 0.35 | 0.38 | 0.41 | 0.44 | 0.47 | 0.23 | 0.25 | 0.28 | 0.30 | 0.32 | 0.34 | 0.35 | 0.37 |
|  | 50 | 0 | 0.15 | 0.17 | 0.19 | 0.20 | 0.21 | 0.22 | 0.24 | 0.24 | 0.15 | 0.16 | 0.17 | 0.18 | 0.18 | 0.19 | 0.20 | 0.20 |
|  |  | $+\frac{1}{2}f_y$ | 0.26 | 0.31 | 0.35 | 0.39 | 0.43 | 0.47 | 0.50 | 0.54 | 0.25 | 0.29 | 0.32 | 0.35 | 0.38 | 0.40 | 0.43 | 0.45 |
|  | 60 | 0 | 0.16 | 0.18 | 0.19 | 0.21 | 0.22 | 0.23 | 0.25 | 0.25 | 0.15 | 0.17 | 0.18 | 0.19 | 0.20 | 0.21 | 0.22 | 0.23 |
|  |  | $+\frac{1}{2}f_y$ | 0.29 | 0.34 | 0.39 | 0.44 | 0.48 | 0.52 | 0.56 | 0.59 | 0.28 | 0.33 | 0.37 | 0.41 | 0.45 | 0.48 | 0.51 | 0.54 |

For values of $e_n/b$ less than that at $f_s = 0$, use compression criteria for splices.
For values of $e_n/b$ between the two listed values, use either Class A or B tension splices.
For values of $e_n/b$ greater than that at $f_s = +\frac{1}{2}f_y$, use only Class B tension splices.

reduction in splice length and a corresponding reduction in cost; any benefit in cost is diminished, however, by higher labor costs in fabricating and handling the unequal lengths of bars that must now be marked and kept track of.

For reference, the eccentricity ratios of $e_n/h$ corresponding to the two stress conditions have been computed and tabulated in Table 12-4. As indicated in the footnotes of the table, Class A splices may only be considered if the steel stress is between zero and $+\frac{1}{2}f_y$.

Except for designs having a very large number of highly repetitive columns that fall in the middle range of loads, there seems to be little incentive to investigate and use the Class A splice conditions for columns. In this text, the provision is largely ignored, primarily because the columns in frames braced against sidesway rarely fall within the higher eccentricity levels. Where a fast check against Table 12-4 reveals that the steel in a particular column does in fact go into tension, a Class A splice probably would not be considered; a "safe" Class B splice would likely be adopted, even if slightly conservative.

Some examples will demonstrate the use of Table A-13 in finding required splice lengths.

## EXAMPLE 12-13.

Determination of splice lengths.

**Given:**

Beam section as shown, under negative moment
Grade 50 steel,
$f_c' = 4000$ psi

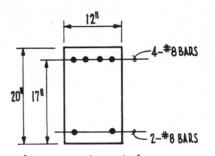

**Find:**

Required splice lengths for both tensile and compressive reinforcement

**Solution:**

Tensile reinforcement is classified as top bars.
Assume $A_s$ provided $= A_s$ required, so Class B splice is required.
For No. 8 bars in tension, Class B splice for top bars is 42 in. long.
For No. 8 bars in compression, splice length is 25 in. long.

### EXAMPLE 12-14.

Determination of splice length.

**Given:**

Column reinforcement, 100% to be lap spliced at the floorline as shown.

Grade 60 steel, $f'_c$ = 4000 psi; $P_n$ = 946 kips, $M_n$ = 247 kip-ft

**Find:**

Splice lengths of No. 9 and No. 8 bars

**Solution:**

Solve for eccentricity ratio $e_n/h$,

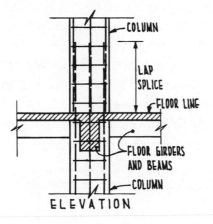

ELEVATION

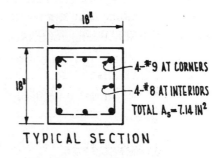

TYPICAL SECTION

$$\frac{e_n}{h} = \frac{M_n/P_n}{h} = \frac{247/946}{18/12} = 0.174$$

Solve for steel ratio $A_s/bh$,

$$\rho = \frac{A_s}{bh} = \frac{7.14}{18 \times 18} = 0.022 \text{ or } 2.2\% \text{ steel}$$

From Table 12-4 it is seen that these bars never enter tension

Use compression lap splice,   $\ell_d$ = 30 in. for No. 8 bars

$\ell_d$ = 34 in. for No. 9 bars

### EXAMPLE 12-15.

Determination of splice lengths.

**Given:**

> Retaining wall of Figure 4-3
> Grade 60 steel, $f_c' = 3000$ psi

**Find:**

> Required splice length of No. 5 dowels to No. 4 vertical bars at back face of wall

**Solution:**

> Code requires that spacing shall not be less than 6 in. or less than 20% of the splice length. These conditions seem to be met by the two sets of bars.
>
> 100% of the bars are spliced at the same level; therefore a Class A splice may not be used.
>
> Splice length for No. 5 bars in a Class B splice is 27 in.
>
> Minimum transverse spacing is therefore $27/5 = 5.2$ in. > 5 in. provided.
>
> Required splice length is therefore 27 in., which is the length indicated.

---

### EXAMPLE 12-16.

Determination of splice length.

**Given:**

> 4 No.6 top bars, bundled, in tension. $f_c' = 4000$, grade 50 steel.
> Fewer than 25% of the bars are spliced at any location.

**Find:**

> Required splice length

**Solution:**

> With less than 25% of the bars spliced at any location, the total $A_s$ is never doubled over the splice length, so the splice is a Class B splice.

From Table A-13, $\ell_d = 30$ in. for top bars

Apply multiplier 1.33 for a four-bar bundle,

$\ell_d = 1.33 \times 30 = 40$ in.

Use 40 in. splice length

---

## EXAMPLE 12-17.

Determination of splice length in columns.

**Given:**

Column 15 in. × 15 in., 4 No. 10 bars longitudinal

Only one bar spliced at any one level, $P_n = 641$ kips, $M_n = 208$ kip-ft.

$f_c' = 4000$ psi, grade 50 steel

**Find:**

Required splice length

**Solution:**

For use in Table 12-4,

$e_n/h = M_n/P_n h = 208 \times 12/641 \times 15 = 0.260$

$A_s = 5.07/15 \times 15 = 0.0225$

From Table 12-4, tensile stress at one side of the column is found by interpolation to be $0.27\, f_y = 13{,}540$ psi in tension.

Since less than half the bars are spliced at any one location, conclude that a Class A splice may be used.

From Table A-13, for a Class A tension splice, $\ell_d = 40$ in.

Use $\ell_d = 40$ in.

---

## OUTSIDE PROBLEMS

Determine the required splice length for beam reinforcement under the given conditions.

| Prob. No. | Concrete Strength $f'_c$ psi | Steel Grade | Type of Splice | % $A_s$ Spliced | Top or Other Location | Bars |
|---|---|---|---|---|---|---|
| 12.45 | 3000 | 40 | Tension | <25% | Top | #6 |
| 12.46 | 3000 | 50 | Tension | 50% | Other | 3 No. 5 bundled |
| 12.47 | 3000 | 60 | Tension | 100% | Top | #7 |
| 12.48 | 3000 | 40 | Compression | | | #9 |
| 12.49 | 3000 | 50 | Compression | | | 4 No. 6 bundled |
| 12.50 | 3000 | 60 | Compression | | | #8 |
| 12.51 | 4000 | 40 | Tension | <25% | Other | #6 |
| 12.52 | 4000 | 50 | Tension | 50% | Top | 3 No. 5 bundled |
| 12.53 | 4000 | 60 | Tension | 100% | Other | #7 |
| 12.54 | 4000 | 40 | Compression | | | #9 |
| 12.55 | 4000 | 50 | Compression | | | 4 No. 6 bundled |
| 12.56 | 4000 | 60 | Compression | | | #8 |
| 12.57 | 5000 | 40 | Tension | <25% | Top | #6 |
| 12.58 | 5000 | 50 | Tension | 50% | Other | 3 No. 5 bundled |
| 12.59 | 5000 | 60 | Tension | 100% | Top | #7 |
| 12.60 | 5000 | 40 | Compression | | | #9 |
| 12.61 | 5000 | 50 | Compression | | | 4 No. 6 bundled |
| 12.62 | 5000 | 60 | Compression | | | #6 |

Determine the required splice lengths for column reinforcement under the given conditions. Splices may be staggered along the column if desired.

| Prob. No. | Concrete Strength $f'_c$ psi | Steel Grade | Column Size | Column Reinf. | $M_n$ kip-ft | $P_n$ kips | % $A_s$ Spliced |
|---|---|---|---|---|---|---|---|
| 12.63 | 3000 | 40 | 12 × 12 | 4 No. 9 | 69 | 330 | <50% |
| 12.64 | 3000 | 60 | 12 × 12 | 4 No. 10 | 121 | 367 | 100% |
| 12.65 | 4000 | 40 | 14 × 14 | 4 No. 10 | 101 | 530 | <50% |
| 12.66 | 4000 | 60 | 14 × 14 | 4 No. 11 | 241 | 602 | 100% |
| 12.67 | 5000 | 40 | 16 × 16 | 4 No. 10 | 137 | 762 | <50% |
| 12.68 | 5000 | 60 | 16 × 16 | 4 No. 11 | 350 | 870 | 100% |

## Elastic Analysis

The elastic equations for bond stress and development length do not appear directly in the 1989 ACI Code. All development lengths are simply stated as empirical formulas, usually nonhomogeneous, based on test data

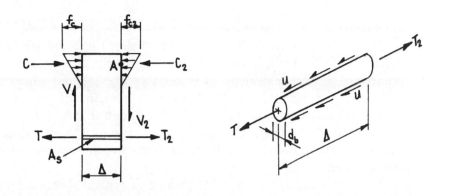

**Figure 12-12**     Elastic bond stress.

without regard for theoretical development. Although the elastic analysis is not overtly recognized, it can offer valuable insight into the working of bond and development length. A highly abbreviated discussion of the elastic analysis of bond is therefore presented in the following paragraphs.

A typical segment of a beam is removed and is shown with its internal resultants of load in Figure 12-12. Note that the segment is shown as $\Delta$ inches thick, with all signs in the positive sense. The tensile reinforcement is removed and shown at larger scale, with the average bond stress u indicated around the bar.

Moments are summed about point A, yielding

$$V\Delta + Tjd - T_2jd = 0 \quad \text{hence } T_2 - T = \frac{V\Delta}{jd} \quad (12\text{-}6)$$

Forces acting on the reinforcement are also summed, yielding

$$T_2 - T = u(\Sigma\pi d_b)\Delta \quad (12\text{-}7)$$

where $\Sigma\pi d_b$ is the perimeter, or the sum of perimeters, of the reinforcement. $\Sigma\pi d_b$ was conventionally designated in older ACI Codes as $\Sigma_0$.

Eq. (12-7) is substituted into Eq. (12-6) and the symbol $\Sigma_0$ is used to indicate sum of perimeters; the result is solved for u:

$$u = \frac{V}{\Sigma_0 jd} \quad (12\text{-}8)$$

where     $V$  = shear on the section
　　　　$u$  = average bond stress around the bar

$jd$  = distance between tensile and compressive resultants.

It should be obvious from Eq. (12-7) that the bond stress is highest where $T_2 - T$ is highest, or where the moment is changing rapidly. Such points are shown in strength of materials to be located where shears are highest, as verified by Eq. (12-8); these points are usually near the supports or near concentrated loads. It may be concluded that high bond stress will occur around the flexural reinforcement wherever high values of shear occur, most commonly at the beam supports and at other points of concentrated loads.

Overall anchorage is now examined; a typical flexural anchorage is shown in Figure 12-13. The embedded length $L_e$ must develop an average bond stress u if the moment at face of support is to be sustained. Moments are summed about the resultant of compressive force C at point A:

$$M = Tjd \qquad (12\text{-}9)$$

where M is the static moment on the section. Forces are now summed along the length of embedment,

$$T = u \, \Sigma_0 L_e \qquad (12\text{-}10)$$

where $\Sigma_o$ is the sum of perimeters, as defined previously.

Eq. (12-10) is substituted into Eq. (12-9) to eliminate $T$, and the result solved for the required embedment length $L_e$:

$$L_e = \frac{M}{u \Sigma_0 jd} \qquad (12\text{-}11)$$

Eq. (12-8) and (12-11) relate the bond stress $u$ and required embedment length $L_e$ to the shear and moment acting on the section. If an allow-

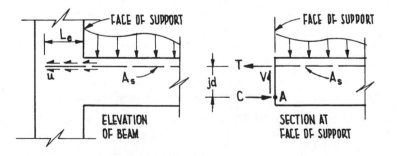

**Figure 12-13**   Anchorage length.

able bond stress $u$ were known, the necessary embedment lengths and bar perimeters could be computed. In past years, the bond stress $u$ was in fact specified by Code and the other unknowns were computed from Eq. (12-8) and Eq. (12-11); in more recent years, the Code has dropped that approach in favor of the more accurate approach presented earlier in this chapter.

Circumstances do arise, however, where the older approach using an allowable stress can be very useful. For example, consider the following problem, which occurs frequently where pad eyes are embedded in the foundations, to be used in subsequent steel erection.

A grade 40 deformed No. 6 reinforcing bar is embedded a distance of 6 in. in a concrete footing. Determine the safe working load that may be applied as a pullout load. ($f_c' = 3000$ psi. The allowable bond stress u may be taken as 350 psi.)

For pullout load $P$, the allowable load is obviously the allowable bond stress $u$ times the embedded contact area (circumference times length):

$$P = u(\pi \times d_b \times L_e) = 350 \times p \times 0.75 \times 6$$
$$= 4.95 \text{ kips}$$

If instead of the allowable bond stress the current Code provisions must be used, the only value that can be computed is the development length at full yield, $\ell_d$, computed as 13 in. (Table A-13). The load $P_y$ in the bar at yield is given by $f_y A_b$, or 17.6 kips. If it is assumed that the pullout load is proportional to the depth of embedment, the reduced pullout load $P_n$ is given by

$$P_n = P_y\left(\frac{6}{13}\right) = 17.6 \times \frac{6}{13}$$
$$= 8.1 \text{ kips}$$

Further, if a factor of safety (F.S.) of 1.67 is assumed to working levels, the allowable load P is then

$$P = \frac{8.1}{\text{F.S.}} = 485 \text{ kips}$$

---

This value of 4.85 computed under the foregoing assumptions compares well with the value of 4.95 kips computed using older criteria. The reliability of the older criterion can be defended further by recognizing that it was used quite successfully for many years before the newer and more accurate criterion was developed. Judiciously applied, the older criterion can still be used with satisfactory results.

The following formulas for computing usable bond stress for deformed bars in tension are extracted from the 1963 ACI Code. The stresses are in psi and apply to bar sizes no larger than No. 11.

|  | Ultimate Stress | Working Stress |
|---|---|---|
| Top bars: | $u = \dfrac{6.7\sqrt{f_c'}}{d_b}$ but $\leq 650$ | $u = \dfrac{3.4\sqrt{f_c'}}{d_b}$ but $\leq 350$ |
| Other bars: | $u = \dfrac{9.5\sqrt{f_c'}}{d_b}$ but $\leq 800$ | $u = \dfrac{4.8\sqrt{f_c'}}{d_b}$ but $\leq 500$ |

For plain bars, the usable bond stresses may be taken as one-half of these values but not more than 160 psi at working stresses nor 250 psi at ultimate stress.

For deformed bars in compression, the following stresses apply.

|  | Ultimate Stress | Working Stress |
|---|---|---|
| All bars: | $u = 13\sqrt{f_c'}$ but $\leq 800$ | $u = 6.5\sqrt{f_c'}$ but $\leq 400$ |

The foregoing older method is now outdated and should not be used where the more recent approach may be applied. The more recent approach, however, can be seen to be somewhat more limited in its range of applications. The older criterion can offer a reassuring check in unusual applications or when the newer criterion is only vaguely applicable.

## REVIEW QUESTIONS

1. What is meant by the "development length" of a reinforcing bar?
2. What is the "bond strength" of concrete, and what are some of the factors that can affect it adversely?
3. How does the bond of concrete to smooth reinforcing bars compare to its bond to deformed bars?
4. In view of the answer to Question 3, about what percentage of the total load on a deformed bar might be attributed to the mechanical bearing of its deformations against the adjacent concrete?

5. How have development lengths of reinforcing bars been determined in recent years?
6. What is the primary advantage in using hooks?
7. What is the primary point of caution in using hooks?
8. When only a part of the required development length of a bar can be embedded in concrete, how can its allowable working load be computed using the strength method? Using the working stress method?
9. On flexural reinforcement in a simple span, where can bond stresses be expected to be highest? Lowest?
10. Why is it necessary to provide hooks for anchoring stirrups?
11. What particular advantages does the 135° hook provide when one is detailing ties and stirrups?
12. Where in the cross section of the beam does the development of stirrups occur?
13. What is "bundling" of reinforcement?
14. When mechanical anchorages or welding is used, how much of the strength of the reinforcement must be developed?

# 13

# INTERMEDIATE LENGTH COLUMNS

The study of reinforced concrete columns is a complex subject, so complex that only the rudiments of column design can be presented in an introductory-level textbook such as this. And, while this introductory study is adequate for small structures braced against sidesway, a great deal more study of columns is recommended before one ventures into more complex rigid frame structures subject to wind and earthquake.

Code (B.6.1) requires that all reinforced concrete columns be designed to sustain the factored ultimate loads. There are no provisions in the Code for designing columns at service levels of stress. Further, there are no requirements (nor are there any restrictions) concerning the control of stresses at day-to-day service levels of load. A column design may therefore include whatever requirements at service levels that the designer considers appropriate, as long as the column section finally selected is capable of sustaining the factored ultimate load.

The procedures outlined in subsequent sections include means to control stresses at service levels of stress. It will be seen that in many if not most cases, the limitations on service stresses will be more stringent than those at ultimate load. As a result, the section finally chosen may be somewhat larger than that required solely for ultimate load, but its performance at service levels will be kept within the desired elastic ranges.

## Configurations and Practices

As indicated in Figure 8-1, many shapes may be used for columns, such as square, rectangular, hollow, circular, Y-shaped, L-shaped, and so on. By far the most common shape is the rectangular shape, which includes the square as a special case. Only the rectangular shape is treated in the succeeding sections; other shapes may be treated similarly.

In designing a column, it is always necessary to make an initial selection for the gross dimensions of the column section, b and h, as shown in Figure 13-1. The reinforcement is then determined for this particular gross size. If the amount of reinforcement required for this gross size is considered to be unacceptable, a new gross size must be selected and the process repeated until an acceptable design is found.

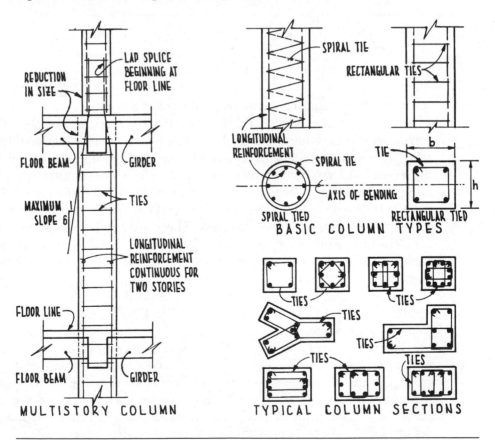

**Figure 13-1**  Column details.

In making the first guess at the size of a column, the rules of thumb given in Chapter 4 can be helpful. Quite often, though, when a group of columns is being designed, the gross size of the column carrying the heaviest load is established first, and all other columns are made the same size. Then only the reinforcement of the other columns need be varied to suit the actual loads.

The rule of thumb for the size of a square column is repeated from Chapter 4:

column width = 12 in. plus 1 in. per story above

Where the column is to be rectangular, it should have roughly the same gross area as this square column, or slightly more. Where significant moments are present, the accuracy of this rule of thumb becomes even worse than usual.

A low concrete building (five stories or less) having a column module of 18 to 20 ft and concrete floors and roof can be expected to have a column loading of 30 to 40 tons per floor. This load is the actual load, not the ultimate load. A nominal column load for small buildings will be seen to be a useful index when guessing sizes of columns.

Older codes specified a minimum dimension of 10 in. for columns, but more recent codes have discontinued this minimum limit. The column size is now controlled by the designer and may be set to match the masonry or concrete walls used elsewhere in the design. Such an arrangement is shown in the wall section of Figure 13-2. While the minimum size is 10 in., the difficulty in casting columns 10 in. square (or less) severely discourages the use of such small columns. A minimum dimension of 12 in. is commonly observed.

For one- and two-story buildings, the loads are so low that a minimum-

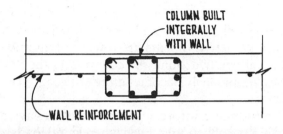

**Figure 13-2**    Integral column and wall.

sized column with minimum longitudinal steel is the usual design. Even these sizes usually provide a capacity much larger than needed. Example 13-5, presented later in this chapter, lists the capacities of minimum-sized columns for various grades of steel and strengths of concrete.

In a typical concrete building, the strength of the concrete specified for columns is usually higher than for the beams and slabs. The primary benefit in using higher strength concrete is in its reduced susceptibility to creep under the high sustained loads. Similarly, the longitudinal reinforcement is almost always the higher-strength steels wherever they are available.

In general, the longitudinal reinforcement in columns is spliced each two stories. Such frequent splicing requires large amounts of additional steel, but the reinforcing bars will not cantilever out of the forms far enough to permit going more than two stories. The splice is almost always placed just above floor level, as indicated earlier in Figure 13-1. The practice of splicing all bars at the same level effectively precludes the use of Class A lap splices in column bars at such locations, since 100% splicing of bars is not permitted with Class A splices (see Table 12-1).

Although Code (10.9.1) allows the steel areas in columns to be as high as 8% of the gross concrete area $bh$, the congestion in such heavily reinforced columns severely limits their use. The congestion is particularly severe where the vertical column bars must pass by the horizontal bars in the two intersecting girders that an interior column usually supports. As a practical matter, the steel areas are generally kept less than 3% of the gross concrete area bh; steel areas above 4% are rare.

To decrease congestion, it is common practice to use the larger bar sizes for longitudinal reinforcement in columns. A larger steel area is thus provided by fewer bars. The use of the larger bar sizes in columns usually does not create problems with bond since the shear in columns can be expected to be quite low.

Two of the more common arrangements of longitudinal reinforcement are shown in Figure 13-3 and repeated in Tables A-9 through A-11 of the Appendix. Where moments are low, the longitudinal steel is usually distributed uniformly around the column. As the magnitude of the moment increases, it obviously becomes more efficient to concentrate the steel at the flexure faces.

Longitudinal reinforcement is almost always arranged symmetrically about the axes of bending. Even where wind loads are not being resisted by the columns and load reversals do not exist, the column reinforcement is

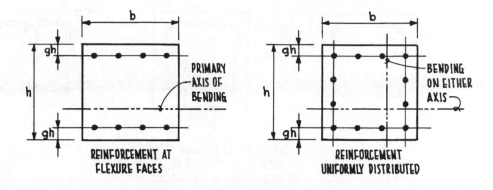

**Figure 13-3**    Common steel arrangements in columns.

still arranged symmetrically. There is thus no chance that the column can be constructed backwards.

Code (7.10.5) requires that the longitudinal bars be tied laterally to prevent their buckling outward under load. Design requirements for the ties themselves are also given by Code:

1. Ties shall be at least No. 3 for longitudinal bar sizes No. 10 or smaller and at least No. 4 for larger bars and bundled bars.
2. Vertical spacing shall not exceed 16 longitudinal bar diameters or 48 tie-bar diameters or the least dimensions of the column.
3. Ties shall be arranged such that every corner bar and alternate longitudinal bars shall have lateral support provided by the corner of a tie having an included angle of no more than 135°; no bar shall be farther than 6 in. clear on each side along the tie from such a laterally supported bar.

Typical tie arrangements are shown in Figure 13-4 for some of the more common configurations.

While seemingly an appendage to the design, the ties used in column design make a critical contribution to the overall strength of the column. Under heavy load, the longitudinal reinforcement has a strong tendency to buckle outward, bursting out of the concrete encasement and destroying the integrity of the column. Column ties are spaced and arranged to prevent such a failure; they force the steel to remain straight and to work monolithically with the concrete in carrying the load. Buckling of the lon-

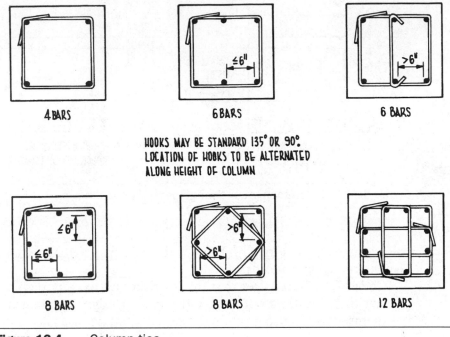

**Figure 13-4**    Column ties.

gitudinal reinforcement, however, remains one of the more serious failure modes to be considered in the design of reinforced concrete columns.

Though not presented in this textbook, longitudinal column reinforcement may also be arranged in a circular pattern. The column itself is usually circular in cross section, but it may be square, simply encasing the circular arrangement of reinforcement in a square cross section. The ties for a circular arrangement of reinforcement are a continuous spiral, wound around the outside of the set of longitudinal bars. Circular columns with spiral ties are useful primarily as extremely heavily loaded columns subject to bending in all directions; such a primary application is outside the scope of this textbook.

## Behavior Under Load

Columns can fail structurally by either of two separate modes: buckling or crushing. Long columns fail by buckling and short columns fail by crushing. However, many concrete columns are neither long nor short, but are somewhere in an intermediate range. For such columns, failure occurs through an indistinct mixture of the two modes.

The "critical" buckling stress of long columns is given by the well-known Euler column formula:

$$\frac{P_c}{A} = \frac{\pi^2 E}{(L/r)^2}$$

(13-1)

where $P_c/A$ = stress just as buckling impends
$E$ = modulus of elasticity
$L$ = column length
$r$ = radius of gyration, $\sqrt{I/A}$

The physical conditions (boundary conditions) assumed for the derivation of the Euler column formula are shown in Figure 13-5. Under axial load, the column fails by lateral displacement, but it will return to its original straight configuration when the load is removed. Note that the length $L$ is measured between the hinge points and that the moment of inertia $I$ is oriented about the axis of bending. Note also that the end conditions are hinges (or points of zero moment, such as inflection points).

It is important to be aware that the end moments are zero in the derivation of the Euler equation. The existence of end moments, even small ones, can seriously reduce the critical load $P_c$. The formula is valid, however, between any two consecutive points of zero moment.

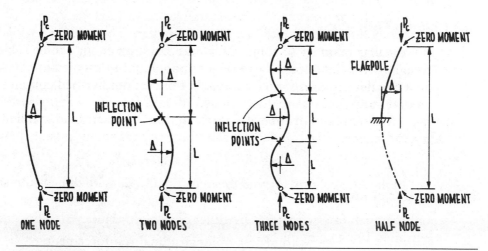

**Figure 13-5**    Euler column conditions.

Equally important in the derivation, it is assumed that the column is perfectly straight; there can be no accidental offset in its alignment that would produce an initial moment. It is also assumed that the load is always concentric on the section; there can be no moments introduced due to eccentric loading. It is further assumed that the moment of inertia $I$ is constant throughout the length of the column; there can be no variations in the dimensions, nor can there be irregularities in reinforcement or in splices.

By this point in the presentation, it should be obvious that very few concrete columns could meet all the conditions required by the Euler formula. Even so, the basic parameters appearing in the Euler formula remain valid indicators for all column performance. The way in which these parameters enter into the semiempirical design of concrete columns is discussed in succeeding sections.

One such parameter that will appear later is the ratio $L/r$, which has a special significance in column design. Called the *slenderness ratio*, it affords an indication of the susceptibility of a column to buckling. The higher the slenderness ratio is, the higher the susceptibility to buckling becomes and the lower the buckling load is.

At the other extreme from long columns are short concrete pedestals. When the ratio of height to least lateral dimension is 3 or less, the member is not subject to Code criteria for columns. The design is performed by direct $P/A + Mc/I$ design procedures.

## ACI Column Formula

For the vast majority of columns that are not long enough to be subject to design by the Euler formula nor short enough to be designed as a pedestal, ACI has developed a simple approach. For rectangular tied columns such as that shown in Figure 13-6, Code (10.3.5.2) requires that the nominal eccentric load $P_n$ shall not exceed 80% of the concentric load at full plasticity, $P_o$, where

$$P_o = 0.85f'_c(A_g - A_{st}) + f_yA_{st} \qquad (13\text{-}2)$$

In the ACI column formula given by Eq. (13-2) with dimensions shown in Figure 13-6,

$$\begin{aligned}
A_g &= \text{gross area of concrete, } bh \\
A_{st} &= \text{total area of longitudinal reinforcement} \\
0.85f'_c &= \text{idealized yield stress of concrete} \\
f_y &= \text{yield stress of reinforcement}
\end{aligned}$$

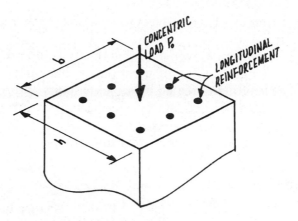

**Figure 13-6**    ACI basic column.

It is emphasized that the load $P_o$ given by Eq. (13-2) is a concentric load. Also, it is the absolute maximum load that the cross section could sustain when it is in a fully plastic state of deformation. There is no extra capacity for moment.

When the nominal axial load $P_n$ is equal to or less than 80% of this peak axial load $P_o$ there is some excess capacity available to take moment. This moment, $M_n$, is viewed by Code (10.3.6) as the "maximum moment that can accompany the axial load." The section is thus designed *first* for axial load; moment can then be added to take up the excess capacity, if any exists.

The nominal ultimate axial load $P_n$ and the nominal ultimate moment $M_n$ are computed as always:

$$P_n = \frac{P_u}{\phi} \quad \text{and} \quad M_n = \frac{M_u}{\phi}$$

where $P_u$ and $M_u$ are the factored ultimate loads and $\phi$ is the strength reduction factor defined and used previously. For columns, $\phi = 0.7$ for both axial load and moment, but as in beams, the strength reduction factor $\phi$ may be varied where $P_n$ is less than $0.10 f'_c bh/\phi$. The same relationship used earlier applies:

$$\phi = \frac{0.9}{1 + 0.2 \dfrac{P_n}{0.10 f'_c bh}} \tag{13-3}$$

This allowable increase in $\phi$ applies only where axial load is very small and moment is very large; under such circumstances, the member begins to act as a beam subject to axial load rather than a column subject to flexure.

It is important to recognize that the ACI column formula has no provisions for length, radius of gyration, or buckling. Buckling criteria and effects of length are treated by separate limits and requirements, imposed and computed independently from the column formula. The column formula provides only the maximum capacity of a section under concentric load, without regard to any other limitations.

## Buckling Criteria

The parameter used by Code (10.11.4) to classify buckling criteria in columns is the slenderness ratio, taken from the Euler column formula:

$$\text{slenderness ratio} = \frac{KL_u}{r}$$

where  $K$  = numerical factor dependent on end conditions of the column and on the type of overall lateral load-carrying system

$L_u$ = unsupported length of the column, taken as the clear distance between attached or supporting members at top and bottom

$r$  = radius of gyration of the column section, taken as 0.3 times the least dimension for rectangular sections

The value of K is given by Code (10.11.2.1) as 1.0 for all buildings or structures braced against sidesway; these are the only structures considered in this text. For these structures, the columns are not used to resist any part of the lateral loads. Another common system, although not considered in this text, is that of an unbraced frame where lateral loads are resisted entirely by flexure on the columns; for such frames the value of K may be much higher. These two systems are shown schematically in Figure 13-7 for comparison.

Although the value of $K$ is known to be 1.0 herein, the symbol $K$ will still be shown wherever the slenderness ratio is referenced. Its inclusion will indicate where the effects of end conditions and sidesway must be considered.

For columns braced against sidesway, Code (10.11.4.1) permits the effects of slenderness to be neglected when

$$\frac{KL_u}{r} < 34 - 12\frac{M_1}{M_2} \qquad (13\text{-}4)$$

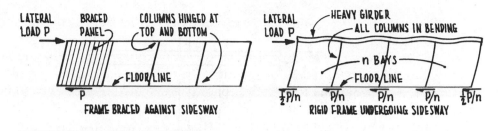

**Figure 13-7**    Braced frame and rigid frame.

where $M_1$ and $M_2$ are the factored end moments from the analysis for vertical loads only (no lateral loads included), and where

$M_1$  =  smaller moment, having a positive sign if the column is bending in single curvature and a negative sign if in double curvature

$M_2$  =  larger moment, always having a positive sign

A sketch of the end moments used in Eq. (13-4) is shown in Figure 13-8.

The least value of the slenderness ratio computed by Eq. (13-4) is seen to occur when $M_1 = M_2$, for which case the slenderness ratio is 22.

All columns considered in this book fall within this intermediate category, where slenderness effects may be neglected. For the other category, where slenderness effects must be considered, Code (10.11.5) prescribes a means to magnify the design moments. The result of such moment magnification is to cause an overdesign of an intermediate column such that it will carry a lighter load (the actual design load) under the more severe conditions.

A final point concerns the length $L_u$. The Euler formula, from which the slenderness ratio is drawn, uses the length between hinge points as its

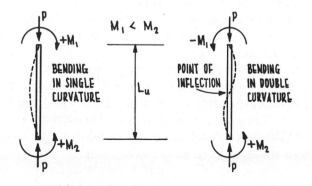

**Figure 13-8**    End moments on a column.

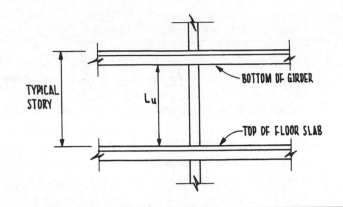

**Figure 13-9**   ACI column length in a building.

buckling length, while the ACI criteria specifies that $L_u$ is the "unsupported" length without regard to end conditions. A typical determination of the length $L_u$ is shown in Figure 13-9. ACI has thus chosen a simple but conservative value for length, in recognition of the myriads of combinations of loads, inflection points, and end rotations that could affect the buckling length of a structural column.

It is again emphasized that the bending in the columns under discussion is developed entirely by vertical loads. There is no bending on these columns due to wind and earthquake. Such bending of columns does not occur in braced frames.

## Ultimate Strength Analysis

In column analysis, a very useful dimension is the eccentricity of load, $e_n$, where

$$e_n = \frac{M_n}{P_n}$$ 
(13-5)

This eccentricity of the load $P_n$ is shown in Figure 13-10 and is used frequently throughout the subsequent discussions and analyses. Note particularly that the eccentricity is measured from the centerline of the section, not from the neutral axis.

In applying the ACI column formula to the analysis of columns, there are other conditions prescribed by ACI that must be met:

1. Longitudinal reinforcement may not be less than 1% or more than 8% of the gross cross-sectional area of concrete, $A_g$.
2. Columns shall be designed for the maximum moment $M_n$ that can accompany the axial load $P_n$. (Where moment varies along the length of the column, the largest value of moment is used.)
3. Strains shall be assumed to vary linearly across the section.
4. Maximum strain in the concrete shall be assumed to be 0.003.
5. When steel is in the elastic range, the tensile stress shall be taken as the modulus $E_s$ times the strain; when in the plastic range, the stress shall be taken at the specified yield stress.
6. Tensile strength of concrete shall be taken as zero.
7. Stress variation in concrete may be assumed to be any shape that provides results in substantial agreement with tests.
8. When $P_n$ is less than the axial load at balanced strain conditions, $P_b$, or less than $0.10 f'_c A_g / \phi$, the ratio of reinforcement $\rho$ provided shall not exceed 75% of the balanced ratio $\rho_b$ that would be required for flexure only.
9. Where there is no computed moment on a column, the column may be designed for an accidental eccentricity no less than $0.1h$ for the column load $P_n$ (not a Code requirement but a recommended practice).

All the foregoing conditions are included in the subsequent analyses. Condition 2, concerning the design for moments, recognizes that the allowable moment on a column is interrelated with the axial force; as axial load is decreased, moment can increase (up to a point). Condition 9 was a Code requirement prior to 1971; it recognizes that there cannot be such a thing as a perfectly concentric load or a perfectly straight column.

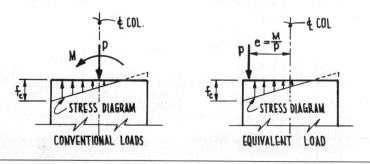

**Figure 13-10**     Eccentricity of load.

It is noted here that when $P_n$ is at its maximum allowable value of $0.8P_o$, there remains a small extra capacity for moment. The magnitude of this small amount of moment is commonly taken to be $P_n(0.1h)$ as suggested in Condition 9. Accordingly, given the computed value for the load $P_n$ when $M_n = 0$ (or $M_n$ is negligible), one may design a column for either of the following cases and achieve essentially the same end result:

- Case 1. Design for the fully plastic load $P_o = P_n/0.8 = 1.25P_n$, with $M_n = 0$.
- Case 2. Design for the computed load $P_n$ plus a moment $M_n = P_n(0.1h)$.

In this text, Case 2 is used exclusively whenever the computed moment $M_n$ is zero or is less than $P_n \times 0.1h$ (see Condition 9).

(Those unwilling to accept the statement that the foregoing Case 1 and Case 2 will produce essentially the same end results should jump ahead to Figure 13-15, where the effects of these two cases appear in an "interaction diagram" between $M_n$ and $P_n$. It can be seen there that for all steel ratios, the break point at $0.80P_n$ coincides almost exactly with the eccentricity value of $0.10h$, demonstrating that the two cases do indeed produce essentially the same end results.)

A typical column section is shown in Figure 13-11. The total area of steel is designated as $A_{st}$ and its steel ratio as $\rho$. That part of the steel distributed to the flanges is $A_s'$ and its steel ratio is $\rho'$. That part of the steel distributed to the web is $A_s''$ and its steel ratio is $\rho''$.

In the equivalent section of Figure 13-11, the web steel $A_s''$ is replaced by an equivalent imaginary strip of steel having a finite width, as shown. All other symbols are the same as used previously.

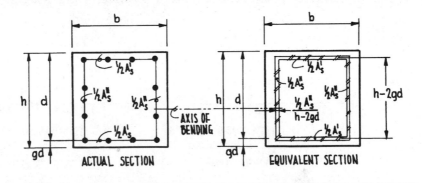

**Figure 13-11**   Column section.

A full definitive analysis for the ultimate loads on a column section is quite long and tedious. For those who are interested, a brief summary of such an analysis is presented in the Appendix. Study of these detailed derivations is not essential to the design procedure; one may skip these derivations and proceed to the next section with no loss in continuity.

## Parametric Design Tables

The equations resulting from the analysis given in the Appendix are unwieldly and complicated. Fortunately, the computer solves complicated expressions as readily as simple ones, with the end result of the computer output being the parametric column design tables, Tables A-9 through A-11, given with the design tables of the Appendix. The following discussions refer to those tables, Tables A-9, A-10, and A-11.

The ratio $R_n$ defined in the design tables is something of an artificial quantity. It is simply a convenient parameter suggested by the 1971 ACI design manual. It is dimensionless, so the values given in the design tables of the Appendix are valid either for Imperial (English) units or for SI units.

It should be observed that the lowest value of $e_n/h$ in the design tables is 0.1. As stated earlier, the assumption of a minimum eccentricity of $0.1h$ is considered to be good practice. In addition, the use of this minimum $e_n/h$ produces the same end result as limiting $P_n$ to $0.80P_o$ when moment is zero. Eccentricities in the tables therefore begin at $e_n/h = 0.1$.

It should also be observed that the design tables do not include values of $R_n$ when $P_n$ is equal to or less than $0.1f'_c bh/\phi$. When $P_n = 0.1bh/\phi$, the ratio $R_n$ is 0.168 for all grades of steel and all strengths of concrete. Such members are more appropriately designed as beams carrying a small axial load than as columns carrying a dominant flexural load. The design procedures for such members are those developed earlier for beams, presented in Chapter 9.

## Examples in the Strength Method

The following examples illustrate the use of the column design tables of the Appendix, Tables A-9 through A-11. In these first few examples, the stress at service levels is not considered; stresses at service levels are treated in a later discussions. Comments are included within the examples to clarify assumptions or identify arbitrary choices. The first example is the design of a simple column to take only axial load, without moment.

### EXAMPLE 13-1.

Strength method, no consideration of service stress.
Design of a column subject only to axial load.
No limitations on dimensions or reinforcement.

**Given:**

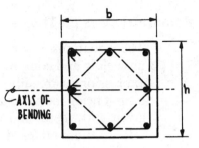

Apartment building, five floors
At a first-floor column, $L_u$ = 9ft 0 in.
$P_{DL}$ = 229 kips, $P_{LL}$ = 201 kips
Grade 60 steel, $f'_c$ = 4000 psi

**Find:**

Suitable column section

**Solution:**

Calculate the nominal ultimate load $P_n$:

$$P_n = \frac{1.4P_{DL} + 1.7P_{LL}}{\phi} = \frac{1.4 \times 229 + 1.7 \times 201}{0.7}$$

$$= 946 \text{ kips}$$

Choose an overall size and gross area (square column). Use the rule of thumb:

$b = h$ = 12 in. + 1 in. × stories above = 12 in. + 1 in. × 4.
   Try $h$ = 16 in., $b$ = 16 in.

Check the buckling criteria:

$$\frac{KL_u}{r} < 34 - 12\frac{M_1}{M_2} \qquad \frac{1 \times 9 \times 12}{0.3 \times 16} = 22.5 < 34 \text{ (O.K.)}$$

Determine $R_n$ and $e_n/h$:

$$R_n = \frac{P_n/bh}{0.85f'_c} = \frac{946,000/16 \times 16}{0.85 \times 4000} = 1.09$$

$$\frac{e_n}{h} = \text{minimum} = 0.1$$

From Table A-10, with steel uniformly distributed;
use $A_s$ = 2.4%

$$= 0.024 \times 16 \times 16$$
$$= 6.14 \text{ in.}^2$$

From Table A-3, select 8 No. 8 bars.

Select the ties: For No. 8 bars, use tie size No. 3.

Spacing:  < 16 bar diameters or  < 16 in.
 < 48 tie diameters or   < 18 in.
 < least dimension or  < 16 in.

Use a column 16 in. × 16 in.; 8 No. 8 longitudinal bars, No. 3 ties at 16 in. o.c. Such a section is shown in the sketch at the beginning of this example.

## EXAMPLE 13-2.

Strength method, no consideration of service stress.
Design of a column subject only to axial load.
One dimension restricted.

**Given:**

Same conditions as in Example 13-1

Height $h$ across axis of bending limited to 12 in.

**Find:**

Suitable column section

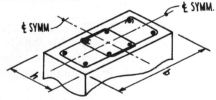

**Solution:**

Calculate the nominal ultimate load $P_n$:

$P_n$ = 946 kips (same as Example 13-1)

Choose an overall size and gross area. Use the rule of thumb: if a square column section is 16 in. × 16 in. in size, a rectangular section should have the same area.

gross area = 16 × 16 = 256 in.$^2$

With one size 12 in., try 12 in. × 22 in., $A_g$ = 264 in.$^2$(O.K.—near enough to 256 in.)

Check the buckling criteria:

$$\frac{KL_u}{r} < 34 - 12\frac{M_1}{M_2} \quad \frac{1 \times 9 \times 12}{0.3 \times 12} = 30 < 34 \text{ (O.K.)}$$

Determine $R_n$ and $e_n/h$:

$$R_n = \frac{P_n/bh}{0.85f_c'} = \frac{946,000/12 \times 22}{0.85 \times 4000} = 1.054$$

$$\frac{e_n}{h} = \text{minimum} = 0.1$$

From Table A-10, steel on two faces only; use 2% steel,

$$A_s = 0.02 \times 12 \times 22 = 5.28 \text{ in.}^2$$

From Table A-3, select 4 No. 8 and 4 No. 7 bars.

Select the ties: for No. 8 bars, use tie size No. 3.

Spacing:  < 16 bar diameters or  < 14 in. (for No. 7 bars)
< 48 tie diameters or   < 18 in.
< least dimension or    < 12 in.

Use 12- × 22-in. column, 4 No. 8 and 4 No. 7 bars, No. 3 ties at 12 in. o.c., with No. 8 bars placed at the four corners. The chosen section is similar to that shown at the beginning of this example.

---

## EXAMPLE 13-3.

Strength method, no consideration of service stress.
Design of a column that is subject to flexure.
Configuration limited to square sections.

### Given:

Same criteria as Example 13-1

At top of column, add   $M_{DL} = 65$ kip-ft
$M_{LL} = 48$ kip-ft

At bottom of column,   $M_{DL} = 33$ kip-ft
$M_{LL} = 24$ kip-ft

### Find:

Suitable column section

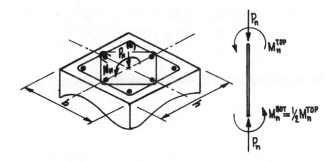

**Solution:**

Calculate the nominal ultimate loads $M_n$ and $P_n$:

$$P_n = 946 \text{ kips (from Example 13-1)}$$

At top:
$$M_n = \frac{1.4 M_{DL} + 1.7 M_{LL}}{\phi} = \frac{1.4 \times 65 + 1.7 \times 48}{\phi}$$

$$= 247 \text{ kip-ft}$$

At bottom: $M_n = 124 \text{ kip-ft}$

For a first trial, choose an overall size: $b = h = 16$ in. (same as Example 13-1).

Check the buckling criteria:

$$\frac{KL_u}{r} < 34 - 12\frac{M_1}{M_2} \qquad \frac{1 \times 9 \times 12}{0.3 \times 16} < 34 - 12 \times \frac{-124}{247}$$

$$22.5 < 40 \text{ (O.K.)}$$

Determine $R_n$ and $e_n/h$:

$$R_n = \frac{P_n/bh}{0.85f_c'} = \frac{946,000/16 \times 16}{0.85 \times 4000} = 1.09$$

$$\frac{e_n}{h} = \frac{M_n/P_n}{h} = \frac{247,000 \times 12/946,000}{16} = 0.196$$

From Table A-10, with steel uniformly distributed, find 4.8% steel (too high).

Since this column section is considered to be unacceptable, the trial size must be increased and the entire procedure repeated for this new trial size. An increment of 4 in. will be added, producing a trial section of 20-in. square. For the new gross size, recalculate $R_n$ and $e_n/h$:

$$R_n = \frac{P_n/bh}{0.85f_c'} = \frac{946,000/20 \times 20}{0.85 \times 4000} = 0.70$$

$$\frac{e_n}{h} = \frac{M_n/P_n}{h} = \frac{247,000 \times 12/946,000}{20} = 0.16$$

From Table A-10, select $\rho$ < 1% steel (too low).

Again the trial size is unacceptable. For the next trial, select 18-in. square, halfway between the first and second trial sizes. For the new gross size, recalculate $R_n$ and $e_n/h$:

$$R_n = \frac{P_n/bh}{0.85f_c'} = \frac{946,000/18 \times 18}{0.85 \times 4000} = 0.86$$

$$\frac{e_n}{h} = \frac{M_n/P_n}{h} = \frac{247,000 \times 12/946,000}{18} = 0.174$$

From Table A-10, find $\rho$ = 2.0% (O.K.—use 2%).

$$A_s = 0.02 \times 18 \times 18 = 6.48 \text{ in.}^2$$

From Table A-3, select 4 No. 9 and 4 No. 8 longitudinal bars.

Select the ties: for No. 9 longitudinal bars, use No. 3 ties.

Spacing:   < 16 bar diameters or   < 16 in. (smallest bar size)
               < 48 tie diameters or   < 18 in.
               < least dimension or   < 18 in.

Use a column 18 in. × 18 in., 4 No. 9 and 4 No. 8 longitudinal bars with the No. 9 bars at the four corners, No. 3 ties at 16 in. o.c.

The final column section is similar to that shown at the beginning of the example.

---

## EXAMPLE 13-4.

Strength method, no consideration of service stress.
Design of a column that is subject to flexure.
Depth of section limited on the axis of bending to a maximum of 11 in.

**Given:**

Same criteria as Example 13-1

At top of column, add  $M_{DL} = 50$ kip-ft
$\qquad\qquad\qquad\quad M_{LL} = 30$ kip-ft

At bottom, add hinge,  $M_{DL} = 0$
$\qquad\qquad\qquad\qquad M_{LL} = 0$

**Find:**

Suitable column section

**Solution:**

Calculate the nominal ultimate loads $P_n$ and $M_n$:

$$P_n \;=946 \text{ kips} \qquad \text{(see Example 13-1)}$$

$$M_n = \frac{1.4M_{DL} + 1.7M_{LL}}{\phi} = \frac{1.4 \times 50 + 1.7 \times 30}{0.7}$$

$$= 173 \text{ kip-ft}$$

Choose an overall size and gross area for a square column. From the rule of thumb, try an area roughly equal to a square section $16 \times 16$ in., having $A_g = 256$ in.$^2$

Try a column size 11 in. $\times$ 24 in., $A_g = 264$ in.$^2$

Check the buckling criteria:

$$\frac{KL_u}{r} < 34 - 12\,\frac{M_1}{M_2} \qquad \frac{1 \times 108}{0.3 \times 11} < 34 - 12 \times \frac{0}{173}$$

$$33 < 34 \quad \text{(O.K.)}$$

Compute $R_n$ and $e_n/h$:

$$R_n = \frac{P_n/bh}{0.85f_c'} = \frac{946,000/11 \times 24}{0.85 \times 4000} = 1.054$$

$$\frac{e_n}{h} = \frac{M_n/P_n}{h} = \frac{173,000 \times 12/946,000}{11} = 0.20$$

From Table A-10 steel on two faces, find $\rho$ of almost 4%. (Too high— try a larger section with lower $\rho$.)

Try a section 11 in. × 30 in. For the new trial size, recompute $R_n$ and $e_n/h$:

$$R_n = \frac{P_n/bh}{0.85f_c'} = \frac{946,000/11 \times 30}{0.85 \times 4000} = 0.84$$

$$\frac{e_n}{h} = \frac{M_n/P_n}{h} = \frac{173,000 \times 12/946,000}{11} = 0.20$$

From Table A-10, with steel at two faces, read $\rho$ of 2% (O.K.—use). Determine the steel area:

$$A_s = \rho bd = 0.02 \times 11 \times 30 = 6.60 \text{ in.}^2$$

From Table A-3, select 4 No. 9 and 4 No. 8 bars.

Select the ties: for No. 9 bars, use tie size No. 3.

Spacing: < 16 bar diameters or  < 16 in. (for No. 8 bars)
< 48 tie diameters or   < 18 in.
< least dimension or   < 11 in.

Use a column 11 in. × 30 in., 4 No. 9 and 4 No. 8 bars, No. 3 ties at 11 in. o.c.

The final section is shown in the sketch at the beginning of the example.

---

A summary of the first four examples follows.

| Example | $P_n$ (kips) | $M_n$ (kip-ft) | $b$ (in.) | $h$ (in.) | $e/h$ | $A_s$ (in.$^2$) |
|---|---|---|---|---|---|---|
| 13-1 | 946 | 0 | 16 | 16 | 0.10 | 6.14 |
| 13-2 | 946 | 0 | 22 | 12 | 0.10 | 5.28 |
| 13-3 | 946 | 247 | 18 | 18 | 0.17 | 6.48 |
| 13-4 | 946 | 173 | 30 | 11 | 0.20 | 6.60 |

In these four examples, the axial load was taken at the same value each time and other factors were then varied. When Example 13-3 is compared to Example 13-1, it is noted that adding a significant amount of moment had only a small effect on the size of the column and its reinforcement. When Example 13-2 is compared to Example 13-4, it is seen that rather drastic changes in one dimension do not change the total area $b \times h$ by a large amount, and even adding a large moment in the shallow direction changes the area of steel only 20%. It is concluded that for braced frames, the axial load dominates to such an extent that other influences become relatively minor.

When axial loads become small, as in one- or two-story buildings, very often the minimum-sized column sections must be used. The next example lists the allowable loads on minimum-sized columns.

## EXAMPLE 13-5.

Strength method, no consideration of service stresses.
Design of columns subject only to axial load.
Minimum dimensions and minimum reinforcement.

Determine the nominal axial load $P_n$ for column sections fabricated using $f_c'$ of 3000, 4000, and 5000 psi, and for grades 40, 50, and 60 steels. Tabulate the results for $e_n/h = 0.1$ for minimum or no moment. Use $A_s$ of 1%, steel uniformly distributed. Read $R_n$ from the Appendix tables and compute $P_n = 0.85 f_c' b h R_n$.

| $f_y$ (ksi) | $f_c'$ (psi) | $R_n$ | $P_n$ (kips) 10 in. × 10 in. $A_g = 100$ in.$^2$ | 12 in. × 12 in. $A_g = 144$ in.$^2$ |
|---|---|---|---|---|
| 40 | 3000 | 0.91 | 232 | 334 |
| 40 | 4000 | 0.87 | 296 | 426 |
| 40 | 5000 | 0.85 | 361 | 520 |
| 50 | 3000 | 0.93 | 237 | 341 |
| 50 | 4000 | 0.89 | 303 | 436 |
| 50 | 5000 | 0.87 | 370 | 532 |
| 60 | 3000 | 0.96 | 245 | 353 |
| 60 | 4000 | 0.91 | 309 | 446 |
| 60 | 5000 | 0.89 | 378 | 545 |

Without incurring load reductions due to length, where $KL_u/r \leq 34$ and $r = 0.3h$,

Maximum length of a 10-in. column: 102 in. = 8 ft 6 in.

Maximum length of a 12-in. column: 122 in. = 10 ft 2 in.

Typical loads per story can be expected to be:

actual $P$ per story = 30 tons or 60 kips

The nominal ultimate load $P_n$ can be expected to be between $1.4/\phi$ and $1.7/\phi$ times the actual load,

$$P_n \text{ per story} = 60 \times 1.4/\phi \text{ to } 60 \times 1.7/\phi$$

$$= 120 \text{ kips to } 146 \text{ kips}$$

---

From an examination of the tabulation, it is seen that the allowable load on a minimum-sized column should be adequate for a two- or three-story building, even where nominal moments are present. For lengths longer than the indicated maximum lengths, Code requires significant reductions in the allowable loads.

## Discussion of the Strength Method

An important assumption was made in the section on strength analysis that is not included in the conditions listed at the beginning of that section: It was tacitly assumed that when the concrete emerges from yield and enters its elastic range (Appendix Figure B-2), it follows the idealized stress-strain curve of Figure 1-4. The assumption is significant, since part of the concrete must therefore behave elastically whenever moment occurs on the section. The accuracy of the ultimate load analysis is thus dependent to some degree on assumptions regarding the elastic behavior of concrete.

With regard to these assumptions about elastic behavior, the Code permits any stress-strain relationship to be used that yields results in substantial agreement with test results. The idealized curve of Figure 1-4 is seen to fulfill this requirement since the results of the foregoing analysis are in agreement with published ACI tables.

As stated earlier, Code requires that the maximum strain in the concrete at failure load shall be 0.003 in./in. The Code makes no further requirements regarding any other strains, such as shrinkage strains or creep strains. In the foregoing analysis, the strain of 0.003 was assumed to include all contributions to strain, whether elastic or inelastic.

The tables of the Appendix, although limited, will provide values of $e_n/h$ and $P_n/bh$ adequate for the design of small structures braced against sidesway. If the project is so large that the tables are inadequate, the project is certainly beyond the intentions of this book. For larger projects, a great deal more study of columns is required than can be presented in an introductory volume such as this.

## OUTSIDE PROBLEMS

Select a column section and its reinforcement for the conditions given below.

| Prob. No. | Concrete Strength $f'_c$ psi | Steel Grade | Axial Load in kips | | Moment in kip-ft | | $L_u$ ft | Limits on $h$ in. | End Conditions |
|---|---|---|---|---|---|---|---|---|---|
| | | | $P_{DL}$ | $P_{LL}$ | $M_{DL}$ | $M_{LL}$ | | | |
| 13.1 | 3000 | 40 | 106 | 135 | 0 | 0 | 11' 8" | None | |
| 13.2 | 3000 | 50 | 124 | 158 | 54 | 68 | 11' 8" | 16 | |
| 13.3 | 3000 | 60 | 148 | 180 | 86 | 74 | 11' 8" | None | |
| 13.4 | 4000 | 40 | 170 | 160 | 0 | 0 | 11' 8" | 16 | |
| 13.5 | 4000 | 50 | 231 | 193 | 102 | 120 | 11' 8" | None | |
| 13.6 | 4000 | 60 | 305 | 225 | 154 | 146 | 11' 8" | 16 | |
| 13.7 | 5000 | 40 | 190 | 206 | 0 | 0 | 11' 8" | None | |
| 13.8 | 5000 | 50 | 254 | 280 | 130 | 155 | 11' 8" | 16 | |
| 13.9 | 5000 | 60 | 405 | 351 | 202 | 195 | 11' 8" | None | |
| 13.10 | 3000 | 40 | 95 | 124 | 0 | 0 | 13' 4" | None | |
| 13.11 | 3000 | 50 | 110 | 143 | 58 | 72 | 13' 4" | 16 | |
| 13.12 | 3000 | 60 | 135 | 162 | 90 | 77 | 13' 4" | None | |
| 13.13 | 4000 | 40 | 161 | 144 | 0 | 0 | 13' 4" | 16 | |
| 13.14 | 4000 | 50 | 224 | 174 | 106 | 121 | 13' 4" | None | |
| 13.15 | 4000 | 60 | 291 | 203 | 159 | 140 | 13' 4" | 16 | |
| 13.16 | 5000 | 40 | 177 | 186 | 0 | 0 | 13' 4" | None | |
| 13.17 | 5000 | 50 | 248 | 172 | 133 | 159 | 13' 4" | 16 | |
| 13.18 | 5000 | 60 | 392 | 316 | 196 | 206 | 13' 4" | None | |
| 13.19 | 3000 | 40 | 91 | 119 | 0 | 0 | 10' 4" | None | |
| 13.20 | 3000 | 50 | 102 | 134 | 59 | 74 | 10' 4" | 15 | |
| 13.21 | 3000 | 60 | 130 | 159 | 92 | 81 | 10' 4" | None | |
| 13.22 | 3000 | 40 | 156 | 139 | 0 | 0 | 10' 4" | 15 | |
| 13.23 | 3000 | 50 | 219 | 168 | 110 | 123 | 10' 4" | None | |
| 13.24 | 3000 | 60 | 280 | 201 | 161 | 140 | 10' 4" | 15 | |
| 13.25 | 3000 | 40 | 172 | 187 | 0 | 0 | 10' 4" | None | |
| 13.26 | 3000 | 50 | 242 | 166 | 135 | 161 | 10' 4" | 15 | |
| 13.27 | 3000 | 60 | 384 | 304 | 201 | 209 | 10' 4" | None | |

## Investigation of a Column Section

The reverse of the design problem can occur occasionally; that is, the size of the column and its reinforcement are known and it is one of the loads that is to be determined. Since the loads $M_n$ and $P_n$ are interrelated, one of the loads must be known, else there are an infinite number of combinations that could be solutions. In most cases it is the axial load $P_n$ that is known and it is desired to find how much moment $M_n$ the section can take in addition to the axial load; the next example is such a case.

### EXAMPLE 13-6.

Strength method.
Investigation of a section to find the service moment, given the axial load.

**Given:**

Section as shown
$P_{DL}$ = 86 kips; $P_{LL}$ = 70 kips
Grade 60 steel; $f_c'$ = 4000 psi

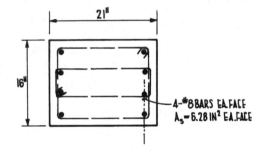

4-#8 BARS EA. FACE
$A_s$ = 6.28 IN² EA. FACE

**Find:**

Service moment $M_{sv}$

**Solution:**

Calculate the nominal ultimate load $P_n$:

$$P_n = \frac{1.4P_{DL} + 1.7P_{LL}}{\phi} = \frac{1.4 \times 86 + 1.7 \times 70}{0.7}$$

$$= 342 \text{ kips}$$

Determine $R_n$ and the steel ratio $\rho$:

$$R_n = \frac{P_n/bh}{0.85f_c'} = \frac{342,000/16 \times 21}{0.85 \times 4000} = 0.30$$

$$\rho = \frac{A_s}{bh} = \frac{6.28}{16 \times 21} = 0.019$$

From Table A-11, read $e_n/h$ = 0.78
Solve for $M_n$ from the $e_n/h$ ratio:

$$\frac{e_n}{h} = \frac{M_n / P_n}{h} \qquad 0.78 = \frac{M_n / 342{,}000}{21}$$

Solve for $M_n$: $M_n$ = 5602 kip-in. = 467 kip-ft

Use a factor of 1.7 to obtain $M_{sv}$:

$$\text{Service moment } M_{sv} = \frac{467 \text{ kip-ft}}{1.7} = 275 \text{ kip-ft}$$

## OUTSIDE PROBLEMS

Determine the service moment $M_{sv}$ for the given column sections. In all cases, the bars are concentrated at the flexure faces.

| Prob. No. | Concrete Strength $f'_c$ psi | Steel Grade | Width $b$ in. | Height $h$ in. | Reinf. Bars (Total) | Ultimate $P_n$ kips |
|---|---|---|---|---|---|---|
| 13.28 | 3000 | 40 | 16 | 16 | 8 No. 08 | 460 |
| 13.29 | 3000 | 50 | 18 | 24 | 10 No. 09 | 644 |
| 13.30 | 3000 | 60 | 20 | 32 | 12 No. 10 | 916 |
| 13.31 | 4000 | 40 | 18 | 18 | 8 No. 09 | 770 |
| 13.32 | 4000 | 50 | 20 | 27 | 10 No. 10 | 1285 |
| 13.33 | 4000 | 60 | 22 | 36 | 12 No. 11 | 1800 |
| 13.34 | 5000 | 40 | 20 | 20 | 10 No. 10 | 1020 |
| 13.35 | 5000 | 50 | 22 | 30 | 12 No. 11 | 1814 |
| 13.36 | 5000 | 60 | 24 | 40 | 14 No. 11 | 2600 |

## Elastic Analysis

In the elastic range, the analysis of a column section is much like that of a beam section except that an axial force has been added and there may be additional reinforcement in the middle of the section. A typical column section is shown in Figure 13-12, with the load components $P_e$ and $M_e$ representing the axial force and moment at working levels. As with the strength method, the web steel is replaced by an imaginary strip of steel having a finite width as shown.

A discontinuity is introduced when the value of $k$ is equal to $1 + g$. For higher values of $k$, all concrete is effective; for lower values of $k$, the concrete in tension is assumed to be cracked and ineffective. Two sets of equations are therefore required to bridge the discontinuity.

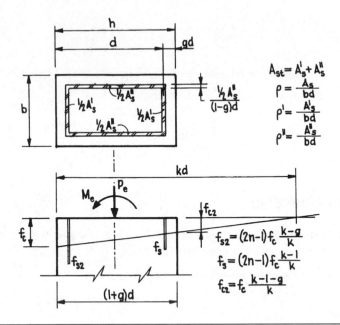

**Figure 13-12**   Elastic column stresses, $k \geq (1+g)$.

For $k$ greater than $(1 + g)$, the stress diagrams of Figure 13-12 are valid. The sum of forces yields an equation for the ratio $R_e$:

$$R_e = \frac{P_e / bh}{f_c} = \frac{2k - 1 - g}{2k(1 + g)}\left[1 + g + (2n - 1)(\rho' + \rho'')\right] \qquad (13\text{-}24)$$

The sum of moments about the centerline yields a similar equation in moments:

$$R_m = \frac{M_e / bh^2}{f_c} = \frac{1}{12k}\left[1 + g + (2n - 1)(3\rho' + \rho'')\frac{(1 - g)^2}{(1 - g)^2}\right] \qquad (13\text{-}25)$$

Dividing Eq. (13-25) by Eq. (13-24) yields a solution for the eccentricity ratio:

$$\frac{e_e}{h} = \frac{R_m}{R_e} \qquad (13\text{-}26)$$

For $k$ less than $(1 + g)$, the state of stress is shown in Figure 13-13. For those stresses, the sum of horizontal forces yields the solution for $R_e$:

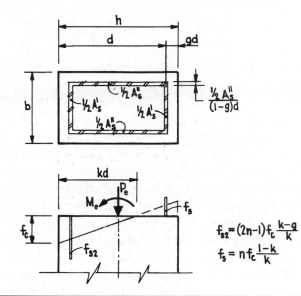

$$f_{s2} = (2n-1)f_c\frac{k-g}{k}$$

$$f_s = nf_c\frac{1-k}{k}$$

**Figure 13-13**   Elastic column stresses, $k \le (1 + g)$.

$$R_e = \frac{P_e/bh}{f_c} = \frac{1}{2k(1+g)}\left[k^2 + (2n-1)\rho'(k-g) + (2n-1) - n\rho''\frac{(1-k)^2}{1-g} - n\rho''(1-k)\right]$$

$$(13\text{-}27)$$

The sum of moments about centerline yields a similar equation in moments:

$$R_m = \frac{M_e/bh^2}{f_c} = \frac{1}{4k(1+g)^2}\left[\frac{k^2}{3}(3+3g-2k) + n\rho'(1-k)(1-g)\right. \qquad (13\text{-}28)$$

$$+ (2n-1)\rho'(k-g)(1-g)$$

$$+ (2n-1)\rho''\frac{(k-g)^2}{3(1-g)}(3-g-2k)$$

$$+ \left. n\rho''\frac{(1-k)^2}{3(1-g)}(1-3g+2k)\right]$$

For this range of $k$, the eccentricity ratio is again given by

$$\frac{e_e}{h} = \frac{R_m}{R_e} \tag{13-29}$$

The actual numerical values of the stress ratio $R_e$ are tabulated in the column tables of the Appendix, Tables A-9 through A-11, at the lower portion of each table. In developing the tables, the value of $k$ was varied to produce the desired value of $e_e/h$ [Eq. (13-26) or (13-29)]; then the value of $R_e$ corresponding to that $e_e/h$ was listed [Eq. (13-24) or (13-27)]. Since all stresses are in the elastic range, the tabulation is seen to be independent of $f_y$, the yield stress of steel.

The tables are set up in terms of two parameters, $e_e/h$ and $R_e$. The stress ratio $R_e$ is analogous to the stress ratio $R_n$ introduced earlier in the strength method; $R_e$, of course, uses elastic load and elastic stress, whereas $R_n$ uses nominal ultimate load and yield stress. For easy reference, both $R_n$ and $R_e$ are defined at the top left corner of each table.

The stress ratio $R_e$ affords a means to determine the compressive stress in the concrete during service conditions. The stress $f_c$ is the maximum working stress in the concrete and includes both axial compression plus flexure. When $R_e$, $P_e$ and the overall dimensions $b$ and $h$ are known, the service stress can be computed.

When investigating the elastic stress as it affects brittle finishes or coverings, bear in mind that creep and shrinkage in concrete affect strain but not stress. When the elastic stress is known, the related elastic strain is found simply by dividing elastic stress by the modulus of elasticity:

$$\text{elastic } E_c = \frac{f_c}{E_c} \tag{13-30}$$

The total strain must then be computed to include both elastic strain and inelastic creep and shrinkage. As discussed in Chapter 3, the inelastic strain may be as much as the elastic strain; hence the total strain may be as much as twice the elastic strain,

$$\text{total } E_c \le 2\frac{f_c}{E_c} \tag{13-31}$$

The additional strain can have severe effects on finishes and coatings.

## Examples in the Elastic Range

The Appendix tables include provisions for the analysis of columns in the elastic range. An indication of the general performance of a column during its day-to-day working life can therefore be obtained. The procedure is directly parallel to that for the plastic range, as illustrated in the following examples.

## EXAMPLE 13-7.

Elastic analysis of columns.
Investigation of service stress.
Column subject to axial load only.

**Given:**

Column of Example 13-1
$P_{DL}$ = 229 kips; $P_{LL}$ = 201 kips
Grade 60 steel; $f_c'$ = 4000 psi

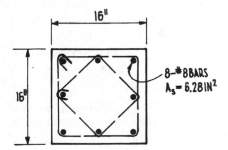

16"

16"

8 - #8 BARS
$A_s$ = 6.28 IN²

**Find:**

Service stress in the concrete

**Solution:**

Calculate the service load $P_{sv}$:

$$\left(\tfrac{1.4}{1.7}P_{DL} + P_{LL}\right)/\phi = 557 \text{ kips}$$

Determine the steel ratio $\rho$

$$\rho = \frac{A_s}{bh} = \frac{6.28}{16 \times 16} = 0.025$$

From Table A-10, using $e_e/h$ = 0.1 to account for any accidental eccentricities, read $R_e$ = 0.89.
Determine the elastic stress:

$$R_e = \frac{P_{sv}/bh}{f_c}; \quad 0.89 = \frac{557,000/16 \times 16}{f_c}$$

Solve for $f_c$:        $f_c$ = 2445 psi.

The elastic stress in concrete is 2445 psi, or 61% of $f_c'$ or 72% of idealized yield.

### EXAMPLE 13-8.

Elastic analysis of columns.
Investigation of service stress.
Column subject to axial load plus flexure.

**Given:**

Column of Example 13-3
$P_{DL}$ = 229 kips; $P_{LL}$ = 201 kips
$M_{DL}$ = 65 kip-ft; $M_{LL}$ = 48 kip-ft
Grade 60 steel; $f'_c$ = 4000 psi

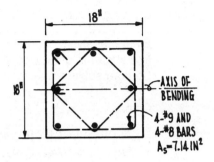

18"

18"

AXIS OF BENDING

4-#9 AND
4-#8 BARS
$A_s$=7.14 IN²

**Find:**

Service stress $f_{sv}$

**Solution:**

Calculate the service loads $M_{sv}$ and $P_{sv}$:

$$M_{sv} = \frac{\frac{1.4}{1.7} M_{DL} + M_{LL}}{\phi} = \frac{\frac{1.4}{1.7} \times 65 + 48}{0.7} = 145 \text{ kip-ft}$$

$$P_{sv} = \frac{\frac{1.4}{1.7} P_{DL} + P_{LL}}{\phi} = \frac{\frac{1.4}{1.7} \times 229 + 201}{0.7} = 557 \text{ kips}$$

Determine the steel ratio $\rho$:

$$\rho = \frac{A_s}{bh} = \frac{7.14}{18 \times 18} = 0.022$$

Calculate the eccentricity ratio $e_e/h$:

$$\frac{e_e}{h} = \frac{M_{sv}/P_{sv}}{h} = \frac{145,000 \times 12/557,000}{18} = 0.174$$

From Table A-10 read $R_e$ = 0.66
Solve for the elastic stress:

$$R_e = \frac{P_{sv}/bh}{f_c}, \quad 0.66 = \frac{557,000/18 \times 18}{f_c}$$

$$f_c = 2605 \text{ psi}$$

The elastic stress in concrete is 2605 psi, or 65% of $f'_c$ or 76% of idealized yield.

---

It should be recognized that the computed stress $f_c$ in Examples 13-7 and 13-8 includes the effects of long-term creep and shrinkage. Until the long-term effects actually happen, the computed value may be considerably in error.

Investigation of elastic stresses can serve a very practical purpose, as demonstrated in the next example.

## EXAMPLE 13-9.

Elastic analysis of columns.
Investigation of elastic stresses.
Column subject to axial load plus flexure.

The column shown below is to be clad with sculptured marble; the marble is to be rigidly attached to the concrete. The modulus of elasticity of the marble is $6 \times 10^6$ psi.

**Given:**

Section as shown
$P_{DL} = 230$ kips; $P_{LL} = 200$ kips
$M_{DL} = 66$ kips; $M_{LL} = 47$ kip-ft
Grade 60 steel; $f'_c = 4000$ psi
$E_c = 57,000^4 = 3,600,000$ psi

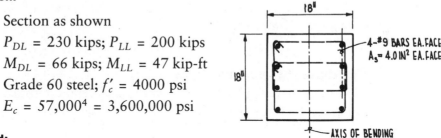

**Find:**

Maximum elastic stress that can occur in the marble at working levels. $(DL + LL)$

**Solution:**

Calculate the elastic loads $P_e$ and $M_e$:

$$P_e = P_{DL} + P_{LL} = 230 + 200 = 430 \text{ kips}$$
$$M_e = M_{DL} + M_{LL} = 66 + 47 = 113 \text{ kip-ft}$$

Determine $\rho$ and $e_e/h$:

$$\rho = \frac{A_s}{bh} = \frac{8.00}{18 \times 18} = 0.0247, \quad \text{say } 2.5\%$$

$$\frac{e_e}{h} = \frac{M_e / P_e}{h} = \frac{113,000 \times 12 / 430,000}{18} = 0.18, \quad \text{say } 0.2$$

From Table A-10, read $R_e = 0.69$.

Solve for $f_c$:

$$R_e = \frac{P_e / bh}{f_c} \qquad 0.69 = \frac{430,000 / 18 \times 18}{f_c}$$

$$f_c = 1923 \text{ psi (maximum).}$$

Determine the elastic strain in the concrete:

$$\varepsilon_c = \frac{f_c}{E_c} = \frac{1923}{3.6 \times 10^6} = 0.000534 \text{ in./in.}$$

---

The column tables of the Appendix include the effects of creep and shrinkage in the concrete. The computed elastic strain of 0.000534 in./in. cannot occur, therefore, until an additional inelastic strain of 0.000534 in./in. has also occurred. The total strain at the surface of the concrete is thus the sum of the two,

$$\varepsilon_c = 0.000534 + 0.000534 = 0.00107 \text{ in./in.}$$

The stress in the marble can now be computed from this total strain. Since the marble is rigidly attached to the concrete, the total strain in the marble $\varepsilon_m$ will be equal to the total strain in the concrete; the stress can therefore be found from Hooke's law.

$$\varepsilon_m = \varepsilon_c = 0.00107 \text{ in./in.}$$
$$f_m = E_m \varepsilon_m = 6 \times 10^6 \times 0.00107 = 6420 \text{ psi}$$

The maximum stress that can occur in the marble is 6420 psi.

It should be recognized that the computed stress in the marble in Example 13-9 is quite high, probably approaching the ultimate strength of the marble. Not all of it is likely to occur, however, since a large part of the inelastic strain in the concrete will have occurred during construction. As construction progresses, the dead load increases and much of the creep that is going to occur will occur progressively during that time. Attaching the marble would, of course, be one of the last items on the construction schedule.

Nonetheless, the elastic strain by itself induces a stress in the marble of 3210 psi (half the total), which is still a significant stress. A great deal of attention will have to be paid to the attachments, or the marble may separate from its concrete backing. Such failures do occur.

## Discussion of the Elastic Analysis

The accuracy of the elastic stresses computed using the column tables of the Appendix is of course subject to question. For example, it was assumed that stress in compressive steel was doubled due to shrinkage and creep. If the actual shrinkage and creep ever reach the assumed values, the stress in the compression steel will double, as assumed. Until that happens, if ever, the computed value of $f_c$ will always be in error, but always on the "safe" side.

As a second point, the accuracy of the idealized stress-strain curve used in the analysis is always subject to question. Although its accuracy may be somewhat uncertain, it is concluded that it is at least accurate enough that it agrees with published test results in flexure. It is therefore considered to be a reasonably accurate indicator of plastic deformation in concrete.

Due to various uncertainties, the accuracy of the elastic analysis is considered by the practice to be somewhat unreliable. The use of the elastic tables for final design of columns is therefore not recommended. It is intended that the elastic tables of the Appendix should be used only to obtain a reasonable indication of stresses and strains at elastic levels; there is at present no other method sanctioned by the Code to investigate these stresses and strains.

## OUTSIDE PROBLEMS

13.37 through 13.45. For the column sections of Problems 13.28 through 13.36, determine the elastic service stress in the concrete.

## Full Range Design of Columns

Like beams, columns may be designed over their full range of elastic and inelastic stresses. The only Code requirement governing such a design is that a column must be designed to sustain the nominal ultimate loads $M_n$

**Table 13-1** Stress Ratios and Service Stresses for Columns.
$f'_c = 4000$ psi; $n = 8.04$; $g = 0.125$

$$R_n = \frac{P_n / bh}{0.85 f'_c}$$

$$e_n = M_n / P_n$$

Steel distributed equally to the four faces

Steel Ratio $A_s/bh$

Stress Ratio $R_n$ and Service Stress $f_{sv}$ for Grade 40 Steel*

| $e_n/h$ | 0.01 | 0.02 | 0.03 | 0.04 | 0.05 | 0.06 | 0.07 | 0.08 |
|---|---|---|---|---|---|---|---|---|
| 0.10 | 0.88:2420 | 0.96:2330 | 1.04:2260 | 1.13:2210 | 1.21:2160 | 1.29:2130 | 1.37:2090 | 1.46:2070 |
| 0.20 | 0.67:2250 | 0.75:2480 | 0.82:2430 | 0.89:2380 | 0.96:2330 | 1.03:2300 | 1.09:2260 | 1.16:2230 |
| 0.30 | 0.51:2770 | 0.58:2700 | 0.65:2630 | 0.71:2570 | 0.77:2520 | 0.83:2480 | 0.89:2440 | 0.95:2400 |
| 0.40 | 0.37:2870 | 0.46:2900 | 0.53:2860 | 0.59:2780 | 0.64:2720 | 0.69:2660 | 0.74:2610 | 0.79:2580 |
| 0.50 | 0.27:2780 | 0.36:2910 | 0.43:2920 | 0.49:2900 | 0.55:2860 | 0.59:2800 | 0.64:2740 | 0.68:2700 |
| 0.60 | 0.20:2580 | 0.29:2840 | 0.36:2900 | 0.41:2910 | 0.46:2900 | 0.51:2880 | 0.56:2840 | 0.60:2790 |
| 0.70 | | 0.24:2730 | 0.30:2850 | 0.35:2890 | 0.40:2890 | 0.44:2880 | 0.49:2860 | 0.53:2850 |
| 0.80 | | 0.20:2630 | 0.26:2780 | 0.30:2840 | 0.35:2870 | 0.39:2870 | 0.43:2860 | 0.47:2850 |
| 0.90 | | 0.17:2530 | 0.22:2720 | 0.27:2800 | 0.31:2830 | 0.35:2850 | 0.38:2850 | 0.42:2840 |
| 1.00 | | | 0.20:2660 | 0.24:2750 | 0.27:2800 | 0.31:2820 | 0.34:2830 | 0.37:2830 |
| 2.00 | | | | | | | | 0.18:2690 |

Stress Ratio $R_n$ and Service Stress $f_{sv}$ for Grade 60 Steel*

| $e_n/h$ | 0.01 | 0.02 | 0.03 | 0.04 | 0.05 | 0.06 | 0.07 | 0.08 |
|---|---|---|---|---|---|---|---|---|
| 0.10 | 0.91:2520 | 1.04:2510 | 1.16:2520 | 1.28:2510 | 1.40:2520 | 1.53:2520 | 1.65:2520 | 1.78:2520 |
| 0.20 | 0.70:2670 | 0.81:2690 | 0.91:2700 | 1.01:2710 | 1.11:2710 | 1.21:2710 | 1.31:2710 | 1.41:2710 |
| 0.30 | 0.54:2890 | 0.63:2920 | 0.72:2920 | 0.81:2920 | 0.89:2900 | 0.97:2900 | 1.06:2900 | 1.14:2890 |
| 0.40 | 0.41:3180 | 0.51:3180 | 0.59:3150 | 0.66:3130 | 0.74:3120 | 0.81:3100 | 0.88:3090 | 0.95:3070 |
| 0.50 | 0.32:3230 | 0.42:3380 | 0.49:3330 | 0.56:3300 | 0.62:3260 | 0.69:3250 | 0.75:3230 | 0.81:3210 |
| 0.60 | 0.25:3180 | 0.35:3390 | 0.42:3450 | 0.48:3400 | 0.54:3370 | 0.60:3340 | 0.65:3330 | 0.71:3310 |
| 0.70 | 0.20:3110 | 0.29:3350 | 0.37:3470 | 0.43:3490 | 0.48:3450 | 0.53:3420 | 0.58:3400 | 0.62:3380 |
| 0.80 | 0.17:3030 | 0.25:3310 | 0.32:3440 | 0.38:3510 | 0.43:3510 | 0.47:3480 | 0.52:3460 | 0.56:3440 |
| 0.90 | | 0.22:3270 | 0.28:3410 | 0.33:3500 | 0.38:3530 | 0.43:3530 | 0.47:3500 | 0.51:3480 |
| 1.00 | | 0.19:3240 | 0.25:3380 | 0.30:3480 | 0.35:3530 | 0.39:3550 | 0.43:3540 | 0.47:3520 |
| 2.00 | | | | | 0.17:3390 | 0.19:3450 | 0.22:3500 | 0.24:3530 |

*Stress ratio $R_n$ precedes the colon; service stress $f_{sv}$ in psi follows the colon.

and $P_n$. There are no requirements nor are there any restrictions concerning the control of stress at elastic or service levels.

The need for some degree of control over service stresses in concrete is readily demonstrated. Typical levels of service stress have been computed for a concrete having an ultimate strength of 4000 psi; these service stresses are shown in Table 13-1. It is noted therein that theoretical service stresses in excess of the idealized yield stress (3400 psi) can occur at several steel ratios and eccentricities. Some prudent level of control over service stresses is therefore indicated.

In the face of the many improbabilities that can occur in the loading of columns, Code permits the service stress in columns to reach consistently higher levels than in beams. Those higher levels of service stress are very evident in Table 13-1. In view of this feature of the Code, a limiting stress in concrete of 60% of idealized yield, as used for beams, has been found to be wastefully conservative; for columns, a higher allowable service stress is considered to be justified. The following logic is used to arrive at one such suitable but higher level for the service stress in columns.

It was noted in earlier discussions that Code permits an overload of 33% when lateral loads are included. Throughout this period of overload, however transient, it is of course desirable that stresses remain elastic; that is, under this 33% overload, stresses should never exceed the yield stress of $0.85f'_c$. Such stresses, if they occur, would produce inelastic deformations while the structure is under its allowable day-to-day conditions of load, a highly questionable feature in a design. Accordingly, adopting a maximum allowable service stress at 75% of the idealized yield will assure that the stresses will remain elastic throughout a 33% overload. This value of $f_{sv} = 0.75(0.85f'_c) = 0.6375f'_c$ is adopted here, and will be used in subsequent examples.

## Examples in Full Range Design

With the foregoing limit on service stresses thus established, the procedure for designing columns becomes one of selecting a column section to suit the allowable service stress, then reviewing the section so chosen for its suitability at ultimate load. If the section is found to be inadequate at ultimate load, a revised section is chosen and the process repeated. When a section is found that suits the limitations both at service levels and at ultimate load, the ties are selected and the design is then complete.

Some examples will illustrate the procedure.

## EXAMPLE 13-10.

Full range design method.
Design of a column subject only to axial load.
No limitations on dimensions or reinforcement.

**Given:**

Apartment building, 5 floors; at a first-floor column, $L_u$ = 9 ft 0 in.
$P_{DL}$ = 229 kips; $P_{LL}$ = 201 kips.
Grade 60 steel; $f'_c$ = 4000 psi; Service stress limited to $0.6375f'_c$

**Find:**

Suitable column section

**Solution:**

1. Design first for service conditions.
   Calculate the service load $P_{sv}$:

   $$P_{sv} = \left(\tfrac{1.4}{1.7} P_{DL} + P_{LL}\right)/\phi = \left(\tfrac{1.4}{1.7} \times 229 + 201\right)/0.7 - 557 \text{ kips}$$

   Choose an overall size and gross area (square column): use the rule of thumb

   $b = h$ = 12 in. plus 1 in. × stories above = 12 + 1 × 4.
   Try a column 16 in. square.

   Check the buckling criteria:

   $$\frac{kL_u}{r} \leq 34 - \frac{M_1}{M_2}; \quad \frac{1 \times 9 \times 12}{0.3 \times 16} - 22.5 < 34 \quad \text{(O.K.)}$$

   Calculate $R_{sv}$ and $e_{sv}/h$:

   $$R_{sv} = \frac{P_{sv}/bh}{f_{sv}} = \frac{557,000}{0.6375 \times 4000 \times 16^2} = 0.85$$

   Use $e_{sv}/h$ at its minimum value of 0.10.

   From Table A-10, with steel uniformly distributed,

   Try $A_s$ = 2.1%, $R_{sv}$ = 0.85, $e_{sv}/h$ = 0.1, 16 in. square

2. Review for ultimate load conditions.
   Calculate the ultimate load $P_n$:

$$P_n = 1.7 \times P_{sv} = 1.7 \times 557 = 947 \text{ kips.}$$

From Table A-10, find $R_n$ for $\rho = 2.1\%$, grade 60 steel, find $R_n = 1.05$:

$$R_n = \frac{P_n / bh}{0.85f_c'} \quad 1.05 = \frac{P_n}{0.85 \times 4000 \times 16^2}$$

allowable $P_n = 914$ kips < 947 kips (no good)

This column size is inadequate at ultimate load. Slightly more capacity is needed.

Revise $A_s$ to 2.4%. From Table A-10, find $R_n = 1.09$

$$R_n = \frac{P_n / bh}{0.85f_c'} \quad 1.09 = \frac{P_n}{0.85 \times 4000 \times 16^2}$$

$$P_n = 949 \text{ kips} > 947 \text{ kips (O.K.)}$$

3.  Complete the design.

    Select longitudinal reinforcing bars:

    $A_s = \rho bh = 0.024 \times 16 \times 16 = 6.14 \text{ in.}^2$

    Use 8 No. 8 bars

Select ties:

    For No. 8 longitudinal bars, use No. 3 ties

    Spacing  < 16 bar diameters or   < 16 in.
             < 48 tie diameters or    < 18 in.
             < least dimension or    < 16 in.

Use:  column 16 in. square
      8 No. 8 bars longitudinal reinforcement
      No. 3 ties @ 16 in. o.c.

The final section is shown in the following sketch.

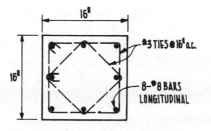

The design of the column section chosen in Example 13-10 is seen to be governed by ultimate load rather than by service stresses. The final service stress was not actually computed in the example but it is known to be slightly less than $0.6375f'_c$. If needed, however, the actual elastic stress at service levels can be verified easily: the stress ratio $R_e$ is found from Table A-10 to be 0.88 for 2.4% reinforcement, for which case the elastic stress is computed to be 2422 psi or $0.62f'_c$. The elastic stress is thus seen to be slightly less than the limiting value of $0.6375f'_c$ and is therefore acceptable.

### EXAMPLE 13-11.

Full range design method.
Design of a column subject only to axial load.
One dimension restricted.

**Given:**

Conditions of Example 13-10
Height $h$ across axis of bending limited to 12 in.

**Find:**

Suitable column section

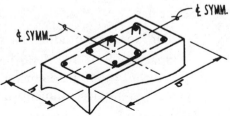

**Solution:**

1. Design first for service conditions
   Calculate the service load $P_{sv}$:

   $P_{sv}$ = 557 kips (same as Example 13-10)

   Choose an overall size and gross area (rectangular column). Use the rule of thumb to find

   Gross area = 16 × 16 = 256 in.² (same as Example 13-10)
   With $h$ = 12 in., solve for $b$ = 22 in., $A_g$ = 264 in.²
   Check the buckling criteria:

$$\frac{KL_u}{r} < 34 - \frac{M_1}{M_2}, \quad \frac{1 \times 9 \times 12}{0.3 \times 12} = 30 < 34 \quad \text{(O.K.)}$$

Compute $RL_{sv}$ and $e_{sv}/h$:

$$RL_{sv} = \frac{P_{sv}/bh}{f_{sv}} = \frac{557,000/12 \times 22}{0.6375 \times 4000} = 0.83$$

use $e_{sv}/h$ at its minimum value of 0.10

From Table A-10, with steel on two faces only,
Try $A_s$ at 2%, $R_{sv}$ = 0.86, $e_{sv}/h$ = 0.10, 12 in. × 22 in.

2.  Review for ultimate load conditions
    Calculate the ultimate load $P_n$:

    $$P_n = 1.7P_{sv} = 1.7 \times 557 = 947 \text{ kips}$$

    From Table A-10, find $R_n$ for $\rho$ = 0.02, $e_e/t$ = 0.10, Grade 60 steel, read $R_n$ = 1.06

    $$R_n = \frac{P_n/bh}{0.85f_c'}, \quad 1.06 = \frac{P_n/12 \times 22}{0.85 \times 4000}$$

    $P_n$ = 951 kips > 947 kips  (O.K.)

    Use 12 in. × 22 in. column, 2% longitudinal reinforcement.

3.  Complete the design
    Select longitudinal reinforcing bars:

    $$A_s = 0.02bh = 0.02 \times 12 \times 22 = 5.28 \text{ in.}^2$$

    use 4 No. 8 bars and 4 No. 7 bars.

Select ties and spacing:

> For No. 8 longitudinal bars, use No. 3 ties
> Spacing:  < 16 bar diameters or  < 14 in. (for No. 7 bars)
>           < 48 tie diameters or  < 18 in.
>           < least dimension or   < 12 in.
> Use column 12 in. × 22 in.
> 4 No. 8 bars at corners, 4 No. 7 bars interior
> No. 3 ties at 12 in. (set of 2 ties)

The final section is similar to that shown at the beginning of the example.

---

**EXAMPLE 13-12.**

---

Full range design method.
Design of a column subject to flexure.
Configuration limited to square sections.

**Given:**

Conditions of Example 13-10

At top of column: Add    $M_{DL}$ = 65 kip-ft
                            $M_{LL}$ = 48 kip-ft

At base of column: Add    $M_{DL}$ = 33 kip-ft
                            $M_{LL}$ = 24 kip-ft

Moment directions are shown in sketch

**Find:**

Suitable square column section

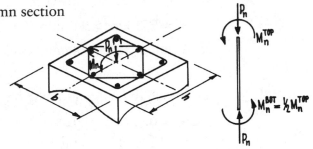

**Solution:**

1. Design first for service conditions
   Calculate the service load $P_{sv}$ and moment $M_{sv}$:

   $P_{sv}$ = 557 kips (same as Example 13-10)

   $$M_{sv} = \frac{\frac{1.4}{1.7} M_{DL} + M_{LL}}{\phi} = \frac{\frac{1.4}{1.7} \times 65 + 48}{0.7}$$

   = 145 kip-ft at top of column

   = 73 kip-ft at base of column

Choose an overall size 16 in. × 16 in. (same as Example 13-10)
Check the buckling criteria:

$$\frac{KL_u}{r} < 34 - 12\frac{M_1}{M_2}, \quad \frac{1 \times 9 \times 12}{0.3 \times 16} < 34 - 12\frac{-73}{145}$$

22.5 < 40 (O.K.)

Determine $R_e$ and $e_e/h$:

$$R_{sv} = \frac{P_{sv}/bh}{f_{sv}} = \frac{557,000/16 \times 16}{0.6375 \times 4000} = 0.85$$

$$e_{sv} = \frac{M_{sv}}{P_{sv}} = \frac{145,000 \times 12}{557,000} = 3.12 \text{ in.}$$

$$\frac{e_{sv}}{h} = \frac{3.12}{16} = 0.195, \text{ use } 0.20$$

From Table A-10, with steel uniformly distributed, find 4.8% steel (too high). This column is unacceptable since the steel ratio is too high. Revise overall size to 18 in. square and try again.
For column size 18 in. × 18 in.:

$$R_{sv} = \frac{P_{sv}/bh}{f_{sv}} = \frac{557,000/18 \times 18}{0.6375 \times 4000} = 0.67$$

$$e_{sv} = \frac{M_{sv}}{P_{sv}} = \frac{145,000 \times 12}{557,000} = 3.12 \text{ in.}$$

$$\frac{e_{sv}}{h} = 0.17$$

From Table A-10, steel uniformly distributed, select 2% steel.
Try column 18 in. × 18 in., 2% steel.

2. Review for ultimate load conditions
   Calculate the ultimate loads $P_n$ and $M_n$:

   $$P_n = P_{sv} \times 1.7 = 557 \times 1.7 = 947 \text{ kips}$$
   $$M_n = M_{sv} \times 1.7 = 145 \times 1.7 = 247 \text{ kip-ft}$$

From Table A-10, find $R_n$ for 2% steel, $e_n/t = 0.17$, read $R_n = 0.88$

$$R_n = \frac{P_n/bh}{0.85f_c'} \quad 0.88 = \frac{P_n/18 \times 18}{0.85 \times 4000}$$

$P_n = 969$ kip-ft $> 947$ kip-ft (O.K.)

Use column 18 in. $\times$ 18 in., 2% steel.

3. Complete the design

Select longitudinal reinforcing bars:

$A_s = 0.02bh = 0.02 \times 18 \times 18 = 6.48$ in.$^2$

Use 4 No. 9 and 4 No. 8 bars, with the No. 9 bars at the corners.

Select the ties: for No. 9 longitudinal bars, use No. 3 ties:

Spacing: < 16 bar diameters or  < 16 in.
< 48 tie diameters or  < 16 in.
< least dimension or  < 18 in.

Use column 18 in. $\times$ 18 in.

4 No. 9 bars at corners

4 No. 8 bars at interior faces

No. 3 ties at 16 in. o.c.

The final section is similar to that shown at the beginning of this example.

## Discussion of the Full Range Design Method

It has already been noted that the accuracy of a column design at elastic or service levels of stress is seriously subject to question. Nonetheless, a design of a column at service levels affords at least some measure of control over service stresses that is not otherwise provided in the strength method of design.

It should be remembered that this textbook is limited to the study of structures braced against sidesway. For such structures, bending moments on columns are minimized and $e/h$ ratios will be small, usually less than 0.3. Even at these low values of $e/h$, however, some of the service stresses given in Table 13-1 are seen to be uncomfortably high and some measure of control over service stresses should be exercised.

Closer examination of Table 13-1 also indicates that service stresses in concrete are significantly higher for columns built with grade 60 steel than for those built with grade 40 steel. Higher allowable stresses (and strains) in steel inherently mean higher strains (and stresses) in concrete. One way to reduce service stresses in concrete, therefore, is simply to design for a reduced stress in steel; that is, design for grade 40 steel rather than grade 60 steel.

Reducing the stress in the longitudinal steel bars will also produce a second beneficial side effect. Under the reduced stress, the tendency of the bars to buckle outward is reduced commensurately, relieving somewhat one of the more serious failure modes in reinforced concrete. Since the tie spacing is the same for all grades of steel, tie spacing will be less critical for the lower-stressed grade 40 steel than for the more highly-stressed grade 60 steel. The reduction in stress thus produces an additional margin against this failure mode.

In summary, the full range design method is seen to offer a degree of control over the service stresses in columns that is not otherwise provided by the strength method. And, as a consequence of the final review at ultimate load, the column so selected must have at least as much capacity at ultimate loading as it would have if it were designed under the strength method alone. The end result is that the full range design method provides a bit more insight and control of column behavior under day-to-day service conditions than is provided by the ACI strength method alone.

## OUTSIDE PROBLEMS

13.46 through 13.72. For the conditions given in Problems 13.1 through 13.27, design the columns using full range design, using an allowable elastic stress of $0.6375f'_c$.

## Interaction Diagrams

A column under a heavy axial load and a light bending load will fail in compression in the concrete. At the other extreme, a column under light axial load and a heavy bending load will fail in tension in the steel. The dividing line between these two failure modes is the balanced strain condition, which occurs when the concrete reaches its ultimate strain of 0.003 in./in., just as the tensile reinforcing steel starts to enter yield.

The concept of the "balanced strain condition" used here for columns

# Table 13-2    Axial Load and Moment on Columns at Balanced Strain Conditions

Steel at Flexure Faces Only

Steel Uniformly Distributed

## Coefficients for Axial Loads $P_o$, $P_b$ and Moment $M_b$*

### Steel at Flexure Faces Only

| $f'_c$ psi | Steel Grade | | Steel Ratio $A_s/bh$ | | | | | | | |
|---|---|---|---|---|---|---|---|---|---|---|
| | | | 0.01 | 0.02 | 0.03 | 0.04 | 0.05 | 0.06 | 0.07 | 0.08 |
| 3000 | 40 | $P_o$ | 2920 | 3300 | 3670 | 4050 | 4420 | 4800 | 5170 | 5550 |
| | | $P_b$ | 1290 | 1270 | 1240 | 1220 | 1190 | 1170 | 1140 | 1120 |
| | | $M_b$ | 470 | 620 | 770 | 920 | 1070 | 1220 | 1370 | 1520 |
| | 50 | $P_o$ | 3020 | 3500 | 3970 | 4450 | 4920 | 5400 | 5870 | 6350 |
| | | $P_b$ | 1200 | 1170 | 1150 | 1120 | 1100 | 1070 | 1050 | 1020 |
| | | $M_b$ | 500 | 690 | 880 | 1070 | 1250 | 1440 | 1630 | 1820 |
| | 60 | $P_o$ | 3120 | 3700 | 4270 | 4850 | 5420 | 6000 | 6570 | 7150 |
| | | $P_b$ | 1120 | 1090 | 1060 | 1040 | 1010 | 990 | 960 | 940 |
| | | $M_b$ | 540 | 760 | 990 | 1210 | 1440 | 1660 | 1890 | 2110 |

### Steel Uniformly Distributed

| $f'_c$ psi | Steel Grade | | Steel Ratio $A_s/bh$ | | | | | | | |
|---|---|---|---|---|---|---|---|---|---|---|
| | | | 0.01 | 0.02 | 0.03 | 0.04 | 0.05 | 0.06 | 0.07 | 0.08 |
| 3000 | 40 | $P_o$ | 2920 | 3300 | 3670 | 4050 | 4420 | 4800 | 5170 | 5550 |
| | | $P_b$ | 1350 | 1380 | 1400 | 1430 | 1460 | 1490 | 1510 | 1540 |
| | | $M_b$ | 420 | 520 | 620 | 730 | 830 | 930 | 1040 | 1140 |
| | 50 | $P_o$ | 3020 | 3500 | 3970 | 4450 | 4920 | 5400 | 5870 | 6350 |
| | | $P_b$ | 1240 | 1250 | 1260 | 1270 | 1280 | 1290 | 1310 | 1320 |
| | | $M_b$ | 440 | 570 | 700 | 830 | 960 | 1090 | 1210 | 1340 |
| | 60 | $P_o$ | 3120 | 3700 | 4270 | 4850 | 5420 | 6000 | 6570 | 7150 |
| | | $P_b$ | 1130 | 1120 | 1110 | 1100 | 1080 | 1070 | 1060 | 1050 |
| | | $M_b$ | 460 | 620 | 770 | 920 | 1070 | 1220 | 1370 | 1530 |

# Table 13-2   Continued

**Left block — Steel Ratio $A_s/bh$**

| $f'_c$ psi | Steel Grade | | 0.01 | 0.02 | 0.03 | 0.04 | 0.05 | 0.06 | 0.07 | 0.08 |
|---|---|---|---|---|---|---|---|---|---|---|
| 4000 | 40 | $P_o$ | 3770 | 4130 | 4500 | 4860 | 5230 | 5600 | 5960 | 6330 |
| | | $P_b$ | 1680 | 1650 | 1620 | 1580 | 1550 | 1510 | 1480 | 1450 |
| | | $M_b$ | 570 | 720 | 870 | 1020 | 1170 | 1320 | 1470 | 1620 |
| | 50 | $P_o$ | 3870 | 4330 | 4800 | 5260 | 5730 | 6200 | 6660 | 7130 |
| | | $P_b$ | 1560 | 1520 | 1490 | 1460 | 1420 | 1390 | 1350 | 1320 |
| | | $M_b$ | 610 | 790 | 980 | 1170 | 1360 | 1540 | 1730 | 1920 |
| | 60 | $P_o$ | 3970 | 4530 | 5100 | 5660 | 6230 | 6800 | 7360 | 7930 |
| | | $P_b$ | 1450 | 1420 | 1380 | 1350 | 1310 | 1280 | 1250 | 1210 |
| | | $M_b$ | 640 | 860 | 1090 | 1310 | 1540 | 1760 | 1990 | 2210 |
| 5000 | 40 | $P_o$ | 4610 | 4970 | 5320 | 5680 | 6040 | 6400 | 6750 | 7110 |
| | | $P_b$ | 2060 | 2010 | 1970 | 1930 | 1890 | 1840 | 1880 | 1760 |
| | | $M_b$ | 670 | 820 | 970 | 1120 | 1270 | 1420 | 1570 | 1720 |
| | 50 | $P_o$ | 4710 | 5170 | 5620 | 6080 | 6540 | 7000 | 7450 | 7910 |
| | | $P_b$ | 1900 | 1860 | 1820 | 1780 | 1730 | 1690 | 1650 | 1610 |
| | | $M_b$ | 710 | 900 | 1080 | 1270 | 1460 | 1650 | 1830 | 2020 |
| | 60 | $P_o$ | 4810 | 5370 | 5920 | 6480 | 7040 | 7600 | 8150 | 8710 |
| | | $P_b$ | 1770 | 1730 | 1690 | 1640 | 1600 | 1560 | 1520 | 1470 |
| | | $M_b$ | 740 | 960 | 1190 | 1410 | 1640 | 1860 | 2090 | 2310 |

**Right block — Steel Ratio $A_s/bh$**

| $f'_c$ psi | Steel Grade | | 0.01 | 0.02 | 0.03 | 0.04 | 0.05 | 0.06 | 0.07 | 0.08 |
|---|---|---|---|---|---|---|---|---|---|---|
| 4000 | 40 | $P_o$ | 3770 | 4130 | 4500 | 4860 | 5230 | 5600 | 5960 | 6330 |
| | | $P_b$ | 1740 | 1760 | 1770 | 1790 | 1810 | 1830 | 1850 | 1870 |
| | | $M_b$ | 520 | 630 | 730 | 830 | 940 | 1040 | 1140 | 1240 |
| | 50 | $P_o$ | 3870 | 4330 | 4800 | 5260 | 5730 | 6200 | 6660 | 7130 |
| | | $P_b$ | 1600 | 1600 | 1600 | 1600 | 1610 | 1610 | 1610 | 1620 |
| | | $M_b$ | 550 | 680 | 800 | 930 | 1060 | 1190 | 1320 | 1450 |
| | 60 | $P_o$ | 3970 | 4530 | 5100 | 5660 | 6230 | 6800 | 7360 | 7930 |
| | | $P_b$ | 1460 | 1440 | 1420 | 1410 | 1390 | 1370 | 1350 | 1330 |
| | | $M_b$ | 570 | 720 | 870 | 1020 | 1170 | 1320 | 1480 | 1630 |
| 5000 | 40 | $P_o$ | 4610 | 4970 | 5320 | 5680 | 6040 | 6400 | 6750 | 7110 |
| | | $P_b$ | 2110 | 2120 | 2130 | 2140 | 2150 | 2160 | 2170 | 2180 |
| | | $M_b$ | 630 | 730 | 830 | 940 | 1040 | 1140 | 1250 | 1350 |
| | 50 | $P_o$ | 4710 | 5170 | 5620 | 6080 | 6540 | 7000 | 7450 | 7910 |
| | | $P_b$ | 1940 | 1940 | 1930 | 1920 | 1920 | 1910 | 1910 | 1900 |
| | | $M_b$ | 650 | 780 | 910 | 1030 | 1160 | 1290 | 1420 | 1550 |
| | 60 | $P_o$ | 4810 | 5370 | 5920 | 6480 | 7040 | 7600 | 8150 | 8710 |
| | | $P_b$ | 1790 | 1760 | 1730 | 1700 | 1670 | 1640 | 1620 | 1590 |
| | | $M_b$ | 670 | 820 | 970 | 1120 | 1270 | 1420 | 1580 | 1730 |

\* $P_o$ in lbs. = coefficient $\times bh$, $b$ and $h$ in inches.
\* $P_l$ in lbs. = coefficient $\times bh$, $b$ and $h$ in inches.
\* $M_b$ in lb.-in. = coefficient $\times bh^2$, $b$ and $h$ in inches.

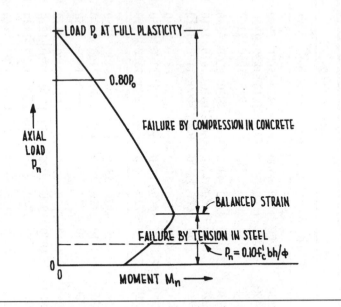

**Figure 13-14**  Interaction between axial load and moment.

is the same as that introduced earlier for beams. (It should not be confused with the "balanced stress condition," however, in which both steel and concrete are deliberately given the same elastic margins to their yield stresses.) In the derivations of Appendix B, the balanced strain condition for columns is represented in Figure B-2 by the line separating sector 4 from sector 5.

The axial load that produces the balanced strain condition is designated $P_b$ and the accompanying moment is designated $M_b$. Values of $P_b$ and $M_b$ may be computed by setting $k = 1/(1 + s)$ and solving the appropriate equations listed in the derivations in the Appendix. The results of such a computation are shown in Table 13-2 for various concrete strengths and steel grades.

From the analysis given earlier and summarized in the equations of Appendix Table B-1, it is seen that there is at failure a fixed interaction between the axial load $P_n$ and the moment $M_n$. The interaction occurs at all levels of load, however, not just at the balanced strain condition. A generalized graph of these interactions is shown in the sketch of Figure 13-14.

Sets of interaction diagrams similar to that of Figure 13-14 are commonly used for design of columns. An example of one such set of diagrams

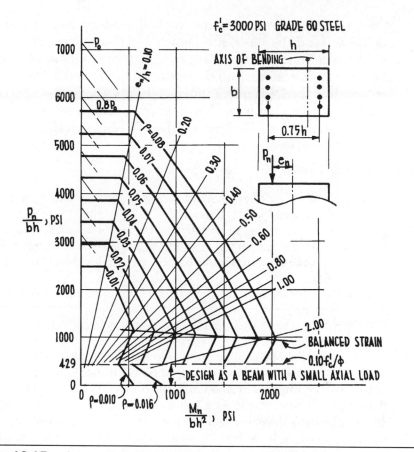

**Figure 13-15**    Interaction diagram.

is given in Figure 13-15. This set of interaction diagrams has been developed entirely from data given in the parametric design Table A-10 in the Appendix, along with the balanced strain conditions given in Table 13-2. Some designers prefer to use these diagrams while others prefer to use parametric design tables such as those in the Appendix; the author has a mild preference for the more compact parametric design tables.

The transition from a column carrying a dominant flexural load to a beam carrying a small axial load is a vague point in the Code. For the columns given in Table 13-2, for example, Code requires that such members be designed with a minimum of 1% reinforcement and a maximum of 8%. Yet if the member were to be designed as a beam carrying a small axial load, Code requires a minimum of 0.33% reinforcement and a maximum of 1.6% (see Tables A-5, A-6, and A-7). Under such a dual set of requirements, it is considered prudent to satisfy both sets of conditions. In this case the reinforcement could be selected between the 1% minimum for a column and the 1.6% maximum for a beam. The final design could then be made either for a beam carrying a small axial load or for a column carrying a large flexural load. The end results will be identical.

## REVIEW QUESTIONS

1. In the working stress method for design of columns, how is the allowable axial working load $P_w$ obtained?
2. Why should the factor of safety be higher for columns than for beams?
3. What purpose do lateral ties serve in a reinforced concrete column?
4. What is the minimum required steel area for longitudinal reinforcement in a column?
5. What is the maximum allowable steel area for longitudinal reinforcement in a column?
6. Why is the steel area in columns usually held to less than 4% of the gross concrete area $bh$?
7. What is the radius of gyration, and how it is computed?
8. What is the slenderness ratio, and how is it used?
9. Describe the state of stress in a column section when it is loaded by the load $P_o$ as defined by the ACI column formula. What is the magnitude of the strain in the concrete under this load?
10. In the strength method, at what axial load does a member change from a column carrying a dominant flexural load to a beam carrying a small axial load?
11. How are end moments on a column accounted for when using the ACI column formula?
12. How is the effect of slenderness of a column accounted for when using the ACI column formula?
13. The ACI column formula is expressed only in terms of axial load. How is a moment accommodated?
14. In the full range design method, a column design cannot produce a section smaller than that produced by the strength design method. Why not?
15. In the full range design method, what is the maximum stress in the concrete at 33% overload?

# 14

# FOUNDATIONS AND SLABS ON GRADE

$P$robably the single most common application of concrete in buildings is its use in foundations and slabs on grade. Regardless of whether a building is constructed of concrete, masonry, timber, or steel, its foundations will almost certainly be constructed of concrete. Any textbook concerning design in concrete must therefore include at least an introduction to the design of simple foundations.

Foundation loadings are more complex than those used for superstructure design. The upward support for the foundation comes as a result of soil pressures induced by the structure. There is a secondary interaction between the structure and its foundation which can sometimes require a complex, time-consuming solution. The solution for these soil–structure interactions is much simpler for shallow foundations than for deep foundations.

The following discussions are limited to simple shallow foundations. Shallow foundations are defined herein as those founded within about 6 ft of the finished floor. They are by far the most common type of foundations.

## Types of Shallow Foundations

Three distinct types of shallow foundations, shown in Figure 14-1, are considered in the following discussions:

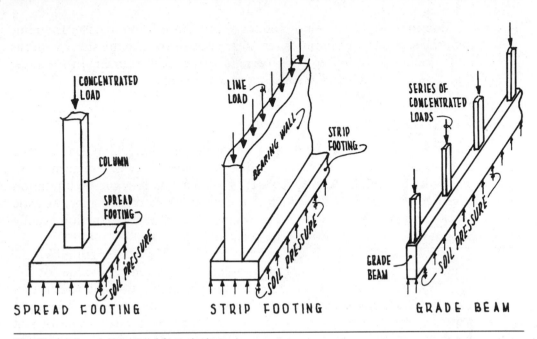

**Figure 14-1**    Types of foundations.

1.  Spread footings under individual columns
2.  Strip footings under bearing walls
3.  Grade beams under repetitive columns

All three types of foundations perform their function by distributing a concentrated load over a larger bearing area.

Spread footings, as shown in Figure 14-1, are subject to flexure as a cantilever in two directions. The column location is kept concentric; any eccentricities will result in a nonuniform distribution of soil pressure under the footing. Where space is restricted at one side, the spread footing may be made rectangular rather than square, but the column is still centered.

Strip footings are subject to flexure only in the outstanding legs; flexural reinforcement is required only in the short direction. There is no stress in the footing in the direction of the wall. The wall itself must be designed to take any variations in loads or settlements along its length.

Grade beams are continuous beams subject to flexure longitudinally along the line of columns they support. They are loaded on the bottom face by the distributed soil pressure. The loading system of the grade beam of Figure 14-1 may look more familiar if the sketch is viewed upside down.

Footings are rarely reinforced for shear. Since forming costs are minimal, the sections are simply made large enough to take the shear without reinforcement. Although Code permits shear reinforcement in footings, that case is not included in subsequent discussions.

## Foundation Loading and Failure Modes

Allowable soil pressure under a footing may be limited by one of two considerations: either by the differential settlement between adjacent footings or by the bearing strength of soil. By far the more common limitation is settlement. Even in sandy soils, the progressive crushing or inelastic deformation of friable sands within the first year after construction usually produces the limiting case for footing pressures.

When settlement of the foundation rather than soil strength governs a design, the size of the contact area is based on a reduced value for the allowable soil pressure; this reduced pressure serves to limit settlements under day-to-day service loads. For the other case (i.e., where soil strength governs), the size of the contact area is based on a peak value of soil pressure, but still at working levels. There is no accepted way to correlate test data on soils to ultimate strength criteria, recognizing that the ultimate strength of the soil can fluctuate drastically due to such things as daily weather changes.

Thus, whether the size of the contact area is limited by settlements or by soil strength, it is seen that all soil criteria are specified at working levels or service levels. Ultimate soil pressure is not used. The working dead and live loads must therefore be carried forward in the calculations to be used in establishing the size of the contact area of the foundation.

## Allowable Soil Pressures

The allowable soil pressure is the increase in soil pressure that may be applied to a soil at a certain stratum. The soil will safely accept this pressure in addition to its regular overburden pressures. Usually, if the overburden is permanently removed (e.g., basements), the allowable soil pressure may be increased proportionately.

Settlements in soil are time dependent, usually taking place over several months or several years. Consequently, when establishing the size of the contact area that will limit settlements, only those loads may be considered that will exist long enough to produce long-term settlements. Certainly, the dead load of the structure meets this requirement but only about half the live load (furnishings, carpets, files, books, etc.) may be considered to be long term. The remaining live load comes and goes, causing only elastic deformations in the soil; long-term inelastic settlements have no time to develop.

When establishing the contact area that will limit settlements, it is common practice to use dead load plus about 50% live load for buildings that serve routine architectural functions. The live load in buildings such as libraries may be much higher, as much as 80% of maximum, while in auditoriums the long-term load may contain only 20% of the maximum live load. For simplicity in the subsequent examples, a value of dead load plus 50% live load will be used.

The allowable soil pressure corresponding to this long-term load is specified as a result of the soils investigation (or by Code). This pressure is commonly termed the "reduced allowable soil pressure" and is used to compute the required size of the contact area.

Once the size of the contact area is established, there is no further need for the reduced allowable soil pressure. When the higher peak loads are applied to the building, a higher soil pressure will obviously result. This higher soil pressure, called the peak pressure, is used for all subsequent calculations for stress; the steel and concrete must of course sustain the peak pressure even if only for a short time.

When settlements control the design, the peak soil pressure is not limited to a maximum value. At the time the allowable soil pressure is being determined, all conditions of load, strength, and settlement are considered. When the reduced allowable soil pressure is finally established, it may thereafter be assumed that the strength has been found adequate for all peak-load conditions; the soil pressure is limited only by settlements.

For the other case of loading (i.e., when the strength of the soil rather than settlement governs the design), the peak load is the only load ever considered. Combinations of loads that produce peak conditions are discussed with the load factors in Chapter 3. In the subsequent examples, the peak loading will be taken as dead load plus 100% live load.

## Footing Rotations

Where soils are so soft that foundations must be designed to limit settlements, the soil can be expected to offer very little resistance to small rotations

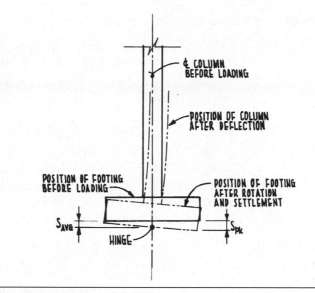

**Figure 14-2**    Footing rotations.

of the footing. Although the soil–structure interaction of such problems is far beyond the scope of this book, it should be apparent that a soil has only a small fraction of the rigidity of concrete. As a consequence, the footing will simply rotate slightly, similar to a raft floating in water, and thereby relieve any moments that occur at the base of the column. Further, the rotation is so small that no significant changes in soil pressure are induced.

Since the footings rotate so easily compared to the concrete columns, columns founded on isolated spread footings should be assumed to be hinged at their bases. The hinge, however, should be assumed to occur at the contact surface at the underside of the footing, as shown in Figure 14-2. The length of the column for calculations is then the distance from the underside of the footing to the underside of the first-floor girders.

Consequently, the effective load delivered to an isolated footing should be taken as the axial load only; moments are simply dissipated. There may, in fact, be some rotations due to the elastic column moments, but the soil pressures can be expected to vary very little due to the small rotations common to an elastic structure. In subsequent discussions, the average soil pressure under a spread footing will be computed as the tributary concentric load divided by the contact area of the footing.

## Spread Footings

Spread footings are rarely, if ever, reinforced for shear. Like slabs, it is far more practical to slightly increase the depth of the footing to increase its shear capacity rather than place numerous stirrups. Since footings are cast directly on soil, formwork is minimal and the extra cost is only that of bulk concrete.

Since shear reinforcement is not used, spread footings can be expected to be quite thick in order to sustain the shear loads. Spans (or overhangs) are comparatively short, so flexure is not usually large enough to require a deeper section than that required for shear. As a general guide, the overall thickness of a spread footing can be expected to be about one-sixth of the maximum dimension, or less, with a minimum thickness of about 12 in. being commonly observed.

When concrete is cast against soil, the minimum cover over the reinforcement is specified by Code (7.7.1) at 3 in. The consequences of foundation failure are so serious, however, that this limit is usually observed as the barest minimum. In subsequent examples, the concrete cover is taken as 4 in. from the centerline of flexural reinforcement, which provides somewhat more than the minimum 3-in. clear cover.

Spread footings are assumed to act as cantilevered beams in two directions. The critical section for bending in a spread footing that supports a concrete column is at the face of the column. For a footing that supports a steel column, the critical section for bending is taken halfway between the face of the steel column and the edge of the base plate. Critical sections are shown in Figure 14-3.

It should be noted that the bending of the footing is assumed to occur uniformly along the full width of the footing as shown in Figure 14-3, as if the column were actually a wall extending all the way across the footing. Although this assumption is admittedly a drastic simplification, its success over the years justifies its continued use. For rectangular footings, the procedure is repeated in the other direction to obtain the reinforcement in that direction.

The same simplifying assumption concerning bend lines applies also when designing the spread footing for beam shear. As for other types of beams, the critical section for beam shear is taken at a distance $d$ from the face of the column, shown as lines AB and CD in Figure 14-3. The ultimate shear stress on the concrete at the critical section is, as usual in beams,

$$v_n = 2\sqrt{f_c'}$$

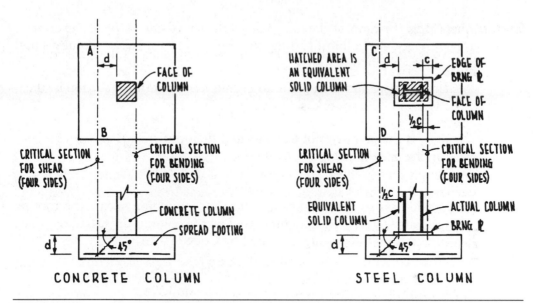

**Figure 14-3**    Critical sections in spread footings.

With no shear reinforcement, the ultimate shear force at the critical section may not be greater than $V_c$, where

$$V_c = v_n b_w d$$

and $b_w$ is the total width of the footing at the critical section.

In addition to the foregoing design criteria for beam shear and flexure, Code imposes an additional limitation due to "punching shear." Punching shear produces the pyramidal punch-out shown in Figure 14-4. Whereas critical beam shear is taken at a distance $d$ from face of support, critical punching shear is taken at a distance $d/2$ from face of support, shown as line ABCD in Figure 14-4.

The ultimate shear stress for punching shear in concrete without shear reinforcement is specified by Code (11.11.2.1):

$$v_n = \left(1 + \frac{2}{\beta_c}\right) 2\sqrt{f'_c} \tag{14-1a}$$

but, as a maximum,

$$v_n \leq 4\sqrt{f'_c} \tag{14-1b}$$

where $\beta_c$ is the ratio of the long side to the short side ($t/b$) of the contact area between the column and the footing ($\beta_c = 1$ for square columns and $\beta_c = \infty$ for continuous walls).

The ultimate punch-out load to be taken by the critical section is, then,

$$V_c = v_n b_o d \tag{14-2}$$

where $b_o$ is the perimeter of the critical punch-out shown in Figure 14-4.
For the line ABCD in Figure 14-4,

$$b_o = 2(b + d) + 2(h + d) \tag{14-3}$$

In addition to flexure, beam shear, and punching shear, the footing must sustain the bearing load presented by the column or wall above. The bearing load on the footing is limited by Code (10.15):

$$P_n = 0.85 f'_c A_g \tag{14-4}$$

where all symbols are as used earlier and the undercapacity factor $\phi$ is 0.7 for bearing.

The allowable bearing load given by Eq. (14-4) may be increased by multiplying by the factor $\sqrt{A_2 / A_1}$, but not by more than 2, where $A_2$ is the

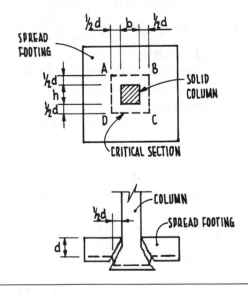

**Figure 14-4**     Punching shear.

area of the top of the footing concentric with $A_1$, and for footings the factor $\sqrt{A_2/A_1}$ will be much greater than 2, with the final result that bearing load will almost always be double that of Eq. (14-4).

For the design of footings, there are thus four conditions to be met: bearing load on top face, punching shear, beam shear, and beam flexure. There are special design tables and graphs for use in designing spread footings; they are useful if one is frequently engaged in such designs. For infrequent use, it is far faster simply to guess the depth of footing and check the four conditions to see if the guess is adequate. That procedure is illustrated in the following examples, first for a square footing and second for a rectangular footing.

### EXAMPLE 14-1.

Design of a Square Spread Footing.
Footing subject only to axial load (no moments).

**Given:**

$P_{DL}$ = 115 kips
$P_{LL}$ = 95 kips
Soil pressure $p_a$ = 3000 psf
Column size 16 × 16 in.
$f'_c$ = 3000 psi
Grade 40 steel

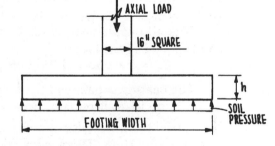

In foundation design, the difference in weight between concrete (145 pcf) and the displaced overburden (115 pcf) is commonly ignored. In subsequent calculations here, it is ignored. If it were to be included, the increase in dead load would be estimated and added to the column dead load; at the conclusion of the calculations the accuracy of this initial estimate would be checked and verified.

The required contact area of the footing is the total load, $P_{DL} + P_{LL}$, divided by the allowable soil pressure, 3000 psf:

$$\text{area} = \frac{P_{DL} + P_{LL}}{P_a} = \frac{115 + 95}{3000} = 70 \text{ ft}^2$$

Use square footing, 8 ft 6 in. or 102 in. The actual soil pressure under this footing size is

$$P_{DL} = \frac{P_{DL}}{\text{area}} = \frac{115}{8.5 \times 8.5} = 1.6 \text{ ksf}$$

$$P_{LL} = \frac{P_{LL}}{\text{area}} = \frac{95}{8.5 \times 8.5} = 1.3 \text{ ksf}$$

The first check is for bearing load on top of the footing. The total load $P_n$ delivered to the top of the footing is, where $\phi = 0.7$ for bearing,

$$P_n = \frac{1.4P_{DL} + 1.7P_{LL}}{\phi} = \frac{1.4 \times 115 + 1.7 \times 95}{0.7}$$

$$= 461 \text{ kips}$$

The allowable bearing load on the footing is

$$P_n = 0.85f_c'A_g = 0.85 \times 3000 \times 16 \times 16 = 653 \text{ kips}$$

This allowable load is multiplied by $\sqrt{A_2 / A_1}$ :

$$\sqrt{A_2 / A_1} = \sqrt{(102 \times 192)/(16 \times 16)} = 6.38 \text{ but} < 2$$

Hence

$$P_n = 2 \times 653 = 1306 \text{ kips} > 461 \text{ kips (O.K.)}$$

The bearing capacity of the footing is more than adequate to carry the column load. It remains now to select the thickness of the footing.

A trial thickness $h$ of about one-sixth of the footing width is assumed, or about 18 in. The effective depth $d$ is then taken at 14 in. This trial thickness of 14 in. is first checked for punching shear. The average ultimate punching shear stress occurs at a distance $d/2$ from the face of the column on all four sides of the column, as shown in the following sketch.

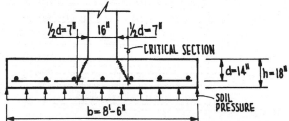

The total punching shear force acting on the section is found by statics:

$$P_{DL} = p_{DL}[b^2 - (t + d)^2] = 1.6(8.5^2 - 2.5^2) = 106 \text{ kips}$$
$$P_{LL} = p_{LL}[b^2 - (t + d)^2] = 1.3(8.5^2 - 2.5^2) = 86 \text{ kips}$$

Hence

$$P_n = \frac{1.4 P_{DL} + 1.7 P_{LL}}{\phi}$$

$$= \frac{1.4 \times 106 + 1.7 \times 86}{0.85} = 347 \text{ kips}$$

The average punching shear stress at ultimate load is given by Code, where $\beta_c = 1$:

$$v_n = 2\sqrt{f_c'}\left(1 + \frac{2}{\beta_c}\right) = 6\sqrt{f_c'}$$

but $v_n$ must not exceed $4\sqrt{f_c'}$, Hence

$$v_n = 4\sqrt{f_c'} = 220 \text{ psi}$$

The capacity of the footing to carry punching shear is, where $b_o = 4(h + d) = 120$ in.,

$$V_c = v_n b_o d = 220 \times 120 \times 14 = 370 \geq 347 \text{ kips (O.K.)}$$

This trial depth of 14 in. is thus seen to be adequate to take the punching shear.

Next the trial depth of 14 in. is checked for its capacity in beam shear. The critical shear force is found at a distance $d$ from the face of the column, as shown in the following sketch.

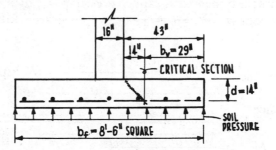

The shear acting on the footing beyond the line of the critical section is

$$V_{DL} = p_{DL} \times b_v \times b_F = 1.6 \times (^{29}/_{12}) \times 8.5 = 33 \text{ kips}$$
$$V_{LL} = p_{LL} \times b_v \times b_F = 1.3 \times (^{29}/_{12}) \times 8.5 = 27 \text{ kips}$$

Hence

$$V_n = \frac{1.4V_{DL} + 1.7V_{LL}}{\phi}$$

$$= \frac{1.4 \times 33 + 1.7 \times 27}{0.85} = 108 \text{ kips}$$

The capacity of the concrete in beam shear is

$$V_{cr} = v_n b_w d = 2\sqrt{f_c'} \times 102 \times 14 = 156 > 108 \text{ kips (O.K.)}$$

The trial depth of 14 in. is thus seen to be adequate to carry the beam shear on the section.

The trial depth of 14 in. is now checked for beam flexure, where $M = wL_c^2/2$ on the cantilever length $L_c$, as shown in the following sketch.

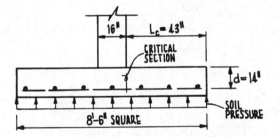

The moment acting on the critical section is

$$M_{DL} = \frac{w_{DL}L_c^2}{2} = \frac{1.6 \times 8.5\left(\frac{43}{12}\right)^2}{2} = 87 \text{ kip-ft}$$

$$M_{LL} = \frac{w_{LL}L_c^2}{2} = \frac{1.3 \times 8.5\left(\frac{43}{12}\right)^2}{2} = 71 \text{ kip-ft}$$

Hence

$$M_n = \frac{1.4M_{DL} + 1.7M_{LL}}{\phi}$$

$$= \frac{1.4 \times 87 + 1.7 \times 71}{0.9} = 269 \text{ kip-ft}$$

The section modulus $Z_c$ is found by the usual means:

$$\frac{M_n}{0.85f_c'} = Z_c \qquad \frac{269{,}000 \times 12}{0.85 \times 3000} = \text{coeff.} \times 102 \times 14 \times 14$$

$$\text{coeff.} = 0.063$$

From Table A-5, select $\rho = 0.050$ ($\rho_{\min}$ for grade 40),

Solve for $A_s$, $A_s = \rho_{bd} = 0.005 \times 102 \times 14 = 7.14$ in.$^2$

Use 12 No. 7 bars each way, evenly spaced ($A_s$ provided = 7.22 in.$^2$). The trial depth of 14 in. is thus seen to be adequate for flexure, beam shear, punching shear, and bearing; it is selected as the final depth of the footing.

The following sketch shows the final design, to include dowels from the footing to the column above.

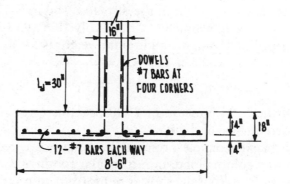

The dowels are chosen rather arbitrarily in this case. Had the applied column load been higher than the allowable bearing capacity of the footing, the excess load would have to be transferred by dowels; the size of the dowels and their embedment would then be designed to carry this excess load. In this footing, however, the footing is adequate to carry the entire load in bearing so theoretically the dowels are not needed. For such cases, the practice is to design the dowels to carry about 10% of the flexural capacity of the column to account for any accidental moments on the footing. This moment is, where $Z_c = bh^2/6$ (approximately),

$$M_n = \left(0.85f_c'Z_c\right)(0.10) = \left(\frac{0.85 \times 3000 \times 16 \times 16^2}{6}\right)(0.10)$$

$$= 174 \text{ kip-in.} = 14.5 \text{ kip-ft}$$

For dowels at the four corners, $\rho' = \rho$ and

$$\frac{M_n}{0.85f_c'} = Z_c \qquad \frac{174{,}000}{0.85 \times 3000} = \text{coeff.} \times 16 \times 16^2$$

$$\text{coeff.} = 0.0167$$

From Table A-5, use minimum $\rho$; hence

$$A_s = \rho_{\min}b_w d = 0.005 \times 16 \times 14 = 1.12 \text{ in.}^2$$

Use No. 7 bars at the four corners.

---

As a point of interest, the difference between the dead load of the concrete and the weight of the displaced soil in Example 10-1 can now be checked to see how much error is involved in ignoring this difference. At a total depth of 18 in., the additional weight is for a concrete weight of 145 pcf and a soil weight of 115 pcf,

$$\text{weight/m}^2 = (145 - 115)(^{18}\!/_{12}) = 45 \text{ psf (error)}$$

Of the total pressure of 2900 psf, this error is less than 2%, and is considered to be negligible.

The next example presents the design of a rectangular footing subjected to a column load. In rectangular footings, Code (15.4.4.2) requires that a major part of the shorter reinforcement must be located in a middle "band." The width of this middle band is equal to the narrower width of the footing, as shown in Figure 14-5.

The portion of the reinforcement to be concentrated in the middle band width is given by Code (15.4.4.2) by the formula

$$\frac{\text{reinforcement in middle band width}}{\text{total reinforcement in short direction}} = \frac{2}{\beta + 1}$$

where $\beta$ is the ratio of the long side to the short side of the footing. For a square footing, $\beta = 1$ and the equation simply requires equal distribution in both directions. The use of this formula is demonstrated in the following example.

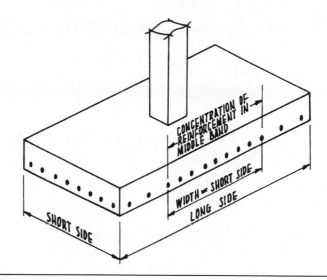

**Figure 14-5**     Middle band in rectangular footing.

### EXAMPLE 14-2.

Design of a Rectangular Footing.
Footing subject to axial load (no moment).

**Given:**

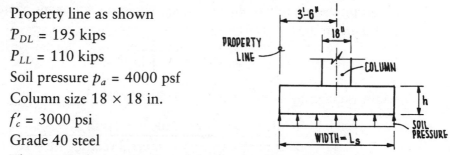

Property line as shown
$P_{DL}$ = 195 kips
$P_{LL}$ = 110 kips
Soil pressure $p_a$ = 4000 psf
Column size 18 × 18 in.
$f'_c$ = 3000 psi
Grade 40 steel

The required contact area of the footing is computed:

$$\text{Area} = \frac{P_{DL} + P_{LL}}{P_a} = \frac{195 + 110}{4} = 76.3 \text{ ft}^2$$

Use a footing 7 ft × 11 ft.

The actual soil pressure is then

$$P_{DL} = \frac{P_{DL}}{\text{area}} = \frac{195}{7 \times 11} = 2.53 \text{ ksf}$$

$$P_{LL} = \frac{P_{LL}}{\text{area}} = \frac{110}{7 \times 11} = 1.43 \text{ ksf}$$

The bearing load on the footing is checked first. The load delivered to the top of the footing is

$$P_n = \frac{1.4P_{DL} + 1.7P_{LL}}{\phi}$$

$$= \frac{1.4 \times 195 + 1.7 \times 110}{0.7} = 657 \text{ kips}$$

The allowable load on the footing is

$$P_n = 0.85 f'_c A_g = 0.85 \times 3000 \times 18 \times 18 = 826 \text{ kips}$$

This allowable load is multiplied by the factor $\sqrt{A_2 / A_1} = \sqrt{7 \times 11 / 1.5 \times 1.5} = 5.85$, but may not be more than doubled. Hence, the allowable load is

$$P_n = 2 \times 826 = 1652 > 660 \text{ kips (O.K.)}$$

The footing is therefore adequate to receive the column load from above.

The footing is checked next for punching shear. Punching shear occurs along the pyramidal punch-out shown in the following sketch.

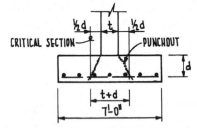

A trial thickness of about one-sixth of the longer side is assumed; hence $h = 11 \text{ ft}/6 = 2 \text{ ft}$ or 24 in. The effective depth $d$ is taken at 20 in., $h + d = 38$ in. = 3.17 ft. The punching load is found by simple statics:

$$P_{DL} = \rho_{DL}[L_L \times L_s - (h + d)^2] = 2.53(7 \times 11 - 3.17^2) = 169 \text{ kips}$$
$$P_{LL} = \rho_{LL}[L_L \times L_s - (h + d)^2] = 1.43(7 \times 11 - 3.17^2) = 96 \text{ kips}$$

Hence

$$P_n = \frac{1.4P_{DL} + 1.7P_{LL}}{\phi}$$

$$= \frac{1.4 \times 169 + 1.7 \times 96}{0.85} = 470 \text{ kips}$$

The punching shear stress in the concrete at ultimate load is

$$\upsilon_n = 2\sqrt{f_c'}\left(1 + \frac{2}{\beta_c}\right) \quad \text{where } \beta_c = 1$$

$$= 2\sqrt{f_c'}(3) = 6\sqrt{f_c'};$$

but also,

$$\upsilon_n \leq 4\sqrt{f_c'},$$

Hence,

$$\upsilon_n = 4\sqrt{f_c'} = 4\sqrt{3000} = 219 \text{ psi}$$

The capacity of the footing in punching shear then is

$$V_c = \upsilon_n b_w d = 219(38 \times 4)20 = 666 > 470 \text{ kips (O.K.)}$$

The trial depth of 20 in. is seen to be adequate for the punching shear.

The trial depth of 20 in. is next checked for beam shear at a distance $d$ from the face of column, first in the long direction, then in the short direction. The locations of the critical sections for beam shear are shown in the following sketch:

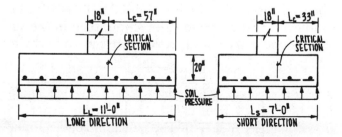

The shear force acting on the footing beyond the critical section in the long direction is found by statics:

$$V_{DL} = \rho_{DL} b_v L_s = 2.53(^{37}\!/_{12})7.0 = 54.6 \text{ kips}$$
$$V_{LL} = \rho_{LL} b_v L_s = 1.43(^{37}\!/_{12})7.0 = 30.9 \text{ kips}$$

Hence

$$v_n = \frac{1.4 V_{DL} + 1.7 V_{LL}}{\phi}$$

$$= \frac{1.4 \times 54.6 + 1.7 \times 30.9}{0.85} = 152 \text{ kips}$$

The capacity of the concrete in beam shear is

$$V_{cr} = v_n b_w d = 2\sqrt{3000} \times 84 \times 20 = 184 > 152 \text{ kips (O.K.)}$$

The section is thus seen to be adequate in the long direction. In the short direction, the shear force is

$$V_{DL} = \rho_{DL} b_v L_L = 2.53(^{13}\!/_{12})11 = 30 \text{ kips}$$
$$V_{LL} = \rho_{LL} b_v L_L = 1.43(^{13}\!/_{12})11 = 17 \text{ kips}$$

Hence

$$V_n = \frac{1.4 V_{DL} + 1.7 V_{LL}}{\phi}$$

$$= \frac{1.4 \times 30 + 1.7 \times 17}{0.85} = 83 \text{ kips}$$

The capacity of the concrete in beam shear is

$$V_{cr} = v_n b_w d = 2\sqrt{3000} \times 132 \times 20 = 289 > 83 \text{ (O.K.)}$$

The trial depth of 20 in. is thus seen to be adequate for beam shear in both directions.

The trial depth of 20 in. is now used to design for beam flexure on the cantilever, first in the long direction, then in the short direction. The dimensions in the two directions are shown in the following sketch.

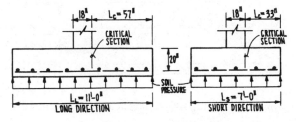

In the long direction, the moment at the critical section is

$$M_{DL} = \frac{w_{DL}L_c^2}{2} = \frac{2.53 \times 7(57/12)^2}{2} = 200 \text{ kip-ft}$$

$$M_{LL} = \frac{w_{LL}L_c^2}{2} = \frac{1.43 \times 7(57/12)^2}{2} = 113 \text{ kip-ft}$$

Hence

$$M_n = \frac{1.4M_{DL} + 1.7M_{LL}}{\phi}$$

$$= \frac{1.4 \times 200 + 1.7 \times 113}{0.9} = 525 \text{ kip-ft}$$

The plastic section modulus is calculated:

$$\frac{M_n}{0.85f_c'} = Z_c \quad \frac{525,000 \times 12}{0.85 \times 3000} = \text{coeff.} \times 84 \times 20^2$$

$$\text{coeff.} = 0.0735$$

From Table A-5, select $\rho = 0.0050$ (minimum $\rho$).
Compute $A_s = \rho bd = 0.005 \times 84 \times 20 = 8.4 \text{ in.}^2$
Use 11 No. 8 bars (long direction).
In the short direction, the moment at the critical section is

$$M_{DL} = \frac{w_{LL}L_c^2}{2} = \frac{2.53 \times 11\left(\frac{33}{12}\right)^2}{2} = 105 \text{ kip-ft}$$

$$M_{LL} = \frac{w_{LL}L_c^2}{2} = \frac{1.43 \times 11\left(\frac{33}{12}\right)^2}{2} = 59 \text{ kip-ft}$$

Hence

$$M_n = \frac{1.4M_{DL} + 1.7M_{LL}}{\phi}$$

$$= \frac{1.4 \times 105 + 1.7 \times 59}{0.9} = 275 \text{ kip-ft}$$

The plastic section modulus is determined:

$$\frac{M_n}{0.85f_c'} = Z_c \qquad \frac{275,000 \times 12}{0.85 \times 3000} = \text{coeff.} \times 132 \times 20^2$$

$$\text{coeff.} = 0.0245$$

From Table A-5, select $\rho = 0.005$ (minimum $\rho$).

Compute $A_s = \rho bd = 0.005 \times 132 \times 20 = 13.2$ in.$^2$

The portion of the total $A_s$ that must be placed in the middle band, 7 ft 0 in. wide,

$$\frac{A_s \text{ in middle band}}{\text{total } A_s (11/7) + 1} = \frac{2}{(11/7) + 1} = 0.78$$

$A_s$ in middle band $= 0.78 \times 13.2 = 10.3$ in.$^2$

Use 19 No. 8 bars, 13 in the middle 7 ft, with 3 additional bars in each end. The following sketch shows the final steel arrangement.

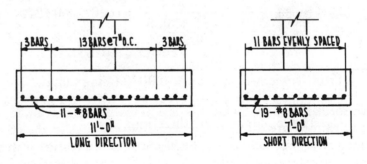

## Strip Footings

A strip footing is assumed to act as a cantilevered slab projecting from under the wall it supports. For strip footings supporting a concrete wall, the critical section for bending is at the face of the wall. If the wall is masonry, an equivalent concrete wall is used which is half the thickness of the masonry wall. The critical section for bending is then at the face of the equivalent concrete wall, as shown in Figure 14-6.

Although Code (15.5.2) permits the critical section for beam shear to

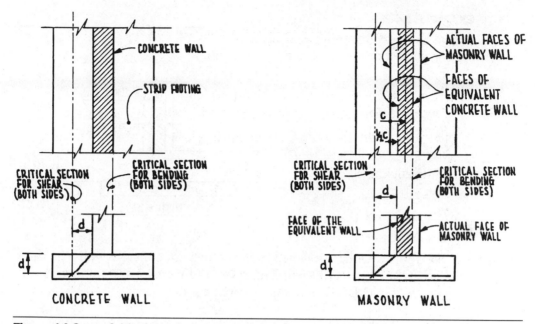

**Figure 14-6**    Critical sections in strip footings.

be measured from the actual face of the masonry wall, it is common practice to measure the critical shear from the face of the equivalent wall. In such an approach, the entire design of the foundation is thus performed for an equivalent wall, affording a consistent treatment of the design. The ultimate strength of the equivalent wall in this approach is taken at twice the ultimate strength of the masonry, $f'_m$.

As with spread footings, shear reinforcement is rarely if ever used in strip footings. The strip footing is simply made thicker to take any excess shear. The thickness of strip footings may be expected to be somewhat less than the thickness of spread footings, but a thickness less than 12 in. is rarely seen.

There is no punching shear on a strip footing. A strip footing is designed only for beam shear and beam moment on the outstanding legs. There are no other special considerations in the design of a strip footing. It is simply a double cantilevered beam. A brief example will illustrate the procedure.

### EXAMPLE 14-3.

Design of strip footing for a masonry wall.
Footing subject only to axial load (no moment).

**Given:**

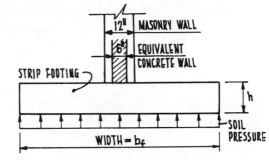

$P_{DL}$ = 8 kips/ft
$P_{LL}$ = 5.5 kips/ft
Soil pressure $\rho_a$ = 2000 psf
Wall thickness 12 in.
$f'_m$ = 1350 psi
$f'_c$ = 3000 psi
Grade 60 steel

The footing will be designed for an equivalent concrete wall 6 in. thick, with $f'_c = 2f'_m = 2700$ psi. The relative dimensions are shown in the sketch above. Determine the required contact area:

$$\text{area} = \frac{P_{DL} + P_{LL}}{P_a} = \frac{8 + 5.5}{2000} = 6.75 \text{ ft}$$

Use a footing width of 7 ft.

Calculate the actual soil pressures:

$$P_{DL} = \frac{P_{DL}}{b_F} = \frac{8}{7} = 1.14 \text{ kips/ft}$$

$$P_{LL} = \frac{P_{LL}}{b_F} = \frac{5.5}{7} = 0.79 \text{ kips/ft}$$

Although not usually necessary for walls, the bearing load on the top of the footing will now be checked. The load delivered to the top of the footing is

$$P_n = \frac{1.4P_{DL} + 1.7P_{LL}}{\phi}$$

$$= \frac{1.4 \times 8 + 1.7 \times 5.5}{0.7} = 29.4 \text{ kips/ft}$$

The allowable bearing load on the footing is

$$P_n = 0.85f'_c A_g = 0.85 \times 3000 \times 12 \times 6 = 184 \text{ kips/ft}$$

The allowable load is multiplied by the factor

$$\sqrt{A_2 / A_1} = \sqrt{7 \times 1 / 0.5 \times 1} = 4; \text{ use 2.}$$

$$P_n = 2 \times 184 = 368 \gg 29.4 \text{ kips/ft (O.K.)}$$

The width of 7 ft is thus seen to be more than adequate for bearing capacity.

A trial thickness is now assumed for the footing. As for spread footings, a thickness of one-sixth to one-eight of the total width in flexure is usually a good start. For a total width of 7 ft 0 in., the estimated overall thickness $h$ is then

$$h = \frac{\text{width}}{6} = \frac{7 \times 12}{6} = 14 \text{ in.}$$

The effective depth $d$ is taken at 10 in. This trial depth of 10 in. for $d$ is used to check the beam shear at a distance $d$ from the face of the equivalent wall. The dimensions are shown in the following sketch.

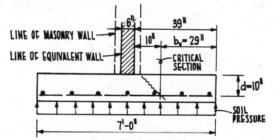

The shear on the section is, per foot of length,

$$V_{DL} = P_{DL}b_vL = 1.14(^{29}\!/_{12}) \times 1 = 2.76 \text{ kips/ft}$$
$$V_{LL} = P_{LL}b_vL = 0.79(^{29}\!/_{12}) \times 1 = 1.91 \text{ kips/ft}$$

Hence

$$V_n = \frac{1.4V_{DL} + 1.7V_{LL}}{\phi}$$

$$= \frac{1.4 \times 2.76 + 1.7 \times 1.91}{0.85} = 8.37 \text{ kips/ft}$$

The capacity of the section is, per foot of length,

$$V_{cr} = v_n b_w d = 2\sqrt{3000} \times 12 \times 10 = 13 > 8.37 \text{ kips/ft (O.K.)}$$

The trial depth of 10 in. is thus seen to be adequate to take the beam shear.

The trial depth of 10 in. for $d$ is now used for the design of the member in flexure. The critical section for flexure is at the face of the equivalent wall, as shown in the following sketch.

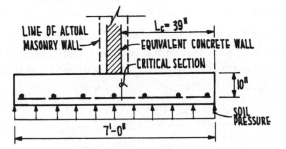

The moment acting on the critical section is

$$M_{DL} = \frac{P_{DL}L_c^2}{2} = \frac{1.14 \times \left(\frac{39}{12}\right)^2}{2} = 6.02 \text{ kip-ft/ft}$$

$$M_{LL} = \frac{P_{LL}L_c^2}{2} = \frac{0.79 \times \left(\frac{39}{12}\right)^2}{2} = 4.17 \text{ kip-ft/ft}$$

Hence

$$M_n = \frac{\left(1.4 M_{DL} + 1.7 M_{LL}\right)}{\phi}$$

$$= \frac{\left(1.4 \times 6.02 + 1.7 \times 4.17\right)}{0.9} = 17.2 \text{ kip-ft/ft}$$

The section modulus $Z_c$ is found by the usual means:

$$\frac{M_n}{0.85 f_c'} = Z_c \qquad \frac{17,200 \times 12}{0.85 \times 3000} = \text{coeff.} \times 12 \times 10^2$$

$$\text{coeff.} = 0.0675$$

From Table A-5, select $\rho = 0.0035$ (minimum).

Solve for $A_s = \rho bd = 0.0035 \times 12 \times 10 = 0.420$ in.$^2$/ft.

Use No. 6 bars at 12 in. o.c.

In the longitudinal direction, there is no computable load on the footing. The only requirement for steel is that for temperature and shrinkage, which in this case is $0.0018bh$. The required area of steel in the cross section is

$A_s = 0.0018bt = 0.0018 \times 84 \times 14 = 2.12$ in.$^2$

Use 7 No. 5 bars, evenly spaced.

A sketch of the final design is shown below. The actual masonry wall is shown rather than the equivalent wall used throughout the foregoing calculations.

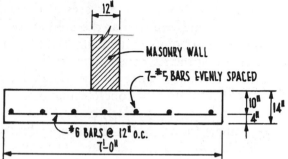

The dowels connecting the wall to its footing are as arbitrarily chosen for strip footings as they are for spread footings. If the wall is reinforced, it is common practice simply to match the wall reinforcement for size and location of dowels. If the wall is unreinforced, dowels may also be omitted.

## Grade Beams

Grade beams are one of the more economical foundations in common use for light buildings. They are used quite often for the foundations of lightweight prefabricated steel buildings as well as for one- or two-story concrete frames. They are not suited for heavy loads, however, and are rarely feasible for buildings having more than two stories.

A grade beam may be designed the same way as any other continuous rectangular beam. Code (8.3) lists approximate coefficients for computing such moments. As noted earlier in this chapter, a grade beam is like any other continuous beam except that loads are upside down.

It may be argued (successfully) that a grade beam is an elastic beam on an elastic foundation. Consequently, when the beam at midspan deflects upward due to soil pressure, the soil pressure decreases somewhat. Approximately, a sinusoidal soil pressure distribution is eventually produced, as shown in Figure 14-7. The proposed use of a uniform pressure distribution and the ACI coefficients can therefore be seen to be in error by an undetermined amount.

Although the foregoing argument is true, the magnitude of error is not prohibitively high. In comparing the sinusoidal pressure distribution of Figure 14-7 with the uniform distribution, it is seen that in all cases the load decreases at midspan and increases at the column points. Thus the moments both at midspan and at the column points decrease while the shears at the column points remain almost constant.

It is evident that the ACI coefficients will yield a conservative value of moment and a reasonable value for shear. Considering the inaccuracies in both the soils analysis and the concrete analysis, the additional error is not considered prohibitive, particularly since it is always on the "safe" side.

In the example following this discussion, the ACI coefficients are used with averaged distributed soil pressures. The soil–structure interaction that would produce the sinusoidal pressure distribution is left to more advanced study.

As with other types of footings, shear reinforcement is rarely used in grade beams. Since falsework and shoring are minimal, it is far more practical to increase slightly the size of the beam rather than to fabricate and place a large number of stirrups. As a consequence, the size of a grade beam is usually much deeper than that of other concrete beams having a similar span.

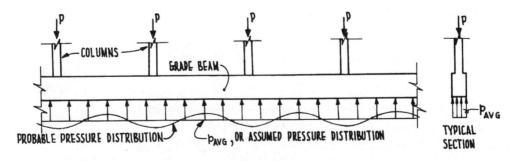

**Figure 14-7**     Soil pressures on a grade beam.

In addition, usually no effort is made to cut off the longitudinal reinforcement in a grade beam. The maximum required area of flexural steel, both top and bottom, is simply made continuous throughout the length of the grade beam. For such a design, only the highest numerical values of positive and negative moments are therefore needed when flexural reinforcement is selected.

Very often, the required width of the grade beam is so narrow that the column is wider than this width; the column would have to overhang the grade beam if it were to be centered on the grade beam. A simple solution is to make the grade beam as wide as the column; the column would then be properly seated on the grade beam. In such cases, the resulting soil pressure may be considerably less than the allowable pressure, but the overall suitability of the design justifies the configuration. The design used in the following example includes such circumstances.

## EXAMPLE 14-4.

Design of a grade beam foundation.
Repetitive column loads at uniform spacing.

### Given:

Grade beam and loads as shown in the sketch

Load on the end column is roughly half the load on an interior column

Column size 12 in. × 12 in. for all columns

$P_{DL}$ = 30 kips and $P_{LL}$ = 24 kips (each column)

Allowable soil pressure $p_a$ = 3000 psf

$f_c'$ = 3000 psi, grade 60 steel

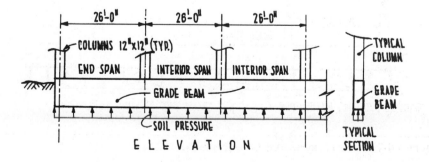

In this example, the difference in weight between the concrete grade beam and the soil it displaces is ignored. The required contact area under the grade beam is computed by simple statics, assuming that the grade beam has a bearing area 26 ft long.

$$\text{area} \quad = \frac{P_{DL} + P_{LL}}{p_a} = \frac{30 + 24}{3000} = 18 \text{ ft}^2 \text{ per bay}$$

$$\text{width} \quad = \frac{\text{area}}{\text{length}} = \frac{18}{26} = 0.69 \text{ ft} = 8\tfrac{1}{2} \text{ in.}$$

If the footing width were to be made 8½ in. as indicated in the foregoing calculations, the column with its width at 12 in. would be eccentric on the grade beam. To keep the column load concentric on the grade beam, the grade beam is simply made 12 in. wide, recognizing that the soil pressure will be somewhat less than the allowable;

$$P_{DL} = \frac{P_{DL}}{\text{area}} = \frac{30}{1 \times 26} = 1.15 \text{ kips/ft}^2$$

$$p_{LL} = \frac{P_{LL}}{\text{area}} = \frac{24}{1 \times 26} = 0.92 \text{ kips/ft}^2$$

Total $p = 1.15 + 0.92 = 2070$ psf $< 3000$ psf allowed

The grade beam will now be checked for beam shear, using the ACI coefficients for maximum shear. At an end span:

$$V_{DL} = 1.15 p_{DL} L_n = \frac{1.15(1.15 \times 25)}{2} = 16.5 \text{ kips}$$

$$V_{LL} = 1.15 p_{LL} L_n = \frac{1.15(0.92 \times 25)}{2} = 13.2 \text{ kips}$$

Hence

$$V_n = \frac{1.4 V_{DL} + 1.7 V_{LL}}{\phi}$$

$$= \frac{1.4 \times 16.5 + 1.7 \times 13.2}{0.85} = 53.6 \text{ kips}$$

A trial depth of about $L/12$ or 26 in. is adopted for the grade beam, with a corresponding value of 22 in. for the effective depth $d$. This tall slender section of $12 \times 26$ in. is now reviewed for its adequacy in shear. The critical section for shear is shown on the following partial shear diagram.

Critical shear of the section in shear is

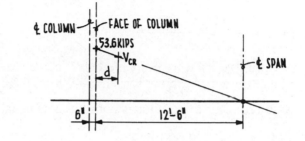

$$V_{cr} = \frac{150 - 22}{150} 53.6 = 45.7 \text{ kips}$$

The capacity of the section in shear is

$$V_c = v_n b_w d = 2\sqrt{3000} \times 12 \times 22 = 29 \text{ kips} < 45.7 \text{ kips } (\text{no good})$$

This section is not adequate for critical shear. A larger section, 38 in. deep, will be tried, effective depth $d = 34$ in. The critical shear on the new section is again computed by ratios:

$$V_{cr} = \frac{150 - 34}{150} 53.6 = 41.5 \text{ kips}$$

The capacity of this deeper section in shear is

$$V_c = v_n b_w d = 2\sqrt{3000} \times 12 \times 34 = 44.7 > 41.5 \text{ (O.K.)}$$

The trial effective depth of 34 in. is seen to be adequate for beam shear.

This trial depth of 34 in. is now reviewed for flexure. The maximum negative moment acting on the section is taken from the ACI coefficients:

$$M_{DL} = \frac{w_{DL} \times L_n^2}{10} = \frac{(1.15 \times 1) \times 25^2}{10} = 71.9 \text{ kip-ft}$$

$$M_{LL} = \frac{w_{LL} \times L_n^2}{10} = \frac{(0.92 \times 1) \times 25^2}{10} = 57.5 \text{ kip-ft}$$

Hence

$$M_n = \frac{1.4 M_{DL} + 1.7 M_{LL}}{\phi}$$

$$= \frac{1.4 \times 71.9 + 1.7 \times 57.5}{0.9} = 220 \text{ kip-ft}$$

The required section modulus is computed as usual:

$$\frac{M_n}{0.85 f'_c} = Z_c \quad \frac{220{,}000 \times 12}{0.85 \times 3000} = \text{coeff.} \times 12 \times 34^2$$

$$\text{coeff.} = 0.075$$

From Table A-5, select $\rho = 0.0035$.

Compute $A_s = \rho bd = 0.0035 \times 12 \times 34 = 1.428 \text{ in.}^2$

Since the maximum positive moment is less than the maximum negative moment, the minimum value of $\rho$ will be required both for positive and negative steel. Use 2 No. 8 bars, top and bottom continuous. Since the distance along the side of the beam between positive and negative reinforcement is more than 18 in., use extra No. 8 bars at mid-depth. The final design of the grade beam is shown in the following sketch.

It should be noted that this tall slender beam may be unstable, tending

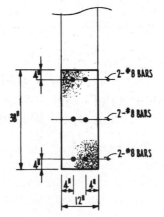

to "roll" under load. Bridging for grade beams is not specifically prescribed by Code, but the criteria for joists (also tall, slender beams) gives a clue to the bridging requirements. For spans 20 ft to 30 ft, extra bracing is required at midspan (i.e., bridging is required at column lines and at midspan). Consequently, bridging is provided for this grade beam at 13 ft on center, at column lines, and midspan. The stiffeners are selected as beams 12 in. wide × 20 in. deep ($d$ = 16 in.), with minimum ratio of steel, $\rho$ = 0.0035.

$A_s = \rho_{\min}bd = 0.0035 \times 12 \times 16 = 0.672$ in.$^2$
Use 2 No. 6 bars top and bottom.

---

## Other Types of Footings

Other types of shallow foundations in common use are usually only variations of the preceding types. A common type is the combined footing shown in Figure 14-8. The combined footing can have almost any reasonable shape as long as the resultant of column loads is collinear with the resultant of a uniform soil pressure.

Another common type of arrangement is the cantilever footing, also shown in Figure 14-8. It is frequently used where it is necessary to place a column so close to the property line that there is no way to keep the column concentric. It can also be a useful temporary foundation when one is underpinning a building for foundation repairs.

Foundations of unreinforced concrete or unreinforced masonry have fallen from use in recent years. They are, however, very satisfactory foundations when properly utilized. As suggested by the sketch in Figure 14-8, there is no tensile stress where the side slopes are kept within the natural shear planes.

Unreinforced foundations can be especially useful when the foundation is submerged in groundwater containing chlorides (salt). The chlorides can attack reinforcement but have little effect otherwise on the concrete. The use of unreinforced footings serves to eliminate the risk of corrosion of reinforcement.

All the footings shown in Figure 14-8 may be designed by the methods presented in the foregoing sections. A great deal of number juggling may be necessary in some cases to match the foundation to the loads. It should be recognized that the combined footing shown in Figure 14-8 acts as an

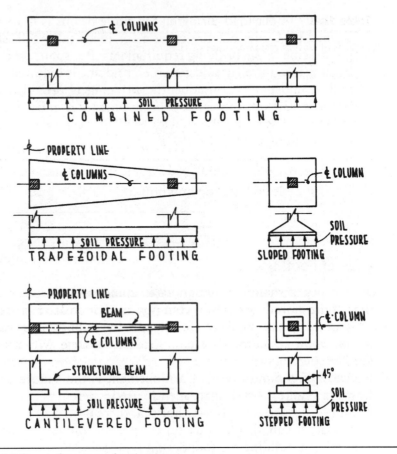

**Figure 14-8**     Common footing types.

elastic beam on an elastic foundation, the same way that a grade beam does.

## Slabs on Grade

The design of concrete slabs on grade is prescribed in ACI 302, Recommended Practice for Concrete Floor and Slab Construction (American Concrete Institute, 1980). The design method for the slab thickness is again empirical, based on wheel load, tire pressure, and concrete strength. A summary of some typical results is presented in Table 14-1.

**Table 14-1**        Minimum Slab Thicknesses*

| | |
|---|---|
| Light foot traffic: residential use or tile covered | 4 in. minimum |
| Foot traffic | |
|    Offices, churches, schools, hospitals | 4 in. minimum (5 in. preferred) |
|    Sidewalks in residential areas | 5 in. minimum |
| Automobile wheels: automobile driveways, garages | 5 in. minimum |
| Forklift and light truck loads; light industrial, commercial | 5 in. minimum |
| Forklift and medium truck loads: abrasive wear, industrial/commercial | 6 in. minimum |

*If trucks are to pass over isolation joints that have no provision for load transfer, such as doorways, the slab should be thickened approximately 50% and tapered to the required thickness at a slope no more than 1 in 10.

In addition to presenting a method of design of slabs on grade, ACI 302 also presents criteria for isolation joints, construction joints, and control joints. Typical isolation joints are shown in Figure 14-9. Isolation joints are used to isolate structural walls and columns from the floor slab such that "hard points" of lateral restraint are not created at the floor line.

Construction joints are placed in the floor slab to permit the casting to be interrupted and to be continued at a later time. When the interruption lasts long enough for the concrete to set, one of the construction joints of Figure 14-10 may be used. The joints shown in Figure 14-10 are taken from ACI 302; there are others in common use.

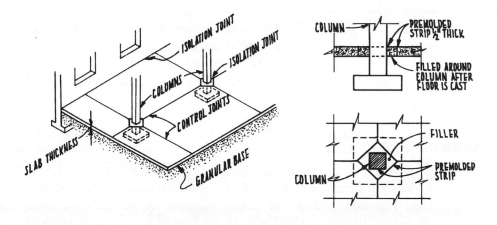

**Figure 14-9**    Isolation joints.

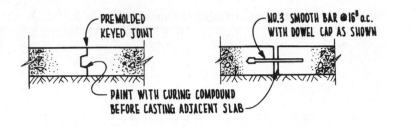

**Figure 14-10** Construction joints.

Control joints create stress raisers that serve to control the location of shrinkage and thermal cracking. Cracking still occurs, but it is much more likely to occur along a control joint than elsewhere. Typical control joints are shown in Figure 14-11. Some common practices in using control joints to control temperature and shrinkage cracking are given in Chapter 3.

Vapor barriers are also discussed in ACI 302. A vapor barrier is a plastic membrane placed under the floor slab, usually specified as polyethylene film. Rather elaborate precautions are required to protect the film from being punctured while the concrete is being cast.

As long as the vapor barrier is intact, it will presumably prevent "vapors" (whatever those are) from percolating upward through the slab. Assuming that it works, it will also prevent "vapors" (whatever those are) from percolating downward. ACI 302 suggests an alternative: A coarse granular fill about 4 in. thick will provide an effective water barrier under a slab.

Slabs on grade are one of the more common features of construction in the industry and are used in all types of buildings. Details such as placement of joints, embedment of pipes, isolation of ducts, reinforcement of edges, prevention of undercutting, minimization of snow and ice hazards,

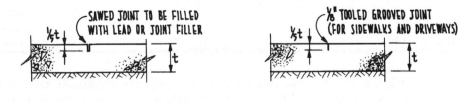

**Figure 14-11** Control joints.

and a wealth of other details are presented in ACI 302. A thorough knowledge of the accepted practices in ACI 302 is recommended to anyone designing in concrete.

## OUTSIDE PROBLEMS

Design an isolated spread footing for the given columns under the given limitations.

| Prob. No. | Column Size (in.) | Axial $P_{DL}$ (kips) | Load $P_{LL}$ (kips) | Soil Pressure (psf) | $f'_c$ (psi) | Steel Grade | Remarks |
|---|---|---|---|---|---|---|---|
| 14.1 | 12 × 12 | 81 | 54 | 4000 | 3000 | 40 | — |
| 14.2 | 10 × 10 | 16 | 17 | 2000 | 3000 | 40 | Property line 1.5 ft from centerline of column |
| 14.3 | 16 × 16 | 122 | 99 | 3000 | 4000 | 50 | — |
| 14.4 | 10 × 10 | 37 | 28 | 2000 | 4000 | 50 | Obstruction 2 ft from centerline column |
| 14.5 | 12 × 12 | 40 | 27 | 2900 | 3000 | 60 | Obstruction 1.5 ft from centerline column |
| 14.6 | 14 × 14 | 79 | 37 | 3000 | 4000 | 60 | — |
| 14.7 | 12 × 12 | 50 | 34 | 3000 | 4000 | 40 | — |
| 14.8 | 12 × 12 | 69 | 67 | 2000 | 3000 | 60 | Property line 3.5 ft from centerline of column |

Design a strip footing for the following wall loads.

| Prob. No. | Wall Size (in.) | Axial $P_{DL}$ (k/ft) | Load $P_{LL}$ (k/ft) | Soil Pressure (psf) | $f'_c$ (psi) | $f'_m$ (psi) | Steel Grade | Remarks |
|---|---|---|---|---|---|---|---|---|
| 14.9 | 8 | 14 | 12 | 4000 | 3000 | 1500 | 40 | Masonry wall |
| 14.10 | 12 | 7 | 9 | 2000 | 3000 | — | 50 | Concrete wall |
| 14.11 | 12 | 12 | 7 | 3000 | 4000 | — | 50 | Concrete wall |
| 14.12 | 16 | 15 | 12 | 3000 | 4000 | 1500 | 60 | Masonry wall |

Design a grade beam foundation for a repetitive column load as shown. The columns are square. The end column load may be assumed to be roughly half that of an interior column load.

| Prob. No. | Column Size (in.) | Bay Length (ft) | Axial $P_{DL}$ (kips) | Load $P_{LL}$ (kips) | Soil Pressure (psf) | $f'_c$ (psi) | Steel Grade |
|---|---|---|---|---|---|---|---|
| 14.13 | 12 | 20 | 32 | 36 | 3000 | 3000 | 40 |
| 14.14 | 12 | 16 | 24 | 32 | 5000 | 4000 | 50 |
| 14.15 | 10 | 25 | 14 | 20 | 4000 | 4000 | 40 |
| 14.16 | 16 | 30 | 46 | 46 | 5000 | 3000 | 60 |

## REVIEW QUESTIONS

1. Why is it so essential to keep a column concentric on its foundation?
2. Define "allowable soil pressure."
3. Why is the size of the contact area determined at service levels rather than at ultimate load?
4. Compare the design of a foundation where strength limits the allowable soil pressure to the design where settlements limit the allowable pressure.
5. For buildings having ordinary architectural functions and occupancies, how much of the live load can be expected to be in place long enough to produce foundation settlements?
6. Why don't large moments form at the base of a column on an isolated footing?
7. When a footing rotates due to deformations of the structure above, what is the effect on soil pressures under the footing?
8. What is the minimum cover over the reinforcement for concrete foundations cast directly against the soil? For concrete foundations cast against forms?
9. In a spread footing carrying a column load, where is the critical section for beam shear? For punching shear? For moment?
10. On a spread footing supporting a column, what is the limiting load that can be delivered to the footing by direct bearing of the column?
11. What are the load conditions to be checked when designing a spread footing for a column load?
12. What are the load conditions to be checked when designing a strip footing for a wall load?
13. Compute the ultimate average shear stress in Eq. (14-1) for a strip footing carrying a wall load, where the ultimate strength of concrete is 4000 psi.
14. How is a strip footing designed to support a masonry wall?
15. Where is the critical section for punching shear on a strip footing supporting a masonry wall?
16. Under what types and configurations of buildings are grade beams usually a feasible foundation?

17. Why does the soil pressure under a grade beam become sinusoidal?

18. Compared to the exacting design of structural steel or timber members, the design of concrete foundations is treated almost as a rought approximation. Why?

19. In soils where corrosion of reinforcement is recognized as a serious hazard, how can a concrete foundation be designed for maximum durability at minimum risk?

20. How thick should a concrete slab be made when it is used as a parking hard stand for an ordinary passenger car?

21. Why are isolation joints sometimes necessary where columns penetrate a floor slab?

22. How may capillarity be interrupted under a floor slab?

# SELECTED
# REFERENCES

1. American Concrete Institute, *Reinforced Concrete Design Handbook*, Publication SP-3. Detroit: ACI, 1969.
2. American Concrete Institute, *Recommended Practice for Concrete Floor and Slab Construction*, ACI-302. Detroit: ACI, 1980.
3. American Concrete Institute, *Building Code Requirements for Reinforced Concrete*, ACI 318-89. Detroit: ACI, 1989.
4. American Concrete Institute, *Recommendations for Design of Beam-Column Joints in Monolithic Reinforced Concrete Structures*, ACI-352. Detroit: ACI, 1985.
5. American Institute of Steel Construction, *Steel Construction Manual*. New York: AISC, 1989.
6. Concrete Reinforcing Steel Institute, CRSI *Handbook*. Chicago: CRSI, 1989.
7. Concrete Reinforcing Steel Institute, CRSI *Manual of Standard Practice*. Chicago: CRSI, 1989.
8. Corps of Engineers, *Handbook for Concrete and Cement*. Vicksburg, Miss.: U.S. Army Waterways Experiment Station, 1986.
9. Ferguson, P.M., *Reinforced Concrete Fundamentals*. New York: Wiley, 1958.
10. Peck, et al. *Foundation Engineering*, 1st ed.: Wiley, 1953.
11. UNESCO, *Reinforced Concrete: An International Manual*. London: Butterworth, 1971.
12. U.S. Bureau of Reclamation, *Concrete Manual*, 8th ed. Denver, Colo.: USBR, 1975.

13. Whitney, C.S., and E. Cohen, "Guide for Ultimate Strength Design of Reinforced Concrete." ACI *Journal, Proceedings*, 53, November 1956.
14. Portland Cement Association, *Design and Control of Concrete Mixes*. Chicago: PCA, 1979.
15. Rogers, P., "Simplified Method of Stirrup Spacing," *Concrete International*, January 1979.
16. Taylor, H.P.J., "The Fundamental Behavior of Reinforced Concrete Beams in Bending and Shear," *Shear in Reinforced Concrete*, Vol. 1 (SP-42). Detroit: American Concrete Institute, 1974.

# A

# DESIGN TABLES

The tables given in this Appendix are adequate for the design problems introduced in this book. For more detailed analysis or for larger projects, the use of more comprehensive design aids is recommended. The tables herein have been reduced as much as possible where such reduction would not incur an undue amount of interpolation.

The beam tables (Tables A-5 through A-7) and the column tables (Tables A-9 through A-11) are dimensionless and homogeneous. The coefficients are therefore the same for both SI units and Imperial units. The remaining tables are in Imperial units; their counterparts in SI units may be found in numerous other handbooks and pamphlets.

The development of the tables is described in the corresponding text presented in earlier chapters of the book. The equations presented therein are complete and may be used to expand or enlarge these tables. General agreement with standard ACI tables has been noted in the relevant discussions in the text.

The column tables are somewhat at variance with the ACI interaction curves where values of $e/t$ are higher than those corresponding to a "balanced" design. The maximum variation has been found to be less than 8%, which is considered to be acceptable for calculations in concrete. Elsewhere, the tables coincide exactly with the ACI interaction curves.

The tables of development lengths, Tables A-12 and A-13, are sharply limited. For larger sizes of bars, for wire fabric, for unusual conditions of cover or imbedment, or under any circumstances where these tables are not exactly applicable, a more comprehensive set of tables should be consulted. For some circumstances, the development of bond prescribed by ACI 318-63 may offer some insight; that method is briefly discussed toward the end of Chapter 12.

**Table A-1**  Allowable Working Stresses in Concrete

| Type of Stress Normal Weight Concrete: 145 pcf | ACI 318-89 | Strength in psi | | |
|---|---|---|---|---|
| | | 3000 | 4000 | 5000 |
| Modulus of Elasticity, $E_c$ (in psi) | $57000\sqrt{f_c'}$ | $3.1 \times 10^6$ | $3.6 \times 10^6$ | $4.0 \times 10^6$ |
| Modular ratio $n = E_s/E_c$ | $509/\sqrt{f_c'}$ | 9.29 | 8.04 | 7.20 |
| Flexure Stress, $f_c$ | | | | |
| Extreme compression fiber | $0.45f_c'$ | 1350 | 1800 | 2250 |
| Shear Stress, $v$ | | | | |
| Beams, walls, one-way slabs | $1.1\sqrt{f_c'}$ | 60 | 70 | 78 |
| Joist floors, stems | $1.2\sqrt{f_c'}$ | 67 | 76 | 85 |
| [1]Two-way slabs and footings | $(1 + 2/\beta_c)\sqrt{f_c'}$ but $\leq 2\sqrt{f_c'}$ | | | |
| Bearing Stress, $f_c$ | | | | |
| On full area | $0.3f_c'$ | 900 | 1200 | 1500 |
| [2]Less than full area (all sides) | $0.3f_c'\sqrt{A_2/A_1}$ but $\leq 0.6f_c'$ | | | |

[1]$\beta_c$ is the ratio of long side to short side of the concentrated load or reaction area.
[2]$A_1$ is the loaded area.
$A_2$ is the maximum area of the portion of the supporting surface that is geometrically similar to and concentric with the loaded area.

Allowable Working Stresses in Reinforcement

| | |
|---|---|
| Modulus of Elasticity of All Steel: | $E_s = 29,000,000$ psi |
| Tensile and Compressive Stresses: | |
| Grade 40 steel ($f_y = 40,000$ psi) | $f_s = 20,000$ psi |
| Grade 50 steel ($f_y = 50,000$ psi) | $f_s = 20,000$ psi |
| Grade 60 steel ($f_y = 60,000$ psi) | $f_s = 22,000$ psi |

**Table A–2**  Area of Steel (in.²) in a Section 1 Ft. Wide

| Spacing (in.) | #3 | #4 | #5 | #6 | #7 | #8 | #9 | #10 | #11 |
|---|---|---|---|---|---|---|---|---|---|
| 2 | 0.66 | 1.18 | 1.84 | 2.65 | 3.61 | 4.71 | | | |
| 2½ | 0.53 | 0.94 | 1.47 | 2.12 | 2.89 | 3.77 | 4.80 | | |
| 3 | 0.44 | 0.79 | 1.23 | 1.77 | 2.41 | 3.14 | 4.00 | 5.07 | 6.25 |
| 3½ | 0.38 | 0.67 | 1.05 | 1.51 | 2.06 | 2.69 | 3.43 | 4.34 | 5.35 |
| 4 | 0.33 | 0.59 | 0.92 | 1.33 | 1.80 | 2.36 | 3.00 | 3.80 | 4.68 |
| 4½ | 0.29 | 0.52 | 0.82 | 1.18 | 1.60 | 2.09 | 2.66 | 3.38 | 4.16 |
| 5 | 0.27 | 0.47 | 0.74 | 1.06 | 1.44 | 1.88 | 2.40 | 3.04 | 3.75 |
| 5½ | 0.24 | 0.43 | 0.67 | 0.96 | 1.31 | 1.71 | 2.18 | 2.76 | 3.41 |
| 6 | 0.22 | 0.39 | 0.61 | 0.88 | 1.20 | 1.57 | 2.00 | 2.53 | 3.12 |
| 6½ | 0.20 | 0.36 | 0.57 | 0.82 | 1.11 | 1.45 | 1.84 | 2.34 | 2.88 |
| 7 | 0.19 | 0.34 | 0.53 | 0.76 | 1.03 | 1.35 | 1.71 | 2.17 | 2.68 |
| 7½ | 0.18 | 0.31 | 0.49 | 0.71 | 0.96 | 1.26 | 1.60 | 2.03 | 2.50 |
| 8 | 0.17 | 0.29 | 0.46 | 0.66 | 0.90 | 1.18 | 1.50 | 1.90 | 2.34 |
| 9 | 0.15 | 0.26 | 0.41 | 0.59 | 0.80 | 1.05 | 1.33 | 1.69 | 2.08 |
| 10 | 0.13 | 0.24 | 0.37 | 0.53 | 0.72 | 0.94 | 1.20 | 1.52 | 1.87 |
| 11 | 0.12 | 0.21 | 0.33 | 0.48 | 0.66 | 0.86 | 1.09 | 1.38 | 1.70 |
| 12 | 0.11 | 0.20 | 0.31 | 0.44 | 0.60 | 0.79 | 1.00 | 1.27 | 1.56 |
| 13 | 0.10 | 0.18 | 0.28 | 0.41 | 0.56 | 0.72 | 0.92 | 1.17 | 1.44 |
| 14 | 0.09 | 0.17 | 0.26 | 0.38 | 0.52 | 0.67 | 0.86 | 1.09 | 1.34 |
| 15 | 0.09 | 0.16 | 0.25 | 0.35 | 0.48 | 0.63 | 0.80 | 1.01 | 1.25 |
| 16 | 0.08 | 0.15 | 0.23 | 0.33 | 0.45 | 0.59 | 0.75 | 0.95 | 1.17 |
| 17 | 0.08 | 0.14 | 0.22 | 0.31 | 0.42 | 0.55 | 0.71 | 0.89 | 1.10 |
| 18 | 0.07 | 0.13 | 0.20 | 0.29 | 0.40 | 0.52 | 0.67 | 0.84 | 1.04 |

## Properties of Reinforcing Bars

| Bar Size | #3 | #4 | #5 | #6 | #7 | #8 | #9 | #10 | #11 |
|---|---|---|---|---|---|---|---|---|---|
| Diameter, in. | 0.375 | 0.500 | 0.625 | 0.750 | 0.875 | 1.000 | 1.125 | 1.270 | 1.410 |
| Area, in.² | 0.11 | 0.20 | 0.31 | 0.44 | 0.60 | 0.79 | 1.00 | 1.27 | 1.56 |
| Perimeter, in. | 1.18 | 1.57 | 1.96 | 2.36 | 2.75 | 3.14 | 3.55 | 3.99 | 4.43 |
| Weight, lb/ft | 0.38 | 0.67 | 1.04 | 1.50 | 2.04 | 2.67 | 3.40 | 4.30 | 5.31 |

**Table A–3**  Steel Area in Square Inches for Combinations of Bar Sizes

**#3**

| No. | Area |
|-----|------|
| 1 | 0.11 |
| 2 | 0.22 |
| 3 | 0.33 |
| 4 | 0.44 |
| 5 | 0.55 |
| 6 | 0.66 |

**#4**

| No. | Area |
|-----|------|
| 1 | 0.20 |
| 2 | 0.39 |
| 3 | 0.59 |
| 4 | 0.79 |
| 5 | 0.98 |
| 6 | 1.18 |

**#5**

| | | #5 plus #4 — Number of Bars | | | |
|-----|------|------|------|------|------|
| No. | Area | 1 | 2 | 3 | 4 |
| 1 | 0.31 | 0.50 | 0.70 | 0.90 | 1.09 |
| 2 | 0.61 | 0.81 | 1.01 | 1.20 | 1.40 |
| 3 | 0.92 | 1.12 | 1.31 | 1.51 | 1.71 |
| 4 | 1.23 | 1.42 | 1.62 | 1.82 | 2.01 |
| 5 | 1.53 | 1.73 | 1.93 | 2.12 | 2.32 |
| 6 | 1.84 | 2.04 | 2.23 | 2.43 | 2.63 |

**#6**

| | | #6 plus #5 — Number of Bars | | | | #6 plus #4 — Number of Bars | | | |
|-----|------|------|------|------|------|------|------|------|------|
| No. | Area | 1 | 2 | 3 | 4 | 1 | 2 | 3 | 4 |
| 1 | 0.44 | 0.75 | 1.06 | 1.36 | 1.67 | 0.64 | 0.83 | 1.03 | 1.23 |
| 2 | 0.88 | 1.19 | 1.50 | 1.80 | 2.11 | 1.08 | 1.28 | 1.47 | 1.67 |
| 3 | 1.33 | 1.63 | 1.94 | 2.25 | 2.55 | 1.52 | 1.72 | 1.91 | 2.11 |
| 4 | 1.77 | 2.07 | 2.38 | 2.69 | 2.99 | 1.96 | 2.16 | 2.36 | 2.55 |
| 5 | 2.21 | 2.52 | 2.82 | 3.13 | 3.44 | 2.14 | 2.60 | 2.80 | 2.99 |
| 6 | 2.65 | 2.96 | 3.26 | 3.57 | 3.88 | 2.85 | 3.04 | 3.24 | 3.44 |

## #7

| # | #7 | #7 plus #6 | | | | #7 plus #5 | | | | #7 plus #4 | | | |
|---|------|------|------|------|------|------|------|------|------|------|------|------|------|
| 1 | 0.60 | 1.04 | 1.48 | 1.93 | 2.37 | 0.80 | 1.21 | 1.52 | 1.83 | 0.91 | 0.99 | 1.19 | 1.39 |
| 2 | 1.20 | 1.64 | 2.09 | 2.53 | 2.97 | 1.40 | 1.82 | 2.12 | 2.43 | 1.51 | 1.60 | 1.79 | 1.99 |
| 3 | 1.80 | 2.25 | 2.69 | 3.13 | 3.57 | 2.00 | 2.42 | 2.72 | 3.03 | 2.11 | 2.20 | 2.39 | 2.59 |
| 4 | 2.41 | 2.85 | .329 | 3.73 | 4.17 | 2.60 | 3.02 | 3.33 | 3.63 | 2.71 | 2.80 | 2.99 | 3.19 |
| 5 | 3.01 | 3.45 | 3.89 | 4.33 | .477 | 3.20 | 3.62 | 3.93 | 4.23 | 3.31 | 3.40 | 3.60 | 3.79 |
| 6 | 3.61 | 4.05 | 4.49 | 4.93 | 5.38 | 3.80 | 4.22 | 4.53 | 4.84 | 3.91 | 4.00 | 4.20 | 4.39 |

## #8

| # | #8 | #8 plus #7 | | | | #8 plus #6 | | | | #8 plus #5 | | | |
|---|------|------|------|------|------|------|------|------|------|------|------|------|------|
| 1 | 0.79 | 1.39 | 1.99 | 2.59 | 3.19 | 1.09 | 1.67 | 2.11 | 2.55 | 1.23 | 1.40 | 1.71 | 2.01 |
| 2 | 1.57 | 2.17 | 2.77 | 3.37 | 3.98 | 1.88 | 2.45 | 2.90 | 3.34 | 2.01 | 2.18 | 2.49 | 2.80 |
| 3 | 2.36 | 2.96 | 3.56 | 4.16 | 4.76 | 2.66 | 3.24 | 3.68 | 4.12 | 2.80 | 2.97 | 3.28 | 3.58 |
| 4 | 3.14 | 3.74 | 4.34 | 4.95 | 5.55 | 3.45 | 4.03 | 4.47 | 4.91 | 3.58 | 3.76 | 4.06 | 4.37 |
| 5 | 3.93 | 4.53 | 5.13 | 5.73 | 6.33 | 4.23 | 4.81 | 5.25 | 5.69 | 4.37 | 4.54 | 4.85 | 5.15 |
| 6 | 4.71 | 5.31 | 5.92 | 6.52 | 7.12 | 5.02 | 5.60 | 6.04 | 6.48 | 5.15 | 5.33 | 5.63 | 5.94 |

## #9

| # | #9 | #9 plus #8 | | | | #9 plus #7 | | | | #9 plus #6 | | |
|---|------|------|------|------|------|------|------|------|------|------|------|------|
| 1 | 1.00 | 1.79 | 2.57 | 3.36 | 4.14 | 1.44 | 1.60 | 2.20 | 2.80 | 1.88 | 2.33 | 2.77 |
| 2 | 2.00 | 2.79 | 3.57 | 4.36 | 5.14 | 2.44 | 2.60 | 3.20 | 3.80 | 2.88 | 3.33 | 3.77 |
| 3 | 3.00 | 3.79 | 4.57 | 5.36 | 6.14 | 3.44 | 3.60 | 4.20 | 4.80 | 3.88 | 4.33 | 4.77 |
| 4 | 4.00 | 4.79 | 5.57 | 6.36 | 7.14 | 4.44 | 4.60 | 5.20 | 5.80 | 4.88 | 5.33 | 5.77 |
| 5 | 5.00 | 5.79 | 6.57 | 7.36 | 8.14 | 5.44 | 5.60 | 6.20 | 6.80 | 5.88 | 6.33 | 6.77 |
| 6 | 6.00 | 6.79 | 7.57 | 8.36 | 9.14 | 6.44 | 6.60 | 7.20 | 7.80 | 6.88 | 7.33 | 7.77 |

## #10

| # | #10 | #10 plus #9 | | | | #10 plus #8 | | | | #10 plus #7 | | |
|---|------|------|------|------|------|------|------|------|------|------|------|------|
| 1 | 1.27 | 2.27 | 3.27 | 4.27 | 5.27 | 1.87 | 2.84 | 3.62 | 4.41 | 2.47 | 3.07 | 3.67 |
| 2 | 2.53 | 3.53 | 4.53 | 5.53 | 6.53 | 3.13 | 4.10 | 4.89 | 5.68 | 3.74 | 4.34 | 4.94 |
| 3 | 3.80 | 4.80 | 5.80 | 6.80 | 7.80 | 4.40 | 5.37 | 6.16 | 6.94 | 5.00 | 5.60 | 6.21 |
| 4 | 5.07 | 6.07 | 7.07 | 8.07 | 9.07 | 5.67 | 6.64 | 7.42 | 8.21 | 6.27 | 6.87 | 7.47 |
| 5 | 6.33 | 7.33 | 8.33 | 9.33 | 10.3 | 6.94 | 7.90 | 8.69 | 9.48 | 7.54 | 8.14 | 8.74 |
| 6 | 7.60 | 8.60 | 9.60 | 10.6 | 11.6 | 8.20 | 9.17 | 9.96 | 10.7 | 8.80 | 9.40 | 10.0 |

## #11

| # | #11 | #11 plus #10 | | | | #11 plus #9 | | | | #11 plus #8 | | |
|---|------|------|------|------|------|------|------|------|------|------|------|------|
| 1 | 1.56 | 2.83 | 4.09 | 5.36 | 6.63 | 2.35 | 3.56 | 4.56 | 5.56 | 3.13 | 3.92 | 4.70 |
| 2 | 3.12 | 4.39 | 5.66 | 6.92 | 8.19 | 3.91 | 5.12 | 6.12 | 7.12 | 4.69 | 5.48 | 6.26 |
| 3 | 4.68 | 5.95 | 7.22 | 8.48 | 9.75 | 5.47 | 6.68 | 7.68 | 8.68 | 6.26 | 7.04 | 7.83 |
| 4 | 6.25 | 7.51 | 8.78 | 10.1 | 11.3 | 7.03 | 8.25 | 9.25 | 10.3 | 7.82 | 8.60 | 9.39 |
| 5 | 7.81 | 9.07 | 10.3 | 11.6 | 12.9 | 8.59 | 9.81 | 10.8 | 11.8 | 9.38 | 10.2 | 11.0 |
| 6 | 9.37 | 10.6 | 11.9 | 13.2 | 14.4 | 10.2 | 11.4 | 12.4 | 13.4 | 10.9 | 11.7 | 12.5 |

**Table A–4**   Design Guidelines. Minimum Width of Beams or Stems at Interior Exposures[1] (Maximum aggregate size ¾ in.)

| Bar Size | With #3 Stirrups[2] Number of Bars in a Single Layer | | | | | | | With No Stirrups No. of Bars | |
|---|---|---|---|---|---|---|---|---|---|
| | 2 | 3 | 4 | 5 | 6 | 7 | 8 | 2 | 3 |
| #4 | 6.1 | 7.6 | 9.1 | 10.6 | 12.1 | 13.6 | 15.1 | 5.00 | 6.50 |
| #5 | 6.3 | 7.9 | 9.6 | 11.2 | 12.8 | 14.4 | 16.1 | 5.25 | 6.88 |
| #6 | 6.5 | 8.3 | 10.0 | 11.8 | 13.5 | 15.3 | 17.0 | 5.50 | 7.25 |
| #7 | 6.7 | 8.6 | 10.5 | 12.4 | 14.2 | 16.1 | 18.0 | 5.75 | 7.63 |
| #8 | 6.9 | 8.9 | 10.9 | 12.9 | 14.9 | 16.9 | 18.9 | 6.00 | 8.00 |
| #9 | 7.3 | 9.5 | 11.8 | 14.0 | 16.3 | 18.6 | 20.8 | 6.25 | 8.38 |
| #10 | 7.7 | 10.2 | 12.8 | 15.3 | 17.8 | 20.4 | 22.9 | 6.50 | 8.75 |
| #11 | 8.0 | 10.8 | 13.7 | 16.5 | 19.3 | 22.1 | 24.9 | 6.75 | 9.13 |

[1]For exterior exposures, bars #6 and larger, add ¾ inch.
[2]For #4 stirrups, add ¾ inch; for #5 stirrups add 1½ inch.

**Clearances at Interior Exposures**

In exterior exposures, for bar sizes #6 and larger, increase the minimum clear concrete cover to 2 inches.

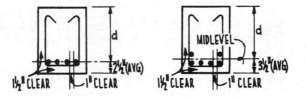

**General Guidelines:**

1. For rectangular beams, width of the beam should be between $0.5d$ and $0.6d$ if there are no constraints.
2. For tee beams, the depth $d$ should be about 5 times the slab thickness and the width $b$ should be about $0.5d$.
3. Increments of overall dimensions for beams should not be less than ½ inch for sizes up to 18 inches and 1 inch for sizes above 18 inches.
4. Increments of overall slab thickness should not be less than ½ inch with a minimum slab thickness of 3½ inches.

5.  Minimum clear cover over reinforcement is $1\frac{1}{2}$ inches, measured from outside of stirrups.
6.  No fewer than 2 reinforcing bars should be used as longitudinal reinforcement in a beam.
7.  Reinforcing bars should be arranged symmetrically about the vertical axis of a beam.
8.  Bar sizes greater than #11 should not be used in beams and no more than two bar sizes should be used in a beam.
9.  Bars should be placed in one layer if possible; if bars of two sizes are placed in multiple layers, the heavier sizes should be placed closest to the outer face.

**Table A–5**  Coefficients for Section Constants. $f'_c = 3000$ psi $(20.6 \text{ N/mm}^2)$; $g = 0.125$, $\beta_1 = 0.85$, $n = 9.29$

| | At Ultimate Loads | | | | | | | | | | | | At Service Loads | | | |
|---|---|---|---|---|---|---|---|---|---|---|---|---|---|---|---|---|
| | Grade 40 Steel $f_y = 40000$ psi | | | | Grade 50 Steel $f_y = 50000$ psi | | | | Grade 60 Steel $f_y = 60000$ psi | | | | Elastic Constants for All Steels | | | |
| $A_s/bd$ | $Z_c$ | $\beta_1 k_n$ | $y_n$ | $f_{sv}$ | $Z_c$ | $\beta_1 k_n$ | $y_n$ | $f_{sv}$ | $Z_c$ | $\beta_1 k_n$ | $y_n$ | $f_{sv}$ | $S_c$ | $k_{sv}$ | $I_{cr}$ | $f_s/f_c$ |
| Compressive Steel Area 0% of Tensile Steel Area | | | | | | | | | | | | | | | | |
| 0.0005 | 0.008 | .008 | .559 | 263 | 0.010 | .010 | .558 | 329 | 0.012 | .012 | .557 | 394 | .045 | .092 | .004 | 91.85 |
| 0.0010 | 0.016 | .016 | .555 | 383 | 0.019 | .020 | .553 | 478 | 0.023 | .024 | .551 | 572 | .061 | .127 | .008 | 63.66 |
| 0.0015 | 0.023 | .024 | .551 | 479 | 0.029 | .029 | .548 | 597 | 0.035 | .035 | .545 | 714 | .073 | .154 | .011 | 51.19 |
| 0.0020 | 0.031 | .031 | .547 | 562 | 0.038 | .039 | .543 | 700 | 0.046 | .047 | .539 | 836 | .082 | .175 | .014 | 43.77 |
| 0.0025 | 0.038 | .039 | .543 | 637 | 0.048 | .049 | .538 | 792 | 0.057 | .059 | .533 | 946 | .091 | .194 | .018 | 38.71 |
| 0.0030 | 0.046 | .047 | .539 | 706 | 0.057 | .059 | .533 | 878 | 0.068 | .071 | .527 | 1047 | .098 | .210 | .020 | 34.98 |
| 0.0035 | 0.053 | .055 | .535 | 771 | 0.066 | .069 | .528 | 957 | 0.079 | .082 | .521 | 1140 | .104 | .225 | .023 | 32.08 |
| 0.0040 | 0.061 | .063 | .531 | 832 | 0.075 | .078 | .523 | 1032 | 0.090 | .094 | .515 | 1228 | .110 | .238 | .026 | 29.75 |
| 0.0050 | 0.075 | .078 | .523 | 946 | 0.093 | .098 | .513 | 1170 | 0.111 | .118 | .504 | 1390 | .119 | .262 | .031 | 26.19 |
| 0.0060 | 0.090 | .094 | .515 | 1051 | 0.111 | .118 | .504 | 1297 | 0.131 | .141 | .492 | 1537 | .128 | .283 | .036 | 23.56 |
| 0.0070 | 0.104 | .110 | .508 | 1148 | 0.128 | .137 | .494 | 1414 | 0.151 | .165 | .480 | 1672 | .136 | .301 | .041 | 21.53 |
| 0.0080 | 0.118 | .125 | .500 | 1240 | 0.145 | .157 | .484 | 1524 | 0.171 | .188 | .468 | 1798 | .142 | .318 | .045 | 19.89 |
| 0.0090 | 0.131 | .141 | .492 | 1327 | 0.161 | .176 | .474 | 1627 | 0.189 | .212 | .457 | 1915 | .148 | .334 | .050 | 18.54 |
| 0.0100 | 0.145 | .157 | .484 | 1410 | 0.177 | .196 | .464 | 1725 | 0.208 | .235 | .445 | 2024 | .154 | .348 | .054 | 17.40 |
| 0.0120 | 0.171 | .188 | .468 | 1564 | 0.208 | .235 | .445 | 1904 | 0.242 | .282 | .421 | 2224 | .164 | .374 | .061 | 15.57 |
| 0.0140 | 0.195 | .220 | .453 | 1705 | 0.237 | .275 | .425 | 2066 | 0.275 | .329 | .398 | 2400 | .172 | .396 | .068 | 14.15 |
| 0.0160 | 0.219 | .251 | .437 | 1836 | 0.265 | .314 | .406 | 2213 | 0.306 | .376 | .374 | 2556 | .179 | .416 | .075 | 13.01 |
| 0.0180 | 0.242 | .282 | .421 | 1957 | 0.291 | .353 | .386 | 2346 | | | | | .186 | .435 | .081 | 12.08 |
| 0.0200 | 0.265 | .314 | .406 | 2069 | 0.315 | .392 | .366 | 2466 | | | | | .192 | .451 | .087 | 11.29 |
| 0.0220 | 0.286 | .345 | .390 | 2173 | 0.338 | .431 | .347 | 2575 | | | | | .197 | .467 | .092 | 10.61 |
| 0.0240 | 0.306 | .376 | .374 | 2270 | | | | | | | | | .202 | .481 | .097 | 10.02 |
| 0.0260 | 0.325 | .408 | .359 | 2359 | | | | | | | | | .206 | .494 | .102 | 9.50 |
| 0.0280 | 0.343 | .439 | .343 | 2442 | | | | | | | | | .211 | .507 | .107 | 9.05 |
| | $0.75\,\rho_b = 0.0278$ | | | | $0.75\,\rho_b = 0.0206$ | | | | $0.75\,\rho_b = 0.0160$ | | | | | | | |
| Compressive Steel Area 20% of Tensile Steel Area | | | | | | | | | | | | | | | | |
| 0.0035 | | | | | | | | | 0.079 | .087 | .519 | 1114 | .107 | .220 | .024 | 32.94 |
| 0.0040 | | | | | 0.076 | .085 | .520 | 1005 | 0.090 | .097 | .514 | 1193 | .113 | .232 | .026 | 30.68 |
| 0.0050 | 0.076 | .086 | .519 | 914 | 0.093 | .100 | .512 | 1126 | 0.111 | .115 | .505 | 1337 | .124 | .254 | .032 | 27.23 |
| 0.0060 | 0.090 | .098 | .514 | 1003 | 0.111 | .115 | .505 | 1237 | 0.131 | .133 | .496 | 1467 | .134 | .273 | .037 | 24.69 |
| 0.0070 | 0.104 | .109 | .508 | 1086 | 0.128 | .129 | .498 | 1339 | 0.152 | .151 | .487 | 1587 | .143 | .290 | .042 | 22.73 |
| 0.0080 | 0.118 | .119 | .503 | 1164 | 0.145 | .143 | .491 | 1434 | 0.172 | .168 | .478 | 1698 | .152 | .305 | .047 | 21.15 |
| 0.0090 | 0.131 | .130 | .498 | 1237 | 0.162 | .157 | .484 | 1523 | 0.191 | .186 | .470 | 1801 | .159 | .319 | 0.51 | 19.85 |
| 0.0100 | 0.145 | .140 | .492 | 1306 | 0.178 | .170 | .477 | 1607 | 0.211 | .203 | .461 | 1898 | .166 | .331 | .056 | 18.75 |
| 0.0120 | 0.172 | .161 | .482 | 1434 | 0.211 | .197 | .464 | 1761 | 0.249 | .237 | .444 | 2075 | .180 | .353 | .064 | 17.00 |
| 0.0140 | 0.198 | .180 | .472 | 1552 | 0.243 | .224 | .450 | 1901 | 0.285 | .271 | .427 | 2232 | .192 | .373 | .072 | 15.64 |
| 0.0160 | 0.225 | .200 | .462 | 1660 | 0.274 | .251 | .437 | 2029 | 0.321 | .305 | .410 | 2373 | .203 | .390 | .080 | 14.56 |
| 0.0180 | 0.250 | .219 | .450 | 1757 | 0.304 | .277 | .421 | 2140 | 0.355 | .339 | .393 | 2500 | .213 | .405 | .087 | 13.67 |
| 0.0200 | 0.274 | .238 | .437 | 1845 | 0.333 | .303 | .406 | 2240 | 0.388 | .373 | .374 | 2608 | .223 | .418 | .094 | 12.92 |
| 0.0220 | 0.298 | .257 | .424 | 1925 | 0.361 | .330 | .390 | 2330 | | | | | .232 | .431 | .100 | 12.28 |
| 0.0240 | 0.322 | .276 | .412 | 1999 | 0.388 | .356 | .374 | 2411 | | | | | .241 | .442 | .107 | 11.73 |
| 0.0260 | 0.344 | .294 | .399 | 2067 | 0.414 | .382 | .359 | 2484 | | | | | .250 | .452 | .113 | 11.24 |
| 0.0280 | 0.367 | .313 | .387 | 2129 | | | | | | | | | .258 | .462 | .119 | 10.82 |
| 0.0320 | 0.409 | .350 | .362 | 2237 | | | | | | | | | .274 | .479 | .130 | 10.09 |
| 0.0360 | 0.449 | .387 | .337 | 2326 | | | | | | | | | .289 | .495 | .141 | 9.50 |

## Table A–5    Continued

| | At Ultimate Loads | | | | | | | | | | | | At Service Loads | | | |
| | Grade 40 Steel $f_y$ = 40000 psi | | | | Grade 50 Steel $f_y$ = 50000 psi | | | | Grade 60 Steel $f_y$ = 60000 psi | | | | Elastic Constants for All Steels | | | |
| $A_s/bd$ | $Z_c$ | $\beta_1 k_n$ | $y_n$ | $f_{sv}$ | $Z_c$ | $\beta_1 k_n$ | $y_n$ | $f_{sv}$ | $Z_c$ | $\beta_1 k_n$ | $y_n$ | $f_{sv}$ | $S_c$ | $k_{sv}$ | $I_{cr}$ | $f_s/f_c$ |
|---|---|---|---|---|---|---|---|---|---|---|---|---|---|---|---|---|
| | | | | | Compressive Steel Area 40% of Tensile Steel Area | | | | | | | | | | | |
| 0.0035 | | | | | | | | | 0.079 | .091 | .517 | 1089 | .109 | .216 | .024 | 33.77 |
| 0.0040 | | | | | 0.076 | .089 | .518 | 979 | 0.090 | .098 | .513 | 1160 | .116 | .227 | .027 | 31.58 |
| 0.0050 | 0.076 | .090 | .517 | 883 | 0.093 | .101 | .512 | 1087 | 0.111 | .113 | .506 | 1289 | .129 | .248 | .033 | 28.24 |
| 0.0060 | 0.090 | .100 | .513 | 961 | 0.111 | .113 | .506 | 1184 | 0.131 | .128 | .499 | 1405 | .140 | .265 | .038 | 25.78 |
| 0.0070 | 0.104 | .108 | .508 | 1032 | 0.128 | .124 | .501 | 1273 | 0.152 | .141 | .492 | 1510 | .151 | .280 | .043 | 23.89 |
| 0.0080 | 0.118 | .116 | .504 | 1098 | 0.145 | .134 | .495 | 1355 | 0.172 | .154 | .485 | 1607 | .161 | .293 | .048 | 22.37 |
| 0.0090 | 0.131 | .124 | .501 | 1160 | 0.162 | .144 | .490 | 1431 | 0.192 | .167 | .479 | 1697 | .170 | .305 | .053 | 21.12 |
| 0.0100 | 0.145 | .131 | .497 | 1217 | 0.179 | .154 | .486 | 1502 | 0.212 | .180 | .473 | 1781 | .179 | .316 | .058 | 20.07 |
| 0.0120 | 0.172 | .145 | .490 | 1323 | 0.213 | .172 | .476 | 1632 | 0.252 | .204 | .461 | 1933 | .195 | .336 | .067 | 18.39 |
| 0.0140 | 0.199 | .157 | .484 | 1419 | 0.246 | .190 | .467 | 1749 | 0.291 | .228 | .449 | 2068 | .211 | .352 | .076 | 17.10 |
| 0.0160 | 0.226 | .170 | .478 | 1506 | 0.279 | .207 | .459 | 1855 | 0.329 | .251 | .437 | 2190 | .226 | .366 | .084 | 16.08 |
| 0.0180 | 0.253 | .181 | .472 | 1585 | 0.312 | .224 | .451 | 1951 | 0.367 | .273 | .426 | 2299 | .240 | .379 | .093 | 15.23 |
| 0.0200 | 0.280 | .192 | .467 | 1659 | 0.344 | .240 | .442 | 2039 | 0.405 | .296 | .415 | 2399 | .253 | .390 | .101 | 14.53 |
| 0.0220 | 0.306 | .202 | .459 | 1725 | 0.376 | .256 | .433 | 2119 | 0.442 | .318 | .404 | 2489 | .266 | .400 | .109 | 13.93 |
| 0.0240 | 0.332 | .213 | .450 | 1784 | 0.407 | .271 | .421 | 2187 | 0.479 | .340 | .393 | 2572 | .279 | .409 | .116 | 13.41 |
| 0.0260 | 0.358 | .222 | .440 | 1838 | 0.438 | .287 | .410 | 2249 | 0.514 | .361 | .379 | 2641 | .292 | .418 | .124 | 12.96 |
| 0.0280 | 0.383 | .232 | .431 | 1887 | 0.467 | .302 | .398 | 2305 | | | | | .304 | .425 | .131 | 12.56 |
| 0.0320 | 0.432 | .250 | .412 | 1973 | 0.515 | .331 | .374 | 2401 | | | | | .328 | .439 | .145 | 11.88 |
| 0.0360 | 0.479 | .268 | .393 | 2045 | 0.581 | .360 | .351 | 2480 | | | | | .351 | .450 | .159 | 11.34 |
| 0.0400 | 0.525 | .285 | .374 | 2105 | | | | | | | | | .374 | .461 | .173 | 10.88 |
| 0.0440 | 0.570 | .301 | .355 | 2155 | | | | | | | | | .397 | .470 | .186 | 10.49 |
| 0.0480 | 0.613 | .317 | .337 | 2197 | | | | | | | | | .419 | .477 | .199 | 10.16 |
| | | | | | Compressive Steel Area 60% of Tensile Steel Area | | | | | | | | | | | |
| 0.0035 | | | | | | | | | 0.079 | .093 | .516 | 1066 | .112 | .212 | .024 | 34.57 |
| 0.0040 | | | | | 0.076 | .092 | .517 | 955 | 0.090 | .100 | .513 | 1130 | .119 | .223 | .027 | 32.45 |
| 0.0050 | 0.076 | .093 | .516 | 855 | 0.093 | .102 | .511 | 1051 | 0.111 | .112 | .506 | 1247 | .133 | .241 | .033 | 29.21 |
| 0.0060 | 0.090 | .101 | .512 | 923 | 0.111 | .112 | .507 | 1137 | 0.131 | .124 | .501 | 1350 | .146 | .257 | .039 | 26.84 |
| 0.0070 | 0.104 | .108 | .509 | 985 | 0.128 | .120 | .502 | 1215 | 0.152 | .135 | .495 | 1443 | .158 | .271 | .044 | 25.02 |
| 0.0080 | 0.118 | .114 | .505 | 1042 | 0.145 | .129 | .498 | 1286 | 0.172 | .145 | .490 | 1527 | .169 | .283 | .050 | 23.56 |
| 0.0090 | 0.131 | .120 | .502 | 1094 | 0.162 | .136 | .494 | 1351 | 0.193 | .154 | .485 | 1605 | .180 | .293 | .055 | 22.36 |
| 0.0100 | 0.145 | .125 | .500 | 1142 | 0.179 | .143 | .491 | 1411 | 0.213 | .164 | .481 | 1677 | .190 | .303 | .060 | 21.35 |
| 0.0120 | 0.172 | .135 | .495 | 1230 | 0.213 | .156 | .484 | 1520 | 0.253 | .181 | .472 | 1806 | .210 | .320 | .070 | 19.75 |
| 0.0140 | 0.200 | .144 | .490 | 1307 | 0.247 | .169 | .478 | 1617 | 0.293 | .197 | .464 | 1920 | .229 | .334 | .080 | 18.53 |
| 0.0160 | 0.227 | .152 | .486 | 1377 | 0.281 | .180 | .473 | 1703 | 0.333 | .213 | .456 | 2021 | .247 | .346 | .089 | 17.55 |
| 0.0180 | 0.254 | .159 | .483 | 1439 | 0.135 | .190 | .467 | 1780 | 0.373 | .227 | .449 | 2112 | .265 | .357 | .098 | 16.76 |
| 0.0200 | 0.281 | .166 | .479 | 1496 | 0.348 | .200 | .462 | 1851 | 0.413 | .241 | .442 | 2194 | .282 | .366 | .107 | 16.09 |
| 0.0220 | 0.309 | .172 | .476 | 1548 | 0.382 | .209 | .458 | 1915 | 0.452 | .255 | .435 | 2269 | .299 | .374 | .116 | 15.53 |
| 0.0240 | 0.336 | .178 | .473 | 1596 | 0.415 | .218 | .453 | 1974 | 0.492 | .268 | .428 | 2337 | .316 | .382 | .125 | 15.05 |
| 0.0260 | 0.363 | .184 | .471 | 1641 | 0.449 | .227 | .449 | 2028 | 0.531 | .281 | .422 | 2400 | .332 | .388 | .134 | 14.62 |
| 0.0280 | 0.390 | .189 | .468 | 1682 | 0.482 | .235 | .445 | 2079 | 0.570 | .293 | .416 | 2458 | .348 | .395 | .142 | 14.25 |
| 0.0320 | 0.444 | .198 | .462 | 1755 | 0.549 | .250 | .437 | 2169 | 0.648 | .317 | .404 | 2562 | .380 | .405 | .159 | 13.63 |
| 0.0360 | 0.497 | .207 | .450 | 1815 | 0.613 | .265 | .421 | 2240 | 0.726 | .340 | .392 | 2651 | .411 | .414 | .176 | 13.12 |
| 0.0400 | 0.549 | .214 | .437 | 1866 | 0.676 | .278 | .406 | 2299 | 0.800 | .362 | .374 | 2718 | .441 | .422 | .192 | 12.70 |
| 0.0440 | 0.600 | .222 | .424 | 1909 | 0.738 | .291 | .390 | 2348 | | | | | .472 | .429 | .208 | 12.35 |
| 0.0480 | 0.651 | .228 | .412 | 1946 | 0.800 | .303 | .374 | 2391 | | | | | .502 | .435 | .224 | 12.05 |

Tabled values are coefficients for the section constants:

$Z_c$ = coeff. $\times bd^2$; $y_n$ = coeff. $\times d$; $S_c$ = coeff. $\times bd^2$; $I_{cr}$ = coeff. $\times bd^3$; $k_n$ and $k_{sv}$ are pure coefficients. Units for $f_{sv}$ are in lb.,sq. in. Values of $I_{cr}$ is for short-term live loads. For long-term loads, use $\frac{1}{2} E_c$ with $I_{cr}$. Coefficients above the interior line apply only to tee shapes. Table is stopped when the sum of tensile and compressive steel exceeds 8% of gross area.

**Table A–6** Coefficients for Section Constants. $f'_c$ = 4000 psi (27.5 N/mm²); $g$ = 0.125, $\beta_1$ = 0.85, $n$ = 8.04

| | At Ultimate Loads | | | | | | | | | | | | At Service Loads | | | |
|---|---|---|---|---|---|---|---|---|---|---|---|---|---|---|---|---|
| | Grade 40 Steel $f_y$ = 40000 psi | | | | Grade 50 Steel $f_y$ = 50000 psi | | | | Grade 60 Steel $f_y$ = 60000 psi | | | | Elastic Constants for All Steels | | | |
| $A_s/bd$ | $Z_c$ | $\beta_1 k_n$ | $y_n$ | $f_{sv}$ | $Z_c$ | $\beta_1 k_n$ | $y_n$ | $f_{sv}$ | $Z_c$ | $\beta_1 k_n$ | $y_n$ | $f_{sv}$ | $S_c$ | $k_{sv}$ | $I_{cr}$ | $f_s/f_c$ |
| Compressive Steel Area 0% of Tensile Steel Area | | | | | | | | | | | | | | | | |
| 0.0005 | 0.006 | .006 | .560 | 282 | 0.007 | .007 | .559 | 352 | 0.009 | .009 | .558 | 422 | .042 | .086 | .004 | 85.76 |
| 0.0010 | 0.012 | .012 | .557 | 409 | 0.015 | .015 | .555 | 511 | 0.017 | .018 | .554 | 612 | .057 | .119 | .007 | 59.53 |
| 0.0015 | 0.017 | .018 | .554 | 511 | 0.022 | .022 | .551 | 638 | 0.026 | .026 | .549 | 763 | .068 | .144 | .010 | 47.92 |
| 0.0020 | 0.023 | .024 | .551 | 600 | 0.029 | .029 | .548 | 748 | 0.035 | .035 | .545 | 894 | .078 | .164 | .013 | 41.00 |
| 0.0025 | 0.029 | .029 | .548 | 680 | 0.036 | .037 | .544 | 847 | 0.043 | .044 | .540 | 1012 | .085 | .181 | .015 | 36.29 |
| 0.0030 | 0.035 | .035 | .545 | 754 | 0.043 | .044 | .540 | 938 | 0.052 | .053 | .536 | 1121 | .092 | .197 | .018 | 32.81 |
| 0.0035 | 0.040 | .041 | .542 | 823 | 0.050 | .051 | .537 | 1023 | 0.060 | .062 | .532 | 1222 | .098 | .211 | .021 | 30.12 |
| 0.0040 | 0.046 | .047 | .539 | 888 | 0.057 | .059 | .533 | 1104 | 0.068 | .071 | .527 | 1317 | .103 | .224 | .023 | 27.94 |
| 0.0050 | 0.057 | .059 | .533 | 1010 | 0.071 | .074 | .526 | 1253 | 0.084 | .088 | .518 | 1493 | .113 | .246 | .028 | 24.62 |
| 0.0060 | 0.068 | .071 | .527 | 1123 | 0.084 | .088 | .518 | 1391 | 0.100 | .106 | .510 | 1654 | .121 | .266 | .032 | 22.18 |
| 0.0070 | 0.079 | .082 | .521 | 1229 | 0.098 | .103 | .511 | 1519 | 0.116 | .124 | .501 | 1803 | .129 | .284 | .037 | 20.28 |
| 0.0080 | 0.090 | .094 | .515 | 1328 | 0.111 | .118 | .504 | 1640 | 0.131 | .141 | .492 | 1943 | .135 | .300 | .041 | 18.76 |
| 0.0090 | 0.100 | .106 | .510 | 1423 | 0.124 | .132 | .496 | 1754 | 0.146 | .159 | .483 | 2075 | .141 | .315 | .044 | 17.50 |
| 0.0100 | 0.111 | .118 | .504 | 1513 | 0.136 | .147 | .489 | 1862 | 0.161 | .176 | .474 | 2199 | .146 | .329 | .048 | 16.43 |
| 0.0120 | 0.131 | .141 | .492 | 1684 | 0.161 | .176 | .474 | 2065 | 0.189 | .212 | .457 | 2430 | .156 | .353 | .055 | 14.72 |
| 0.0140 | 0.151 | .165 | .480 | 1842 | 0.185 | .206 | .460 | 2251 | 0.217 | .247 | .439 | 2639 | .164 | .375 | .062 | 13.40 |
| 0.0160 | 0.171 | .188 | .468 | 1990 | 0.208 | .235 | .445 | 2423 | 0.242 | .282 | .421 | 2830 | .171 | .395 | .068 | 12.34 |
| 0.0180 | 0.189 | .212 | .457 | 2129 | 0.230 | .265 | .430 | 2582 | 0.267 | .318 | .404 | 3004 | .178 | .412 | .073 | 11.46 |
| 0.0200 | 0.208 | .235 | .445 | 2260 | 0.251 | .294 | .415 | 2731 | 0.291 | .353 | .386 | 3164 | .184 | .429 | .079 | 10.72 |
| 0.0240 | 0.242 | .282 | .421 | 2501 | 0.291 | .353 | .386 | 2998 | | | | | .194 | .458 | .089 | 9.53 |
| 0.0280 | 0.275 | .329 | .398 | 2717 | 0.327 | .412 | .357 | 3229 | | | | | .203 | .483 | .098 | 8.62 |
| 0.0320 | 0.306 | .376 | .374 | 2911 | | | | | | | | | .210 | .505 | .106 | 7.89 |
| 0.0360 | 0.334 | .424 | .351 | 3085 | | | | | | | | | .216 | .525 | .114 | 7.29 |
| | 0.75 $r_b$ = 0.0371 | | | | 0.75 $r_b$ = 0.0275 | | | | 0.75 $r_b$ = 0.0214 | | | | | | | |
| Compressive Steel Area 20% of Tensile Steel Area | | | | | | | | | | | | | | | | |
| 0.0035 | | | | | | | | | 0.060 | .071 | .527 | 1207 | .100 | .207 | .021 | 30.79 |
| 0.0040 | | | | | 0.058 | .070 | .528 | 1089 | 0.068 | .078 | .523 | 1291 | .106 | .219 | .023 | 28.68 |
| 0.0050 | 0.058 | .071 | .527 | 991 | 0.071 | .081 | .522 | 1219 | 0.084 | .092 | .516 | 1447 | .117 | .240 | .028 | 25.46 |
| 0.0060 | 0.069 | .080 | .522 | 1087 | 0.084 | .093 | .516 | 1339 | 0.100 | .106 | .510 | 1589 | .126 | .258 | .033 | 23.09 |
| 0.0070 | 0.079 | .089 | .518 | 1176 | 0.098 | .104 | .511 | 1449 | 0.116 | .120 | .503 | 1720 | .135 | .275 | .037 | 21.26 |
| 0.0080 | 0.090 | .098 | .514 | 1259 | 0.111 | .115 | .505 | 1552 | 0.131 | .133 | .496 | 1842 | .143 | .289 | .042 | 19.78 |
| 0.0090 | 0.100 | .106 | .510 | 1338 | 0.124 | .125 | .500 | 1650 | 0.147 | .146 | .489 | 1956 | .150 | .302 | .046 | 18.57 |
| 0.0100 | 0.111 | .114 | .505 | 1413 | 0.136 | .136 | .495 | 1742 | 0.162 | .159 | .483 | 2064 | .157 | .314 | .050 | 17.54 |
| 0.0120 | 0.131 | .130 | .498 | 1553 | 0.162 | .157 | .484 | 1913 | 0.191 | .186 | .470 | 2262 | .169 | .336 | .057 | 15.90 |
| 0.0140 | 0.152 | .145 | .490 | 1682 | 0.187 | .177 | .474 | 2069 | 0.220 | .211 | .457 | 2442 | .180 | .355 | .065 | 14.63 |
| 0.0160 | 0.172 | .161 | .482 | 1802 | 0.211 | .197 | .464 | 2213 | 0.249 | .237 | .444 | 2606 | .191 | .371 | .072 | 13.61 |
| 0.0180 | 0.192 | .175 | .475 | 1914 | 0.235 | .218 | .454 | 2346 | 0.276 | .263 | .431 | 2756 | .200 | .386 | .078 | 12.77 |
| 0.200 | 0.212 | .190 | .467 | 2019 | 0.259 | .238 | .444 | 2470 | 0.303 | .288 | .418 | 2894 | .210 | .400 | .084 | 12.07 |
| 0.0240 | 0.250 | .219 | .450 | 2205 | 0.304 | .277 | .421 | 2686 | 0.355 | .339 | .393 | 3138 | .227 | .424 | .096 | 10.95 |
| 0.0280 | 0.286 | .248 | .431 | 2366 | 0.347 | .317 | .398 | 2868 | 0.404 | .390 | .365 | 3334 | .242 | .444 | .107 | 10.09 |
| 0.0320 | 0.322 | .276 | .412 | 2507 | 0.388 | .356 | .374 | 3023 | | | | | .257 | .461 | .118 | 9.40 |
| 0.0360 | 0.356 | .304 | .393 | 2629 | 0.426 | .395 | .351 | 3154 | | | | | .270 | .476 | .128 | 8.84 |
| 0.0400 | 0.388 | .331 | .374 | 2735 | | | | | | | | | .284 | .490 | .138 | 8.37 |
| 0.0480 | 0.449 | .387 | .337 | 2907 | | | | | | | | | .309 | .513 | .155 | 7.63 |

**Table A–6**     *Continued*

| | At Ultimate Loads | | | | | | | | | | | | At Service Loads | | | |
|---|---|---|---|---|---|---|---|---|---|---|---|---|---|---|---|---|
| | Grade 40 Steel $f_y = 40000$ psi | | | | Grade 50 Steel $f_y = 50000$ psi | | | | Grade 60 Steel $f_y = 60000$ psi | | | | Elastic Constants for All Steels | | | |
| $A_s/bd$ | $Z_c$ | $\beta_1 k_n$ | $y_n$ | $f_{sv}$ | $Z_c$ | $\beta_1 k_n$ | $y_n$ | $f_{sv}$ | $Z_c$ | $\beta_1 k_n$ | $y_n$ | $f_{sv}$ | $S_c$ | $k_{sv}$ | $I_{cr}$ | $f_s/f_c$ |
| Compressive Steel Area 40% of Tensile Steel Area | | | | | | | | | | | | | | | | |
| 0.0035 | | | | | | | | | 0.061 | .076 | .524 | 1188 | .102 | .204 | .021 | 31.44 |
| 0.0040 | | | | | 0.058 | .075 | .525 | 1069 | 0.069 | .082 | .521 | 1265 | .109 | .215 | .024 | 29.39 |
| 0.0050 | 0.058 | .078 | .524 | 966 | 0.071 | .086 | .520 | 1186 | 0.085 | .095 | .515 | 1405 | .120 | .234 | .029 | 26.26 |
| 0.0060 | 0.069 | .085 | .520 | 1050 | 0.085 | .095 | .515 | 1291 | 0.100 | .106 | .509 | 1531 | .131 | .251 | .034 | 23.97 |
| 0.0070 | 0.079 | .093 | .516 | 1127 | 0.098 | .104 | .510 | 1387 | 0.116 | .117 | .504 | 1646 | .141 | .266 | .038 | 22.20 |
| 0.0080 | 0.090 | .100 | .513 | 1199 | 0.111 | .113 | .506 | 1477 | 0.131 | .128 | .499 | 1752 | .150 | .279 | .043 | 20.78 |
| 0.0090 | 0.100 | .106 | .509 | 1266 | 0.124 | .121 | .502 | 1560 | 0.147 | .138 | .494 | 1851 | .158 | .291 | .047 | 19.61 |
| 0.0100 | 0.111 | .112 | .506 | 1329 | 0.136 | .129 | .498 | 1638 | 0.162 | .148 | .489 | 1944 | .167 | .302 | .051 | 18.62 |
| 0.0120 | 0.131 | .124 | .501 | 1444 | 0.162 | .144 | .490 | 1782 | 0.192 | .167 | .479 | 2114 | .182 | .321 | .060 | 17.04 |
| 0.0140 | 0.152 | .135 | .495 | 1549 | 0.187 | .159 | .483 | 1912 | 0.222 | .186 | .470 | 2266 | .196 | .337 | .068 | 15.83 |
| 0.0160 | 0.172 | .145 | .490 | 1645 | 0.213 | .172 | .476 | 2030 | 0.252 | .204 | .461 | 2404 | .209 | .351 | .075 | 14.86 |
| 0.0180 | 0.193 | .154 | .485 | 1734 | 0.238 | .186 | .470 | 2138 | 0.281 | .222 | .452 | 2529 | .222 | .364 | .083 | 14.07 |
| 0.0200 | 0.213 | .164 | .481 | 1816 | 0.262 | .199 | .463 | 2238 | 0.310 | .239 | .443 | 2645 | .235 | .375 | .090 | 13.40 |
| 0.0240 | 0.253 | .181 | .472 | 1964 | 0.312 | .224 | .451 | 2417 | 0.367 | .273 | .426 | 2849 | .258 | .395 | .104 | 12.34 |
| 0.0280 | 0.293 | .197 | .464 | 2095 | 0.361 | .248 | .439 | 2574 | 0.424 | .307 | .409 | 3024 | .280 | .411 | .117 | 11.54 |
| 0.0320 | 0.332 | .213 | .450 | 2203 | 0.407 | .271 | .421 | 2701 | 0.479 | .340 | .393 | 3176 | .302 | .425 | .130 | 10.90 |
| 0.0360 | 0.370 | .227 | .435 | 2296 | 0.452 | .294 | .404 | 2808 | 0.531 | .372 | .372 | 3294 | .322 | .437 | .142 | 10.37 |
| 0.0400 | 0.407 | .241 | .421 | 2378 | 0.497 | .316 | .386 | 2899 | | | | | .343 | .447 | .154 | 9.94 |
| 0.0480 | 0.479 | .268 | .393 | 2510 | 0.581 | .360 | .351 | 3043 | | | | | .382 | .465 | .177 | 9.26 |
| 0.0560 | 0.548 | .293 | .365 | 2609 | | | | | | | | | .420 | .479 | .200 | 8.74 |
| Compressive Steel Area 60% of Tensile Steel Area | | | | | | | | | | | | | | | | |
| 0.0035 | | | | | | | | | 0.061 | .080 | .523 | 1170 | .104 | .201 | .021 | 32.06 |
| 0.0040 | | | | | 0.058 | .079 | .523 | 1049 | 0.069 | .085 | .520 | 1239 | .111 | .211 | .024 | 30.07 |
| 0.0050 | 0.058 | .082 | .522 | 942 | 0.071 | .089 | .518 | 1154 | 0.085 | .096 | .514 | 1366 | .124 | .229 | .029 | 27.04 |
| 0.0060 | 0.069 | .089 | .518 | 1016 | 0.085 | .097 | .514 | 1248 | 0.100 | .106 | .509 | 1479 | .136 | .245 | .034 | 24.82 |
| 0.0070 | 0.079 | .095 | .515 | 1084 | 0.098 | .105 | .510 | 1333 | 0.116 | .115 | .505 | 1581 | .147 | .258 | .039 | 23.11 |
| 0.0080 | 0.090 | .101 | .512 | 1145 | 0.111 | .112 | .507 | 1410 | 0.131 | .124 | .501 | 1674 | .157 | .270 | .044 | 21.74 |
| 0.0090 | 0.100 | .106 | .509 | 1203 | 0.124 | .118 | .503 | 1482 | 0.147 | .132 | .497 | 1760 | .167 | .281 | .049 | 20.62 |
| 0.0100 | 0.111 | .111 | .507 | 1256 | 0.136 | .125 | .500 | 1549 | 0.162 | .140 | .493 | 1840 | .176 | .290 | .053 | 19.67 |
| 0.0120 | 0.131 | .120 | .502 | 1353 | 0.162 | .136 | .494 | 1671 | 0.193 | .154 | .485 | 1985 | .194 | .307 | .062 | 18.16 |
| 0.0140 | 0.152 | .128 | .498 | 1439 | 0.188 | .147 | .489 | 1778 | 0.223 | .168 | .478 | 2113 | .211 | .321 | .071 | 17.00 |
| 0.0160 | 0.172 | .135 | .495 | 1517 | 0.213 | .156 | .484 | 1875 | 0.253 | .181 | .472 | 2228 | .227 | .333 | .079 | 16.08 |
| 0.0180 | 0.193 | .142 | .491 | 1587 | 0.239 | .166 | .480 | 1963 | 0.283 | .193 | .466 | 2331 | .243 | .344 | .087 | 15.33 |
| 0.0200 | 0.213 | .148 | .488 | 1652 | 0.264 | .174 | .475 | 2043 | 0.313 | .205 | .460 | 2425 | .258 | .354 | .095 | 14.70 |
| 0.0240 | 0.254 | .159 | .483 | 1766 | 0.315 | .190 | .467 | 2184 | 0.373 | .227 | .449 | 2591 | .288 | .370 | .111 | 13.70 |
| 0.0280 | 0.295 | .169 | .478 | 1864 | 0.365 | .205 | .460 | 2305 | 0.433 | .248 | .438 | 2732 | .317 | .383 | .126 | 12.94 |
| 0.0320 | 0.336 | .178 | .473 | 1950 | 0.415 | .218 | .453 | 2411 | 0.492 | .268 | .428 | 2855 | .344 | .394 | .141 | 12.35 |
| 0.0360 | 0.377 | .186 | .469 | 2026 | 0.465 | .231 | .447 | 2504 | 0.551 | .287 | .419 | 2962 | .372 | .404 | .155 | 11.86 |
| 0.0400 | 0.417 | .194 | .466 | 2093 | 0.516 | .243 | .441 | 2587 | 0.609 | .305 | .410 | 3056 | .399 | .412 | .170 | 11.46 |
| 0.0480 | 0.497 | .207 | .450 | 2201 | 0.613 | .265 | .421 | 2716 | 0.726 | .340 | .392 | 3215 | .451 | .426 | .198 | 10.84 |

Tabled values are coefficients for the section constants:

$Z_c = $ coeff. $\times bd^2$; $y_n = $ coeff. $\times d$; $S_c = $ coeff. $\times bd^2$; $I_{cr} = $ coeff. $\times bd^3$; $k_n$ and $k_{sv}$ are pure coefficients. Units for $f_{sv}$ are in lb.sq. in. Value of $I_{cr}$ is for short-term live loads. For long-term loads, use $\frac{1}{2} E_c$ with $I_{cr}$. Coefficients above the interior line apply only to tee shapes. Table is stopped when the sum of tensile and compressive steel exceeds 8% of gross area.

**Table A–7**   Coefficients for Section Constants. $f_c' = 5000$ psi (34.4 N/sqmm); $g = 0.125$, $\beta_1 = 0.80$, $n = 7.2$

| | At Ultimate Loads | | | | | | | | | | | | At Service Loads | | | |
|---|---|---|---|---|---|---|---|---|---|---|---|---|---|---|---|---|
| | Grade 40 Steel $f_y = 40000$ psi | | | | Grade 50 Steel $f_y = 50000$ psi | | | | Grade 60 Steel $f_y = 60000$ psi | | | | Elastic Constants for All Steels | | | |
| $A_s/bd$ | $Z_c$ | $\beta_1 k_n$ | $y_n$ | $f_{sv}$ | $Z_c$ | $\beta_1 k_n$ | $y_n$ | $f_{sv}$ | $Z_c$ | $\beta_1 k_n$ | $y_n$ | $f_{sv}$ | $S_c$ | $k_{sv}$ | $I_{cr}$ | $f_s/f_c$ |
| Compressive Steel Area 0% of Tensile Steel Area | | | | | | | | | | | | | | | | |
| 0.0005 | 0.005 | .005 | .560 | 297 | 0.006 | .006 | .560 | 371 | 0.007 | .007 | .559 | 445 | .040 | .081 | .003 | 81.30 |
| 0.0010 | 0.009 | .009 | .558 | 431 | 0.012 | .012 | .557 | 538 | 0.014 | .014 | .555 | 645 | .054 | .113 | .006 | 56.49 |
| 0.0015 | 0.014 | .014 | .555 | 538 | 0.017 | .018 | .554 | 671 | 0.021 | .021 | .552 | 804 | .065 | .137 | .009 | 45.51 |
| 0.0020 | 0.019 | .019 | .553 | 631 | 0.023 | .024 | .551 | 787 | 0.028 | .028 | .548 | 942 | .074 | .156 | .012 | 38.97 |
| 0.0025 | 0.023 | .024 | .551 | 715 | 0.029 | .029 | .548 | 891 | 0.035 | .035 | .545 | 1066 | .081 | .173 | .014 | 34.51 |
| 0.0030 | 0.028 | .028 | .548 | 793 | 0.035 | .035 | .545 | 987 | 0.041 | .042 | .541 | 1180 | .088 | .187 | .016 | 31.22 |
| 0.0035 | 0.032 | .033 | .546 | 865 | 0.040 | .041 | .542 | 1077 | 0.048 | .049 | .538 | 1287 | .094 | .201 | .019 | 28.66 |
| 0.0040 | 0.037 | .038 | .544 | 934 | 0.046 | .047 | .539 | 1162 | 0.055 | .056 | .534 | 1387 | .099 | .213 | .021 | 26.61 |
| 0.0050 | 0.046 | .047 | .539 | 1062 | 0.057 | .059 | .533 | 1320 | 0.068 | .071 | .527 | 1574 | .108 | .235 | .025 | 23.47 |
| 0.0060 | 0.055 | .056 | .534 | 1181 | 0.068 | .071 | .527 | 1465 | 0.081 | .085 | .520 | 1746 | .116 | .254 | .029 | 21.15 |
| 0.0070 | 0.064 | .066 | .530 | 1292 | 0.079 | .082 | .521 | 1602 | 0.094 | .099 | .513 | 1905 | .123 | .271 | .033 | 19.36 |
| 0.0080 | 0.072 | .075 | .525 | 1398 | 0.090 | .094 | .515 | 1730 | 0.107 | .113 | .506 | 2056 | .130 | .287 | .037 | 17.91 |
| 0.0090 | 0.081 | .085 | .520 | 1498 | 0.100 | .106 | .510 | 1852 | 0.119 | .127 | .499 | 2198 | .135 | .301 | .041 | 16.72 |
| 0.0100 | 0.090 | .094 | .515 | 1594 | 0.111 | .118 | .504 | 1968 | 0.131 | .141 | .492 | 2333 | .141 | .314 | .044 | 15.71 |
| 0.0120 | 0.107 | .113 | .506 | 1776 | 0.131 | .141 | .492 | 2187 | 0.155 | .169 | .478 | 2584 | .150 | .338 | .051 | 14.09 |
| 0.0140 | 0.123 | .132 | .497 | 1946 | 0.151 | .165 | .480 | 2390 | 0.178 | .198 | .464 | 2816 | .158 | .359 | .057 | 12.83 |
| 0.0160 | 0.139 | .151 | .487 | 2106 | 0.171 | .188 | .468 | 2579 | 0.200 | .226 | .450 | 3030 | .165 | .378 | .063 | 11.82 |
| 0.0180 | 0.155 | .169 | .478 | 2257 | 0.189 | .212 | .457 | 2756 | 0.222 | .254 | .435 | 3229 | .172 | .396 | .068 | 10.99 |
| 0.0200 | 0.171 | .188 | .468 | 2401 | 0.208 | .235 | .445 | 2923 | 0.242 | .282 | .421 | 3415 | .178 | .412 | .073 | 10.29 |
| 0.0240 | 0.200 | .226 | .450 | 2669 | 0.242 | .282 | .421 | 3230 | 0.281 | .339 | .393 | 3749 | .188 | .440 | .083 | 9.16 |
| 0.0280 | 0.229 | .264 | .431 | 2914 | 0.275 | .329 | .398 | 3504 | | | | | .196 | .465 | .091 | 8.29 |
| 0.0320 | 0.256 | .301 | .412 | 3139 | 0.306 | .376 | .374 | 3750 | | | | | .204 | .486 | .099 | 7.60 |
| 0.0360 | 0.281 | .339 | .393 | 3345 | | | | | | | | | .210 | .506 | .106 | 7.03 |
| 0.0400 | 0.306 | .376 | .374 | 3535 | | | | | | | | | .216 | .524 | .113 | 6.55 |
| | 0.75 $r_b = 0.0437$ | | | | 0.75 $r_b = 0.0324$ | | | | 0.75 $r_b = 0.0252$ | | | | | | | |
| Compressive Steel Area 20% of Tensile Steel Area | | | | | | | | | | | | | | | | |
| 0.0035 | | | | | | | | | 0.049 | .059 | .533 | 1282 | .095 | .198 | .019 | 29.21 |
| 0.0040 | | | | | 0.047 | .059 | .533 | 1158 | 0.055 | .065 | .530 | 1372 | .101 | .209 | .021 | 27.21 |
| 0.0050 | 0.047 | .060 | .532 | 1054 | 0.058 | .068 | .528 | 1296 | 0.068 | .077 | .524 | 1537 | .111 | .229 | .026 | 24.16 |
| 0.0060 | 0.056 | .068 | .528 | 1155 | 0.068 | .078 | .524 | 1422 | 0.081 | .088 | .518 | 1688 | .120 | .247 | .030 | 21.92 |
| 0.0070 | 0.064 | .075 | .525 | 1249 | 0.079 | .087 | .519 | 1540 | 0.094 | .099 | .513 | 1828 | .129 | .263 | .034 | 20.18 |
| 0.0080 | 0.073 | .082 | .521 | 1338 | 0.090 | .096 | .515 | 1650 | 0.107 | .110 | .508 | 1958 | .136 | .277 | .038 | 18.78 |
| 0.0090 | 0.081 | .089 | .518 | 1422 | 0.100 | .104 | .510 | 1753 | 0.119 | .121 | .502 | 2081 | .143 | .290 | .042 | 17.63 |
| 0.0100 | 0.090 | .096 | .515 | 1502 | 0.111 | .113 | .506 | 1852 | 0.131 | .131 | .497 | 2197 | .149 | .302 | .046 | 16.65 |
| 0.0120 | 0.107 | .109 | .508 | 1651 | 0.131 | .130 | .498 | 2036 | 0.156 | .152 | .486 | 2413 | .161 | .323 | .053 | 15.09 |
| 0.0140 | 0.123 | .122 | .502 | 1789 | 0.152 | .147 | .489 | 2205 | 0.180 | .173 | .476 | 2610 | .172 | .341 | .059 | 13.89 |
| 0.0160 | 0.140 | .134 | .496 | 1918 | 0.172 | .163 | .481 | 2361 | 0.203 | .194 | .465 | 2791 | .182 | .358 | .066 | 12.92 |
| 0.0180 | 0.156 | .146 | .489 | 2038 | 0.192 | .179 | .473 | 2507 | 0.226 | .215 | .455 | 2958 | .191 | .372 | .072 | 12.12 |
| 0.200 | 0.172 | .158 | .483 | 2152 | 0.211 | .195 | .465 | 2643 | 0.249 | .235 | .445 | 3113 | .200 | .386 | .078 | 11.46 |
| 0.0240 | 0.204 | .182 | .472 | 2362 | 0.250 | .227 | .449 | 2892 | 0.293 | .276 | .424 | 3393 | .216 | .409 | .089 | 10.39 |
| 0.0280 | 0.235 | .205 | .457 | 2546 | 0.286 | .259 | .431 | 3108 | 0.335 | .317 | .404 | 3637 | .230 | .429 | .099 | 9.56 |
| 0.0320 | 0.265 | .228 | .442 | 2710 | 0.322 | .291 | .412 | 3294 | 0.375 | .357 | .382 | 3842 | .244 | .447 | .109 | 8.91 |
| 0.0360 | 0.294 | .250 | .427 | 2856 | 0.356 | .322 | .393 | 3458 | | | | | .257 | .462 | .118 | 8.37 |
| 0.0400 | 0.322 | .273 | .412 | 2986 | 0.388 | .353 | .374 | 3601 | | | | | .269 | .476 | .127 | 7.92 |
| 0.0480 | 0.375 | .317 | .382 | 3207 | | | | | | | | | .292 | .499 | .144 | 7.21 |
| 0.0560 | 0.425 | .361 | .352 | 3381 | | | | | | | | | .314 | .519 | .160 | 6.67 |

## Table A–7    *Continued*

| | At Ultimate Loads | | | At Service Loads | | |
| --- | --- | --- | --- | --- | --- | --- |
| | Grade 40 Steel $f_y$ = 40000 psi | Grade 50 Steel $f_y$ = 50000 psi | Grade 60 Steel $f_y$ = 60000 psi | Elastic Constants for All Steels | | |
| $A_s/bd$ | $Z_c$  $\beta_1k_n$  $y_n$  $f_{sv}$ | $Z_c$  $\beta_1k_n$  $y_n$  $f_{sv}$ | $Z_c$  $\beta_1k_n$  $y_n$  $f_{sv}$ | $S_c$  $k_{sv}$  $I_{cr}$  $f_s/f_c$ | | |

**Compressive Steel Area 40% of Tensile Steel Area**

| $A_s/bd$ | Grade 40 | Grade 50 | Grade 60 | Service |
| --- | --- | --- | --- | --- |
| 0.0035 | | | 0.049 .065 .530 1270 | .097 .195 .019 29.74 |
| 0.0040 | | 0.047 .065 .530 1144 | 0.056 .070 .527 1351 | .103 .206 .022 27.80 |
| 0.0050 | 0.047 .067 .529 1035 | 0.058 .074 .526 1268 | 0.069 .081 .522 1500 | .114 .225 .026 24.84 |
| 0.0060 | 0.056 .074 .526 1124 | 0.069 .082 .522 1380 | 0.081 .090 .517 1635 | .124 .241 .031 22.66 |
| 0.0070 | 0.064 .080 .522 1206 | 0.079 .089 .518 1483 | 0.094 .099 .513 1758 | .134 .255 .035 20.98 |
| 0.0080 | 0.073 .086 .520 1283 | 0.090 .096 .514 1579 | 0.106 .108 .508 1873 | .142 .268 .039 19.63 |
| 0.0090 | 0.081 .092 .517 1354 | 0.100 .103 .511 1668 | 0.119 .117 .504 1980 | .150 .280 .043 18.51 |
| 0.0100 | 0.090 .097 .514 1422 | 0.111 .110 .507 1752 | 0.131 .125 .500 2080 | .158 .290 .047 17.58 |
| 0.0120 | 0.106 .107 .509 1546 | 0.131 .123 .501 1907 | 0.156 .141 .492 2264 | .172 .309 .055 16.07 |
| 0.0140 | 0.123 .116 .505 1659 | 0.152 .135 .495 2048 | 0.180 .156 .484 2430 | .185 .325 .062 14.92 |
| 0.0160 | 0.140 .125 .500 1763 | 0.172 .147 .489 2176 | 0.204 .171 .477 2581 | .198 .340 .069 14.00 |
| 0.0180 | 0.156 .133 .496 1859 | 0.193 .158 .484 2294 | 0.228 .186 .470 2720 | .210 .352 .076 13.24 |
| 0.0200 | 0.172 .141 .492 1948 | 0.213 .169 .478 2404 | 0.252 .200 .462 2848 | .221 .363 .082 12.60 |
| 0.0240 | 0.205 .156 .485 2110 | 0.253 .190 .468 2602 | 0.299 .228 .448 3077 | .243 .383 .095 11.59 |
| 0.0280 | 0.237 .170 .478 2254 | 0.292 .210 .458 2777 | 0.345 .256 .435 3277 | .263 .400 .107 10.81 |
| 0.0320 | 0.270 .183 .471 2383 | 0.332 .229 .448 2932 | 0.391 .282 .421 3452 | .283 .414 .119 10.20 |
| 0.0360 | 0.301 .195 .461 2494 | 0.370 .248 .435 3064 | 0.436 .309 .408 3607 | .302 .426 .130 9.70 |
| 0.0400 | 0.332 .207 .450 2592 | 0.407 .266 .421 3178 | 0.479 .335 .393 3738 | .320 .437 .141 9.28 |
| 0.0480 | 0.392 .230 .427 2755 | 0.479 .302 .393 3363 | | .356 .455 .163 8.62 |
| 0.0560 | 0.451 .251 .404 2884 | 0.548 .337 .365 3505 | | .391 .470 .183 8.12 |

**Compressive Steel Area 60% of Tensile Steel Area**

| $A_s/bd$ | Grade 40 | Grade 50 | Grade 60 | Service |
| --- | --- | --- | --- | --- |
| 0.0035 | | | 0.049 .069 .528 1256 | .099 .192 .019 30.25 |
| 0.0040 | | 0.047 .069 .528 1128 | 0.056 .074 .526 1330 | .105 .202 .022 28.36 |
| 0.0050 | 0.048 .072 .527 1015 | 0.058 .077 .524 1240 | 0.069 .083 .521 1466 | .117 .220 .026 25.49 |
| 0.0060 | 0.056 .078 .524 1094 | 0.069 .084 .520 1341 | 0.081 .092 .517 1587 | .128 .235 .031 23.38 |
| 0.0070 | 0.065 .083 .521 1166 | 0.079 .091 .517 1432 | 0.094 .099 .513 1696 | .138 .249 .035 21.75 |
| 0.0080 | 0.073 .088 .518 1233 | 0.090 .097 .514 1516 | 0.106 .107 .509 1797 | .148 .260 .040 20.45 |
| 0.0090 | 0.081 .093 .516 1294 | 0.100 .103 .511 1593 | 0.119 .114 .506 1891 | .157 .271 .044 19.37 |
| 0.0100 | 0.090 .097 .514 1352 | 0.111 .108 .508 1666 | 0.131 .120 .502 1978 | .166 .280 .048 18.47 |
| 0.0120 | 0.106 .105 .510 1457 | 0.131 .118 .503 1797 | 0.156 .133 .496 2135 | .183 .297 .056 17.03 |
| 0.0140 | 0.123 .113 .506 1551 | 0.152 .128 .499 1915 | 0.181 .145 .490 2276 | .198 .311 .064 15.92 |
| 0.0160 | 0.140 .119 .503 1635 | 0.172 .136 .494 2021 | 0.205 .156 .485 2402 | .213 .324 .072 15.04 |
| 0.0180 | 0.156 .125 .500 1712 | 0.193 .144 .490 2117 | 0.229 .166 .479 2517 | .228 .334 .079 14.32 |
| 0.0200 | 0.173 .131 .497 1783 | 0.213 .152 .487 2206 | 0.254 .176 .474 2621 | .242 .344 .087 13.72 |
| 0.0240 | 0.205 .141 .492 1909 | 0.254 .166 .480 2362 | 0.302 .195 .465 2807 | .269 .361 .101 12.76 |
| 0.0280 | 0.238 .150 .488 2019 | 0.295 .178 .473 2498 | 0.350 .213 .456 2966 | .295 .374 .115 12.03 |
| 0.0320 | 0.271 .158 .484 2115 | 0.335 .190 .467 2617 | 0.398 .230 .448 3105 | .320 .386 .128 11.46 |
| 0.0360 | 0.303 .165 .480 2200 | 0.376 .201 .462 2722 | 0.445 .246 .440 3228 | .345 .396 .141 10.99 |
| 0.0400 | 0.336 .172 .477 2277 | 0.416 .212 .457 2816 | 0.493 .261 .432 3336 | .369 .404 .154 10.60 |
| 0.0480 | 0.401 .183 .471 2408 | 0.496 .231 .447 2977 | 0.587 .291 .417 3521 | .417 .419 .180 9.99 |

Tabled values are coefficients for the section constants:

$Z_c$ = coeff. $\times bd^2$; $y_n$ = coeff. $\times d$; $S_c$ = coeff. $\times bd^2$; $I_{cr}$ = coeff. $\times bd^3$; $k_n$ and $k_{sv}$ are pure coefficients. Units for $f_{sv}$ are in lb.sq. in. Value of $I_{cr}$ is for short-term live loads. For long-term loads, use $\frac{1}{2} E_c$ with $I_{cr}$. Coefficients above the interior line apply only to tee shapes. Table is stopped when the sum of tensile and compressive steel exceeds 8% of gross area.

**Table A–8**  (Imperial Units) Coefficients for Section Constants at Balanced Stress Conditions*

| | | | Steel Yield and Concrete Stresses, psi | Tensile Steel Ratio | At Full Plastic Rotation to Ultimate Load | | | For Elastic Rotations Up to and Including the Elastic Limit | | | | |
|---|---|---|---|---|---|---|---|---|---|---|---|---|
| | $f'_c$ | $f_{sv}$ | $f_y$ | $\rho$ | $Z_c$ | $\beta_1 k_n$ | $y_n$ | $S_c$ | $k_{sv}$ | $y_{sv}$ | $I_{cr}$ | $f_s/f_c$ |
| | | | | Compressive Steel Area 0% of Tensile Steel Area | | | | | | | | |
| | | 1553 | 40000 | 0.0119 | 0.169 | 0.186 | 0.470 | 0.163 | 0.372 | 0.439 | 0.0606 | 15.69 |
| | 3000 | 1545 | 50000 | 0.0082 | 0.148 | 0.161 | 0.482 | 0.144 | 0.321 | 0.455 | 0.0461 | 19.61 |
| | | 1539 | 60000 | 0.0060 | 0.132 | 0.142 | 0.492 | 0.128 | 0.283 | 0.468 | 0.0363 | 23.53 |
| | | 2078 | 40000 | 0.0173 | 0.182 | 0.203 | 0.461 | 0.176 | 0.406 | 0.427 | 0.0713 | 11.76 |
| | 4000 | 2067 | 50000 | 0.0120 | 0.161 | 0.177 | 0.474 | 0.156 | 0.354 | 0.445 | 0.0551 | 14.71 |
| | | 2058 | 60000 | 0.0089 | 0.144 | 0.157 | 0.484 | 0.140 | 0.313 | 0.458 | 0.0439 | 17.65 |
| | | 2605 | 40000 | 0.0230 | 0.193 | 0.217 | 0.454 | 0.185 | 0.433 | 0.418 | 0.0803 | 9.41 |
| | 5000 | 2591 | 50000 | 0.0161 | 0.172 | 0.190 | 0.468 | 0.166 | 0.379 | 0.436 | 0.0629 | 11.76 |
| | | 2579 | 60000 | 0.0120 | 0.155 | 0.169 | 0.478 | 0.150 | 0.338 | 0.450 | 0.0506 | 14.12 |
| | | | | Compressive Steel Area 20% of Tensile Steel Area | | | | | | | | |
| | | 1548 | 40000 | 0.0139 | 0.197 | 0.180 | 0.473 | 0.191 | 0.372 | 0.439 | 0.0696 | 15.69 |
| | 3000 | 1541 | 50000 | 0.0092 | 0.165 | 0.160 | 0.483 | 0.161 | 0.321 | 0.455 | 0.0510 | 19.61 |
| | | 1536 | 60000 | 0.0066 | 0.143 | 0.143 | 0.491 | 0.139 | 0.283 | 0.468 | 0.0392 | 23.53 |
| | | 2068 | 40000 | 0,0210 | 0,221 | 0,198 | 0.464 | 0.214 | 0.406 | 0.427 | 0.0842 | 11.76 |
| | 4000 | 2059 | 50000 | 0.0139 | 0.185 | 0.176 | 0.475 | 0.180 | 0.354 | 0.445 | 0.0623 | 14.71 |
| | | 2052 | 60000 | 0.0099 | 0.160 | 0.158 | 0.484 | 0.156 | 0.313 | 0.458 | 0.0483 | 17.65 |
| | | 2583 | 40000 | 0.0289 | 0.241 | 0.217 | 0.454 | 0.233 | 0.433 | 0.418 | 0.0972 | 9.41 |
| | 5000 | 2578 | 50000 | 0.0190 | 0.202 | 0.188 | 0.469 | 0.196 | 0.379 | 0.436 | 0.0725 | 11.76 |
| | | 2570 | 60000 | 0.0136 | 0.175 | 0.169 | 0.478 | 0.170 | 0.338 | 0.450 | 0.0565 | 14.12 |
| | | | | Compressive Steel Area 40% of Tensile Steel Area | | | | | | | | |
| | | 1541 | 40000 | 0.0169 | 0.238 | 0.175 | 0.475 | 0.232 | 0.372 | 0.439 | 0.0824 | 15.69 |
| | 3000 | 1536 | 50000 | 0.0105 | 0.187 | 0.159 | 0.483 | 0.183 | 0.321 | 0.455 | 0.0573 | 19.61 |
| | | 1532 | 60000 | 0.0072 | 0.156 | 0.144 | 0.490 | 0.153 | 0.283 | 0.468 | 0.0426 | 23.53 |
| | | 2056 | 40000 | 0.0268 | 0.281 | 0.192 | 0.466 | 0.273 | 0.406 | 0.427 | 0.1042 | 11.76 |
| | 4000 | 2050 | 50000 | 0.0164 | 0.217 | 0.175 | 0.475 | 0.212 | 0.354 | 0.445 | 0.0722 | 14.71 |
| | | 2046 | 60000 | 0.0112 | 0.180 | 0.159 | 0.483 | 0.176 | 0.313 | 0.458 | 0.0537 | 17.65 |
| | | 2561 | 40000 | 0.0387 | 0.322 | 0.218 | 0.453 | 0.314 | 0.433 | 0.418 | 0.1256 | 9.41 |
| | 5000 | 2566 | 50000 | 0.0232 | 0.245 | 0.186 | 0.470 | 0.239 | 0.379 | 0.436 | 0.0863 | 11.76 |
| | | 2560 | 60000 | 0.0157 | 0.201 | 0.169 | 0.478 | 0.196 | 0.338 | 0.450 | 0.0642 | 14.12 |
| | | | | Compressive Steel Area 60% of Tensile Steel Area | | | | | | | | |
| | | 1534 | 40000 | 0.0214 | 0.301 | 0.171 | 0.477 | 0.294 | 0.372 | 0.439 | 0.1021 | 15.69 |
| | 3000 | 1531 | 50000 | 0.0122 | 0.217 | 0.158 | 0.484 | 0.212 | 0.321 | 0.455 | 0.0656 | 19.60 |
| | | 1529 | 60000 | 0.0080 | 0.173 | 0.145 | 0.490 | 0.170 | 0.283 | 0.468 | 0.0469 | 23.53 |
| | | 2042 | 40000 | 0.0369 | 0.386 | 0.188 | 0.469 | 0.378 | 0.406 | 0.427 | 0.1394 | 11.76 |
| | 4000 | 2041 | 50000 | 0.0200 | 0.264 | 0.174 | 0.475 | 0.258 | 0.354 | 0.445 | 0.0863 | 14.71 |
| | | 2039 | 60000 | 0.0128 | 0.205 | 0.160 | 0.482 | 0.201 | 0.313 | 0.458 | 0.0608 | 17.65 |
| | | | 40000 | Balanced stress condition is not feasible | | | | | | | | |
| | 5000 | 2552 | 50000 | 0.0298 | 0.312 | 0.184 | 0.471 | 0.306 | 0.379 | 0.436 | 0.1078 | 11.76 |
| | | 2551 | 60000 | 0.0186 | 0.237 | 0.169 | 0.478 | 0.232 | 0.338 | 0.450 | 0.0748 | 14.12 |
| | | | | Compressive Steel Area 80% of Tensile Steel Area | | | | | | | | |
| | | 1525 | 40000 | 0.0293 | 0.409 | 0.167 | 0.479 | 0.402 | 0.372 | 0.439 | 0.1363 | 15.69 |
| | 3000 | 1526 | 50000 | 0.0146 | 0.258 | 0.157 | 0.484 | 0.253 | 0.321 | 0.455 | 0.0772 | 19.61 |
| | | 1526 | 60000 | 0.0090 | 0.193 | 0.146 | 0.490 | 0.190 | 0.283 | 0.468 | 0.0522 | 23.53 |
| | | | 40000 | Balanced stress condition is not feasible | | | | | | | | |
| | 4000 | 2032 | 50000 | 0.0256 | 0.336 | 0.173 | 0.476 | 0.331 | 0.354 | 0.445 | 0.1084 | 14.71 |
| | | 2033 | 60000 | 0.0151 | 0.240 | 0.161 | 0.482 | 0.236 | 0.313 | 0.458 | 0.0704 | 17.65 |
| | | | 40000 | Balanced stress condition is not feasible | | | | | | | | |
| | 5000 | 2538 | 50000 | 0.0414 | 0.433 | 0.182 | 0.472 | 0.426 | 0.379 | 0.436 | 0.1463 | 11.76 |
| | | 2541 | 60000 | 0.0229 | 0.290 | 0.170 | 0.478 | 0.285 | 0.338 | 0.450 | 0.0902 | 14.12 |

*Tabled values are coefficients for the section constants:
$Z_c$ = coeff. $\times bd^2$; $S_c$ = coeff. $\times bd^2$; $I_{cr}$ = coeff. $\times bd^3$; $y_n$ = coeff. $\times d$; $y_{sv}$ = coeff. $\times d$; $k_n$ and $k_{sv}$ are pure coefficients.
Balanced stress condition is considered not to be feasible if total area of tensile and compressive reinforcement would exceed 8% of gross area.

## Table A–3    Steel Area in Square Inches for Combinations of Bar Sizes

| No. | Area | \#5 plus \#4 — 1 | 2 | 3 | 4 | (2nd group) 1 | 2 | 3 | 4 | (3rd group) 1 | 2 | 3 | 4 |
|---|---|---|---|---|---|---|---|---|---|---|---|---|---|
| 1 | 0.11 | | | | | | | | | | | | |
| 2 | 0.22 | | | | | | | | | | | | |
| 3 | 0.33 | | | | | | | | | | | | |
| 4 | 0.44 | | | | | | | | | | | | |
| 5 | 0.55 | | | | | | | | | | | | |
| 6 | 0.66 | | | | | | | | | | | | |

**\#4**

| No. | Area |
|---|---|
| 1 | 0.20 |
| 2 | 0.39 |
| 3 | 0.59 |
| 4 | 0.79 |
| 5 | 0.98 |
| 6 | 1.18 |

**\#5** — **\#5 plus \#4**

| No. | Area | 1 | 2 | 3 | 4 |
|---|---|---|---|---|---|
| 1 | 0.31 | 0.50 | 0.70 | 0.90 | 1.09 |
| 2 | 0.61 | 0.81 | 1.01 | 1.20 | 1.40 |
| 3 | 0.92 | 1.12 | 1.31 | 1.51 | 1.71 |
| 4 | 1.23 | 1.42 | 1.62 | 1.82 | 2.01 |
| 5 | 1.53 | 1.73 | 1.93 | 2.12 | 2.32 |
| 6 | 1.84 | 2.04 | 2.23 | 2.43 | 2.63 |

**\#6** — **\#6 plus \#5** — **\#6 plus \#4**

| No. | Area | 1 | 2 | 3 | 4 | 1 | 2 | 3 | 4 |
|---|---|---|---|---|---|---|---|---|---|
| 1 | 0.44 | 0.75 | 1.06 | 1.36 | 1.67 | 0.64 | 0.83 | 1.03 | 1.23 |
| 2 | 0.88 | 1.19 | 1.50 | 1.80 | 2.11 | 1.08 | 1.28 | 1.47 | 1.67 |
| 3 | 1.33 | 1.63 | 1.94 | 2.25 | 2.55 | 1.52 | 1.72 | 1.91 | 2.11 |
| 4 | 1.77 | 2.07 | 2.38 | 2.69 | 2.99 | 1.96 | 2.16 | 2.36 | 2.55 |
| 5 | 2.21 | 2.52 | 2.82 | 3.13 | 3.44 | 2.14 | 2.60 | 2.80 | 2.99 |
| 6 | 2.65 | 2.96 | 3.26 | 3.57 | 3.88 | 2.85 | 3.04 | 3.24 | 3.44 |

**\#7** — **\#7 plus \#6** — **\#7 plus \#5** — **\#7 plus \#4**

| No. | Area | 1 | 2 | 3 | 4 | 1 | 2 | 3 | 4 | 1 | 2 | 3 | 4 |
|---|---|---|---|---|---|---|---|---|---|---|---|---|---|
| 1 | 0.60 | 1.04 | 1.48 | 1.93 | 2.37 | 0.91 | 1.21 | 1.52 | 1.83 | 0.80 | 0.99 | 1.19 | 1.39 |
| 2 | 1.20 | 1.64 | 2.09 | 2.53 | 2.97 | 1.51 | 1.82 | 2.12 | 2.43 | 1.40 | 1.60 | 1.79 | 1.99 |
| 3 | 1.80 | 2.25 | 2.69 | 3.13 | 3.57 | 2.11 | 2.42 | 2.72 | 3.03 | 2.00 | 2.20 | 2.39 | 2.59 |
| 4 | 2.41 | 2.85 | .329 | 3.73 | 4.17 | 2.71 | 3.02 | 3.33 | 3.63 | 2.60 | 2.80 | 2.99 | 3.19 |
| 5 | 3.01 | 3.45 | 3.89 | 4.33 | .477 | 3.31 | 3.62 | 3.93 | 4.23 | 3.20 | 3.40 | 3.60 | 3.79 |
| 6 | 3.61 | 4.05 | 4.49 | 4.93 | 5.38 | 3.91 | 4.22 | 4.53 | 4.84 | 3.80 | 4.00 | 4.20 | 4.39 |

**\#8** — **\#8 plus \#7** — **\#8 plus \#6** — **\#8 plus \#5**

| No. | Area | 1 | 2 | 3 | 4 | 1 | 2 | 3 | 4 | 1 | 2 | 3 | 4 |
|---|---|---|---|---|---|---|---|---|---|---|---|---|---|
| 1 | 0.79 | 1.39 | 1.99 | 2.59 | 3.19 | 1.23 | 1.67 | 2.11 | 2.55 | 1.09 | 1.40 | 1.71 | 2.01 |
| 2 | 1.57 | 2.17 | 2.77 | 3.37 | 3.98 | 2.01 | 2.45 | 2.90 | 3.34 | 1.88 | 2.18 | 2.49 | 2.80 |
| 3 | 2.36 | 2.96 | 3.56 | 4.16 | 4.76 | 2.80 | 3.24 | 3.68 | 4.12 | 2.66 | 2.97 | 3.28 | 3.58 |
| 4 | 3.14 | 3.74 | 4.34 | 4.95 | 5.55 | 3.58 | 4.03 | 4.47 | 4.91 | 3.45 | 3.76 | 4.06 | 4.37 |
| 5 | 3.93 | 4.53 | 5.13 | 5.73 | 6.33 | 4.37 | 4.81 | 5.25 | 5.69 | 4.23 | 4.54 | 4.85 | 5.15 |
| 6 | 4.71 | 5.31 | 5.92 | 6.52 | 7.12 | 5.15 | 5.60 | 6.04 | 6.48 | 5.02 | 5.33 | 5.63 | 5.94 |

**\#9** — **\#9 plus \#8** — **\#9 plus \#7** — **\#9 plus \#6**

| No. | Area | 1 | 2 | 3 | 4 | 1 | 2 | 3 | 4 | 1 | 2 | 3 | 4 |
|---|---|---|---|---|---|---|---|---|---|---|---|---|---|
| 1 | 1.00 | 1.79 | 2.57 | 3.36 | 4.14 | 1.60 | 2.20 | 2.80 | 3.41 | 1.44 | 1.88 | 2.33 | 2.77 |
| 2 | 2.00 | 2.79 | 3.57 | 4.36 | 5.14 | 2.60 | 3.20 | 3.80 | 4.41 | 2.44 | 2.88 | 3.33 | 3.77 |
| 3 | 3.00 | 3.79 | 4.57 | 5.36 | 6.14 | 3.60 | 4.20 | 4.80 | 5.41 | 3.44 | 3.88 | 4.33 | 4.77 |
| 4 | 4.00 | 4.79 | 5.57 | 6.36 | 7.14 | 4.60 | 5.20 | 5.80 | 6.41 | 4.44 | 4.88 | 5.33 | 5.77 |
| 5 | 5.00 | 5.79 | 6.57 | 7.36 | 8.14 | 5.60 | 6.20 | 6.80 | 7.41 | 5.44 | 5.88 | 6.33 | 6.77 |
| 6 | 6.00 | 6.79 | 7.57 | 8.36 | 9.14 | 6.60 | 7.20 | 7.80 | 8.41 | 6.44 | 6.88 | 7.33 | 7.77 |

**\#10** — **\#10 plus \#9** — **\#10 plus \#8** — **\#10 plus \#7**

| No. | Area | 1 | 2 | 3 | 4 | 1 | 2 | 3 | 4 | 1 | 2 | 3 | 4 |
|---|---|---|---|---|---|---|---|---|---|---|---|---|---|
| 1 | 1.27 | 2.27 | 3.27 | 4.27 | 5.27 | 2.05 | 2.84 | 3.62 | 4.41 | 1.87 | 2.47 | 3.07 | 3.67 |
| 2 | 2.53 | 3.53 | 4.53 | 5.53 | 6.53 | 3.32 | 4.10 | 4.89 | 5.68 | 3.13 | 3.74 | 4.34 | 4.94 |
| 3 | 3.80 | 4.80 | 5.80 | 6.80 | 7.80 | 4.59 | 5.37 | 6.16 | 6.94 | 4.40 | 5.00 | 5.60 | 6.21 |
| 4 | 5.07 | 6.07 | 7.07 | 8.07 | 9.07 | 5.85 | 6.64 | 7.42 | 8.21 | 5.67 | 6.27 | 6.87 | 7.47 |
| 5 | 6.33 | 7.33 | 8.33 | 9.33 | 10.3 | 7.12 | 7.90 | 8.69 | 9.48 | 6.94 | 7.54 | 8.14 | 8.74 |
| 6 | 7.60 | 8.60 | 9.60 | 10.6 | 11.6 | 8.39 | 9.17 | 9.96 | 10.7 | 8.20 | 8.80 | 9.40 | 10.0 |

**\#11** — **\#11 plus \#10** — **\#11 plus \#9** — **\#11 plus \#8**

| No. | Area | 1 | 2 | 3 | 4 | 1 | 2 | 3 | 4 | 1 | 2 | 3 | 4 |
|---|---|---|---|---|---|---|---|---|---|---|---|---|---|
| 1 | 1.56 | 2.83 | 4.09 | 5.36 | 6.63 | 2.56 | 3.56 | 4.56 | 5.56 | 2.35 | 3.13 | 3.92 | 4.70 |
| 2 | 3.12 | 4.39 | 5.66 | 6.92 | 8.19 | 4.12 | 5.12 | 6.12 | 7.12 | 3.91 | 4.69 | 5.48 | 6.26 |
| 3 | 4.68 | 5.95 | 7.22 | 8.48 | 9.75 | 5.68 | 6.68 | 7.68 | 8.68 | 5.47 | 6.26 | 7.04 | 7.83 |
| 4 | 6.25 | 7.51 | 8.78 | 10.1 | 11.3 | 7.25 | 8.25 | 9.25 | 10.3 | 7.03 | 7.82 | 8.60 | 9.39 |
| 5 | 7.81 | 9.07 | 10.3 | 11.6 | 12.9 | 8.81 | 9.81 | 10.8 | 11.8 | 8.59 | 9.38 | 10.2 | 11.0 |
| 6 | 9.37 | 10.6 | 11.9 | 13.2 | 14.4 | 10.4 | 11.4 | 12.4 | 13.4 | 10.2 | 10.9 | 11.7 | 12.5 |

# Table A-9 — Stress Ratios for Columns. $f'_c = 3000$ psi; $n = 9.29$; $g = 0.125$

$$R_n = \frac{P_n/bh}{0.85 f'_c}$$

$$R_e = \frac{P_e/bh}{f'_c}$$

$$e_n = M_n/P_n$$

$$e_e = M_e/P_e$$

Steel at Flexure Faces Only

Steel Uniformly Distributed

## Steel at Flexure Faces Only

### Grade 40 Steel — Allowable $R_n$ at ultimate load

| $e_n/b$ | Steel ratio $A_s/bh$ | | | | | | | |
|---|---|---|---|---|---|---|---|---|
| | 0.01 | 0.02 | 0.03 | 0.04 | 0.05 | 0.06 | 0.07 | 0.08 |
| 0.10 | 0.91 | 1.03 | 1.15 | 1.26 | 1.38 | 1.50 | 1.61 | 1.72 |
| 0.20 | 0.72 | 0.83 | 0.93 | 1.03 | 1.13 | 1.23 | 1.33 | 1.42 |
| 0.30 | 0.58 | 0.68 | 0.77 | 0.86 | 0.95 | 1.03 | 1.12 | 1.20 |
| 0.40 | 0.46 | 0.57 | 0.65 | 0.73 | 0.81 | 0.88 | 0.96 | 1.03 |
| 0.50 | 0.35 | 0.48 | 0.56 | 0.64 | 0.71 | 0.77 | 0.84 | 0.90 |
| 0.60 | 0.26 | 0.40 | 0.50 | 0.56 | 0.62 | 0.69 | 0.75 | 0.81 |
| 0.70 | 0.20 | 0.33 | 0.43 | 0.50 | 0.56 | 0.62 | 0.67 | 0.72 |
| 0.80 | 0.17 | 0.28 | 0.37 | 0.45 | 0.51 | 0.56 | 0.61 | 0.66 |
| 0.90 | | 0.24 | 0.32 | 0.40 | 0.46 | 0.51 | 0.56 | 0.60 |
| 1.00 | | 0.20 | 0.28 | 0.35 | 0.42 | 0.47 | 0.52 | 0.56 |
| 2.00 | | | | | 0.19 | 0.23 | 0.26 | 0.29 |
| 3.00 | | | | | | | | 0.19 |
| 4.00 | | | | | | | | |

### 50 Steel — Allowable $R_n$ at ultimate load

| $e_n/b$ | Steel ratio $A_s/bh$ | | | | | | | |
|---|---|---|---|---|---|---|---|---|
| | 0.01 | 0.02 | 0.03 | 0.04 | 0.05 | 0.06 | 0.07 | 0.08 |
| 0.10 | 0.95 | 1.09 | 1.24 | 1.39 | 1.53 | 1.68 | 1.83 | 1.97 |
| 0.20 | 0.75 | 0.89 | 1.01 | 1.14 | 1.26 | 1.38 | 1.51 | 1.63 |
| 0.30 | 0.60 | 0.72 | 0.84 | 0.95 | 1.06 | 1.16 | 1.27 | 1.37 |
| 0.40 | 0.49 | 0.61 | 0.71 | 0.81 | 0.90 | 1.00 | 1.09 | 1.18 |
| 0.50 | 0.38 | 0.52 | 0.61 | 0.70 | 0.79 | 0.87 | 0.95 | 1.04 |
| 0.60 | 0.30 | 0.45 | 0.54 | 0.62 | 0.70 | 0.77 | 0.85 | 0.92 |
| 0.70 | 0.24 | 0.38 | 0.48 | 0.56 | 0.63 | 0.70 | 0.76 | 0.83 |

## Steel Uniformly Distributed

### Grade 40 Steel — Allowable $R_n$ at ultimate load

| $e_n/b$ | Steel ratio $A_s/bh$ | | | | | | | |
|---|---|---|---|---|---|---|---|---|
| | 0.01 | 0.02 | 0.03 | 0.04 | 0.05 | 0.06 | 0.07 | 0.08 |
| 0.10 | 0.91 | 1.02 | 1.14 | 1.25 | 1.36 | 1.47 | 1.59 | 1.70 |
| 0.20 | 0.70 | 0.80 | 0.90 | 0.99 | 1.08 | 1.18 | 1.27 | 1.36 |
| 0.30 | 0.54 | 0.64 | 0.72 | 0.80 | 0.88 | 0.96 | 1.04 | 1.11 |
| 0.40 | 0.41 | 0.51 | 0.59 | 0.67 | 0.74 | 0.80 | 0.87 | 0.94 |
| 0.50 | 0.31 | 0.41 | 0.50 | 0.57 | 0.63 | 0.69 | 0.75 | 0.81 |
| 0.60 | 0.23 | 0.34 | 0.42 | 0.48 | 0.55 | 0.60 | 0.66 | 0.71 |
| 0.70 | 0.19 | 0.28 | 0.35 | 0.42 | 0.47 | 0.53 | 0.58 | 0.63 |
| 0.80 | | 0.24 | 0.30 | 0.36 | 0.42 | 0.47 | 0.52 | 0.56 |
| 0.90 | | 0.20 | 0.27 | 0.32 | 0.37 | 0.42 | 0.46 | 0.51 |
| 1.00 | | 0.18 | 0.24 | 0.29 | 0.33 | 0.38 | 0.42 | 0.46 |
| 2.00 | | | | | | 0.18 | 0.21 | 0.23 |
| 3.00 | | | | | | | | 0.19 |
| 4.00 | | | | | | | | |

### 50 Steel — Allowable $R_n$ at ultimate load

| $e_n/b$ | Steel ratio $A_s/bh$ | | | | | | | |
|---|---|---|---|---|---|---|---|---|
| | 0.01 | 0.02 | 0.03 | 0.04 | 0.05 | 0.06 | 0.07 | 0.08 |
| 0.10 | 0.94 | 1.08 | 1.22 | 1.36 | 1.50 | 1.64 | 1.78 | 1.92 |
| 0.20 | 0.72 | 0.84 | 0.96 | 1.08 | 1.19 | 1.30 | 1.42 | 1.53 |
| 0.30 | 0.56 | 0.67 | 0.77 | 0.87 | 0.96 | 1.06 | 1.15 | 1.25 |
| 0.40 | 0.43 | 0.54 | 0.63 | 0.72 | 0.80 | 0.88 | 0.97 | 1.05 |
| 0.50 | 0.33 | 0.45 | 0.54 | 0.61 | 0.69 | 0.76 | 0.83 | 0.90 |
| 0.60 | 0.26 | 0.37 | 0.46 | 0.53 | 0.60 | 0.66 | 0.73 | 0.79 |
| 0.70 | 0.21 | 0.31 | 0.39 | 0.47 | 0.53 | 0.59 | 0.64 | 0.70 |

## Allowable $R_n$ at ultimate load

**Grade**

| $e_n/b$ | | | | | | | | | | | | | | | | |
|---|---|---|---|---|---|---|---|---|---|---|---|---|---|---|---|---|
| 0.80 | 0.18 | 0.27 | 0.34 | 0.41 | 0.47 | 0.53 | 0.58 | 0.63 | 0.20 | 0.32 | 0.43 | 0.50 | 0.57 | 0.63 | 0.69 | 0.76 |
| 0.90 |  | 0.23 | 0.30 | 0.36 | 0.42 | 0.48 | 0.53 | 0.57 | 0.18 | 0.28 | 0.38 | 0.46 | 0.52 | 0.58 | 0.64 | 0.69 |
| 1.00 |  | 0.21 | 0.27 | 0.33 | 0.38 | 0.43 | 0.48 | 0.52 |  | 0.24 | 0.34 | 0.42 | 0.48 | 0.53 | 0.59 | 0.64 |
| 2.00 |  |  |  |  | 0.19 | 0.21 | 0.24 | 0.27 |  |  |  |  | 0.20 | 0.23 | 0.28 | 0.32 |
| 3.00 |  |  |  |  |  |  |  | 0.18 |  |  |  |  |  |  | 0.18 | 0.20 |
| 4.00 |  |  |  |  |  |  |  |  |  |  |  |  |  |  |  | 0.17 |

**Grade 60 Steel**

| $e_n/b$ | | | | | | | | | | | | | | | | |
|---|---|---|---|---|---|---|---|---|---|---|---|---|---|---|---|---|
| 0.10 | 0.98 | 1.15 | 1.33 | 1.51 | 1.69 | 1.86 | 2.04 | 2.22 | 0.96 | 1.12 | 1.29 | 1.46 | 1.62 | 1.79 | 1.96 | 2.13 |
| 0.20 | 0.78 | 0.94 | 1.09 | 1.24 | 1.39 | 1.54 | 1.69 | 1.84 | 0.74 | 0.88 | 1.02 | 1.15 | 1.29 | 1.42 | 1.55 | 1.69 |
| 0.30 | 0.62 | 0.77 | 0.90 | 1.04 | 1.17 | 1.30 | 1.42 | 1.55 | 0.57 | 0.70 | 0.81 | 0.93 | 1.04 | 1.15 | 1.26 | 1.37 |
| 0.40 | 0.51 | 0.64 | 0.77 | 0.88 | 1.00 | 1.11 | 1.22 | 1.33 | 0.45 | 0.57 | 0.67 | 0.77 | 0.86 | 0.96 | 1.05 | 1.15 |
| 0.50 | 0.42 | 0.55 | 0.66 | 0.77 | 0.87 | 0.97 | 1.07 | 1.17 | 0.36 | 0.47 | 0.57 | 0.65 | 0.74 | 0.82 | 0.90 | 0.98 |
| 0.60 | 0.33 | 0.48 | 0.58 | 0.68 | 0.77 | 0.86 | 0.95 | 1.04 | 0.29 | 0.40 | 0.49 | 0.57 | 0.64 | 0.71 | 0.79 | 0.86 |
| 0.70 | 0.28 | 0.43 | 0.52 | 0.61 | 0.69 | 0.77 | 0.86 | 0.94 | 0.24 | 0.34 | 0.43 | 0.50 | 0.57 | 0.63 | 0.70 | 0.76 |
| 0.80 | 0.24 | 0.37 | 0.47 | 0.55 | 0.63 | 0.70 | 0.78 | 0.85 | 0.20 | 0.30 | 0.38 | 0.45 | 0.51 | 0.57 | 0.63 | 0.68 |
| 0.90 | 0.21 | 0.32 | 0.43 | 0.50 | 0.57 | 0.64 | 0.71 | 0.78 | 0.17 | 0.26 | 0.33 | 0.40 | 0.46 | 0.51 | 0.57 | 0.62 |
| 1.00 | 0.19 | 0.29 | 0.39 | 0.46 | 0.53 | 0.59 | 0.66 | 0.72 |  | 0.23 | 0.30 | 0.36 | 0.42 | 0.47 | 0.52 | 0.57 |
| 2.00 |  |  | 0.19 | 0.23 | 0.28 | 0.32 | 0.37 | 0.41 |  |  |  | 0.18 | 0.21 | 0.24 | 0.27 | 0.30 |
| 3.00 |  |  |  |  | 0.19 | 0.22 | 0.24 | 0.27 |  |  |  |  |  |  | 0.18 | 0.20 |
| 4.00 |  |  |  |  |  |  | 0.19 | 0.21 |  |  |  |  |  |  |  |  |

## Elastic stress ratio $R_e$ for any elastic stress

**For All Steel**

| $e_e/b$ | | | | | | | | | | | | | | | | |
|---|---|---|---|---|---|---|---|---|---|---|---|---|---|---|---|---|
| 0.10 | 0.74 | 0.85 | 0.97 | 1.09 | 1.20 | 1.31 | 1.43 | 1.54 | 0.76 | 0.90 | 1.03 | 1.16 | 1.29 | 1.42 | 1.55 | 1.69 |
| 0.20 | 0.54 | 0.63 | 0.71 | 0.79 | 0.88 | 0.97 | 1.05 | 1.14 | 0.56 | 0.67 | 0.78 | 0.88 | 0.98 | 1.09 | 1.19 | 1.30 |
| 0.30 | 0.38 | 0.45 | 0.52 | 0.59 | 0.66 | 0.73 | 0.80 | 0.86 | 0.42 | 0.51 | 0.61 | 0.70 | 0.79 | 0.87 | 0.97 | 1.05 |
| 0.40 | 0.27 | 0.34 | 0.40 | 0.46 | 0.51 | 0.57 | 0.62 | 0.68 | 0.31 | 0.40 | 0.48 | 0.56 | 0.63 | 0.71 | 0.78 | 0.86 |
| 0.50 | 0.21 | 0.27 | 0.32 | 0.37 | 0.42 | 0.46 | 0.51 | 0.56 | 0.24 | 0.32 | 0.39 | 0.46 | 0.52 | 0.59 | 0.65 | 0.72 |
| 0.60 | 0.17 | 0.22 | 0.26 | 0.31 | 0.35 | 0.39 | 0.43 | 0.47 | 0.19 | 0.27 | 0.33 | 0.39 | 0.45 | 0.51 | 0.56 | 0.62 |
| 0.70 | 0.14 | 0.19 | 0.23 | 0.27 | 0.30 | 0.34 | 0.37 | 0.41 | 0.16 | 0.23 | 0.28 | 0.34 | 0.39 | 0.44 | 0.49 | 0.54 |
| 0.80 | 0.12 | 0.16 | 0.20 | 0.23 | 0.27 | 0.30 | 0.33 | 0.36 | 0.14 | 0.20 | 0.25 | 0.30 | 0.35 | 0.39 | 0.44 | 0.48 |
| 0.90 | 0.11 | 0.14 | 0.18 | 0.21 | 0.24 | 0.27 | 0.30 | 0.32 | 0.13 | 0.18 | 0.23 | 0.27 | 0.31 | 0.35 | 0.40 | 0.44 |
| 1.00 | 0.09 | 0.13 | 0.16 | 0.19 | 0.21 | 0.24 | 0.27 | 0.29 | 0.11 | 0.16 | 0.20 | 0.24 | 0.28 | 0.32 | 0.36 | 0.40 |
| 2.00 | 0.05 | 0.06 | 0.08 | 0.10 | 0.11 | 0.12 | 0.14 | 0.15 | 0.06 | 0.08 | 0.11 | 0.13 | 0.15 | 0.17 | 0.19 | 0.21 |
| 3.00 | 0.03 | 0.04 | 0.05 | 0.06 | 0.07 | 0.08 | 0.09 | 0.10 | 0.04 | 0.06 | 0.07 | 0.09 | 0.10 | 0.12 | 0.13 | 0.14 |
| 4.00 | 0.02 | 0.03 | 0.04 | 0.05 | 0.06 | 0.06 | 0.07 | 0.08 | 0.03 | 0.04 | 0.05 | 0.07 | 0.08 | 0.09 | 0.10 | 0.11 |

There is no entry in the table for values of $R_n$ less than 0.168. Such members may be designed either as a beam carrying a small axial load or as a column carrying a dominant flexural load. The design procedures for a beam carrying a small axial load are simpler and are much preferred (see Chapter 9).

# Table A-10  Stress Ratios for Columns. $f'_c = 4000$ psi; $n = 8.04$; $g = 0.125$

$$R_n = \frac{P_n/bh}{0.85 f'_c}$$

$$R_e = \frac{P_e/bh}{f_c}$$

$$e_n = M_n/P_n$$

$$e_e = M_e/P_e$$

## Steel at Flexure Faces Only

Steel ratio $A_s/bh$

### Grade 40 Steel — Allowable $R_n$ at ultimate load

| $e_n/h$ | 0.01 | 0.02 | 0.03 | 0.04 | 0.05 | 0.06 | 0.07 | 0.08 |
|---|---|---|---|---|---|---|---|---|
| 0.10 | 0.88 | 0.97 | 1.05 | 1.14 | 1.22 | 1.31 | 1.39 | 1.48 |
| 0.20 | 0.69 | 0.77 | 0.85 | 0.92 | 1.00 | 1.07 | 1.14 | 1.21 |
| 0.30 | 0.54 | 0.62 | 0.70 | .76 | 0.83 | 0.89 | 0.95 | 1.02 |
| 0.40 | 0.41 | 0.52 | 0.58 | 0.65 | 0.70 | 0.76 | 0.82 | 0.87 |
| 0.50 | 0.30 | 0.42 | 0.50 | 0.56 | 0.61 | 0.66 | 0.71 | 0.76 |
| 0.60 | 0.22 | 0.34 | 0.43 | 0.49 | 0.54 | 0.59 | 0.63 | 0.68 |
| 0.70 | 0.17 | 0.27 | 0.36 | 0.43 | 0.48 | 0.52 | 0.57 | 0.61 |
| 0.80 |  | 0.23 | 0.30 | 0.37 | 0.43 | 0.48 | 0.51 | 0.55 |
| 0.90 |  | 0.19 | 0.26 | 0.32 | 0.38 | 0.43 | 0.47 | 0.50 |
| 1.00 |  |  | 0.23 | 0.29 | 0.34 | 0.39 | 0.43 | 0.47 |
| 2.00 |  |  |  |  |  | 0.18 | 0.20 | 0.23 |
| 3.00 |  |  |  |  |  |  |  |  |
| 4.00 |  |  |  |  |  |  |  |  |

### 50 Steel — Allowable $R_n$ at ultimate load

| $e_n/h$ | 0.01 | 0.02 | 0.03 | 0.04 | 0.05 | 0.06 | 0.07 | 0.08 |
|---|---|---|---|---|---|---|---|---|
| 0.10 | 0.90 | 1.01 | 1.12 | 1.23 | 1.34 | 1.45 | 1.55 | 1.66 |
| 0.20 | 0.71 | 0.81 | 0.91 | 1.00 | 1.10 | 1.19 | 1.28 | 1.37 |
| 0.30 | 0.56 | 0.66 | 0.75 | 0.83 | 0.91 | 0.99 | 1.07 | 1.15 |
| 0.40 | 0.45 | 0.54 | 0.63 | 0.70 | 0.77 | 0.85 | 0.92 | 0.99 |
| 0.50 | 0.34 | 0.46 | 0.54 | 0.61 | 0.67 | 0.74 | 0.80 | 0.86 |
| 0.60 | 0.25 | 0.38 | 0.47 | 0.53 | 0.59 | 0.65 | 0.71 | 0.76 |
| 0.70 | 0.21 | 0.32 | 0.41 | 0.48 | 0.53 | 0.58 | 0.64 | 0.69 |

## Steel Uniformly Distributed

Steel ratio $A_s/bh$

### Grade 40 Steel — Allowable $R_n$ at ultimate load

| $e_n/h$ | 0.01 | 0.02 | 0.03 | 0.04 | 0.05 | 0.06 | 0.07 | 0.08 |
|---|---|---|---|---|---|---|---|---|
| 0.10 | 0.88 | 0.96 | 1.04 | 1.13 | 1.21 | 1.29 | 1.37 | 1.46 |
| 0.20 | 0.67 | 0.75 | 0.82 | 0.89 | 0.96 | 1.03 | 1.09 | 1.16 |
| 0.30 | 0.51 | 0.58 | 0.65 | 0.71 | 0.77 | 0.83 | 0.89 | 0.95 |
| 0.40 | 0.37 | 0.46 | 0.53 | 0.59 | 0.64 | 0.69 | 0.74 | 0.79 |
| 0.50 | 0.27 | 0.36 | 0.43 | 0.49 | 0.55 | 0.59 | 0.64 | 0.68 |
| 0.60 | 0.20 | 0.29 | 0.36 | 0.41 | 0.46 | 0.51 | 0.56 | 0.60 |
| 0.70 |  | 0.24 | 0.30 | 0.35 | 0.40 | 0.44 | 0.49 | 0.53 |
| 0.80 |  | 0.20 | 0.26 | 0.30 | 0.35 | 0.39 | 0.43 | 0.47 |
| 0.90 |  | 0.17 | 0.22 | 0.27 | 0.31 | 0.35 | 0.38 | 0.42 |
| 1.00 |  |  | 0.20 | 0.24 | 0.27 | 0.31 | 0.34 | 0.37 |
| 2.00 |  |  |  |  |  |  |  | 0.18 |
| 3.00 |  |  |  |  |  |  |  |  |
| 4.00 |  |  |  |  |  |  |  |  |

### 50 Steel — Allowable $R_n$ at ultimate load

| $e_n/h$ | 0.01 | 0.02 | 0.03 | 0.04 | 0.05 | 0.06 | 0.07 | 0.08 |
|---|---|---|---|---|---|---|---|---|
| 0.10 | 0.90 | 1.00 | 1.10 | 1.21 | 1.31 | 1.42 | 1.52 | 1.63 |
| 0.20 | 0.69 | 0.78 | 0.87 | 0.95 | 1.04 | 1.12 | 1.21 | 1.29 |
| 0.30 | 0.52 | 0.61 | 0.69 | 0.76 | 0.83 | 0.90 | 0.98 | 1.05 |
| 0.40 | 0.40 | 0.49 | 0.56 | 0.63 | 0.69 | 0.75 | 0.81 | 0.87 |
| 0.50 | 0.30 | 0.40 | 0.47 | 0.53 | 0.59 | 0.64 | 0.70 | 0.75 |
| 0.60 | 0.23 | 0.32 | 0.40 | 0.46 | 0.51 | 0.56 | 0.61 | 0.65 |
| 0.70 | 0.18 | 0.27 | 0.34 | 0.39 | 0.45 | 0.49 | 0.54 | 0.58 |

**Grade 40 (continued) — Allowable $R_n$ at ultimate load**

| $e_n/h$ | | | | | | | | | | | | | | | | |
|---|---|---|---|---|---|---|---|---|---|---|---|---|---|---|---|
| 0.80 | 0.17 | 0.26 | 0.35 | 0.43 | 0.48 | 0.53 | 0.58 | 0.62 | 0.23 | 0.29 | 0.34 | 0.39 | 0.44 | 0.48 | 0.52 |
| 0.90 | | 0.23 | 0.31 | 0.38 | 0.44 | 0.48 | 0.53 | 0.57 | 0.19 | 0.25 | 0.30 | 0.35 | 0.39 | 0.43 | 0.47 |
| 1.00 | | 0.20 | 0.27 | 0.34 | 0.40 | 0.44 | 0.49 | 0.53 | 0.17 | 0.22 | 0.27 | 0.31 | 0.35 | 0.39 | 0.43 |
| 2.00 | | | | | 0.18 | 0.22 | 0.25 | 0.28 | | | | | 0.17 | 0.19 | 0.21 |
| 3.00 | | | | | | | | 0.18 | | | | | | | |
| 4.00 | | | | | | | | | | | | | | | |

## Allowable $R_n$ at ultimate load

**Grade 60 Steel**

| $e_n/h$ | | | | | | | | | | | | | | | | |
|---|---|---|---|---|---|---|---|---|---|---|---|---|---|---|---|---|
| 0.10 | 0.93 | 1.06 | 1.19 | 1.32 | 1.45 | 1.59 | 1.72 | 1.85 | 0.91 | 1.04 | 1.16 | 1.28 | 1.40 | 1.53 | 1.65 | 1.78 |
| 0.20 | 0.73 | 0.85 | 0.97 | 1.08 | 1.19 | 1.30 | 1.41 | 1.52 | 0.70 | 0.81 | 0.91 | 1.01 | 1.11 | 1.21 | 1.31 | 1.41 |
| 0.30 | 0.58 | 0.69 | 0.80 | 0.90 | 0.99 | 1.09 | 1.19 | 1.28 | 0.54 | 0.63 | 0.72 | 0.81 | 0.89 | 0.97 | 1.06 | 1.14 |
| 0.40 | 0.46 | 0.57 | 0.67 | 0.76 | 0.84 | 0.93 | 1.02 | 1.10 | 0.41 | 0.51 | 0.59 | 0.66 | 0.74 | 0.81 | 0.88 | 0.95 |
| 0.50 | 0.37 | 0.49 | 0.57 | 0.65 | 0.73 | 0.81 | 0.88 | 0.96 | 0.32 | 0.42 | 0.49 | 0.56 | 0.62 | 0.69 | 0.75 | 0.81 |
| 0.60 | 0.29 | 0.42 | 0.50 | 0.58 | 0.65 | 0.72 | 0.78 | 0.85 | 0.25 | 0.35 | 0.42 | 0.48 | 0.54 | 0.60 | 0.65 | 0.71 |
| 0.70 | 0.24 | 0.36 | 0.45 | 0.51 | 0.58 | 0.64 | 0.70 | 0.76 | 0.20 | 0.29 | 0.37 | 0.43 | 0.48 | 0.53 | 0.58 | 0.62 |
| 0.80 | 0.20 | 0.30 | 0.40 | 0.46 | 0.52 | 0.58 | 0.64 | 0.69 | 0.17 | 0.25 | 0.32 | 0.38 | 0.43 | 0.47 | 0.52 | 0.56 |
| 0.90 | 0.18 | 0.27 | 0.35 | 0.42 | 0.48 | 0.53 | 0.58 | 0.64 | | 0.22 | 0.28 | 0.33 | 0.38 | 0.43 | 0.47 | 0.51 |
| 1.00 | | 0.24 | 0.31 | 0.39 | 0.44 | 0.49 | 0.54 | 0.59 | | 0.19 | 0.25 | 0.30 | 0.35 | 0.39 | 0.43 | 0.47 |
| 2.00 | | | | 0.19 | 0.22 | 0.26 | 0.29 | 0.32 | | | | | 0.17 | 0.19 | 0.22 | 0.24 |
| 3.00 | | | | | | 0.17 | 0.19 | 0.22 | | | | | | | | |
| 4.00 | | | | | | | | | | | | | | | | |

## Elastic stress ratio $R_e$ for any elastic stress

**For All Steel**

| $e_e/h$ | | | | | | | | | | | | | | | | |
|---|---|---|---|---|---|---|---|---|---|---|---|---|---|---|---|---|
| 0.10 | 0.74 | 0.86 | 0.97 | 1.09 | 1.20 | 1.31 | 1.42 | 1.54 | 0.72 | 0.82 | 0.92 | 1.02 | 1.12 | 1.22 | 1.31 | 1.41 |
| 0.20 | 0.55 | 0.64 | 0.73 | 0.82 | 0.91 | 1.00 | 1.09 | 1.18 | 0.53 | 0.60 | 0.67 | 0.75 | 0.82 | 0.89 | 0.97 | 1.04 |
| 0.30 | 0.40 | 0.49 | 0.57 | 0.65 | 0.72 | 0.80 | 0.88 | 0.96 | 0.37 | 0.43 | 0.49 | 0.55 | 0.61 | 0.67 | 0.73 | 0.79 |
| 0.40 | 0.29 | 0.37 | 0.44 | 0.51 | 0.58 | 0.64 | 0.71 | 0.77 | 0.26 | 0.32 | 0.37 | 0.42 | 0.47 | 0.52 | 0.57 | 0.62 |
| 0.50 | 0.22 | 0.30 | 0.36 | 0.42 | 0.48 | 0.53 | 0.59 | 0.65 | 0.20 | 0.25 | 0.30 | 0.34 | 0.38 | 0.42 | 0.46 | 0.50 |
| 0.60 | 0.18 | 0.25 | 0.30 | 0.36 | 0.41 | 0.46 | 0.51 | 0.55 | 0.16 | 0.21 | 0.25 | 0.28 | 0.32 | 0.36 | 0.39 | 0.43 |
| 0.70 | 0.15 | 0.21 | 0.26 | 0.31 | 0.35 | 0.40 | 0.44 | 0.49 | 0.13 | 0.17 | 0.21 | 0.24 | 0.28 | 0.31 | 0.34 | 0.37 |
| 0.80 | 0.13 | 0.19 | 0.23 | 0.27 | 0.31 | 0.35 | 0.39 | 0.43 | 0.11 | 0.15 | 0.18 | 0.21 | 0.24 | 0.27 | 0.30 | 0.33 |
| 0.90 | 0.12 | 0.16 | 0.21 | 0.25 | 0.28 | 0.32 | 0.36 | 0.39 | 0.10 | 0.13 | 0.16 | 0.19 | 0.22 | 0.24 | 0.27 | 0.29 |
| 1.00 | 0.10 | 0.15 | 0.19 | 0.22 | 0.26 | 0.29 | 0.32 | 0.36 | 0.09 | 0.12 | 0.15 | 0.17 | 0.20 | 0.22 | 0.24 | 0.26 |
| 2.00 | 0.05 | 0.08 | 0.10 | 0.12 | 0.13 | 0.15 | 0.17 | 0.19 | 0.04 | 0.06 | 0.07 | 0.09 | 0.10 | 0.11 | 0.12 | 0.14 |
| 3.00 | 0.03 | 0.05 | 0.07 | 0.08 | 0.09 | 0.10 | 0.12 | 0.13 | 0.03 | 0.04 | 0.05 | 0.06 | 0.07 | 0.08 | 0.08 | 0.09 |
| 4.00 | 0.03 | 0.04 | 0.05 | 0.06 | 0.07 | 0.08 | 0.09 | 0.10 | 0.02 | 0.03 | 0.04 | 0.04 | 0.05 | 0.06 | 0.06 | 0.07 |

There is no entry in the table for values of $R_n$ less than 0.168. Such members may be designed either as a beam carrying a small axial load or as a column carrying a dominant flexural load. The design procedures for a beam carrying a small axial load are simpler and are much preferred (see Chapter 9).

## Table A-11   Stress Ratios for Columns. $f'_c$ = 5000 psi; $n$ = 7.2; $g$ = 0.125

$$R_n = \frac{P_n/bh}{0.85f'_c}$$

$$R_e = \frac{P_e/bh}{f_c}$$

$$e_n = M_n/P_n$$

$$e_e = M_e/P_e$$

**Steel at Flexure Faces Only**

**Allowable $R_n$ at ultimate load** — Steel ratio $A_s/bh$

### Grade 40 Steel

| $e_n/h$ | 0.01 | 0.02 | 0.03 | 0.04 | 0.05 | 0.06 | 0.07 | 0.08 |
|---|---|---|---|---|---|---|---|---|
| 0.10 | 0.86 | 0.93 | 0.99 | 1.06 | 1.13 | 1.19 | 1.26 | 1.33 |
| 0.20 | 0.67 | 0.74 | 0.80 | 0.86 | 0.91 | 0.97 | 1.03 | 1.08 |
| 0.30 | 0.52 | 0.59 | 0.65 | 0.70 | 0.75 | 0.81 | 0.86 | 0.91 |
| 0.40 | 0.38 | 0.48 | 0.54 | 0.59 | 0.64 | 0.68 | 0.73 | 0.77 |
| 0.50 | 0.27 | 0.38 | 0.46 | 0.51 | 0.55 | 0.59 | 0.63 | 0.67 |
| 0.60 | 0.19 | 0.30 | 0.38 | 0.44 | 0.48 | 0.52 | 0.56 | 0.60 |
| 0.70 |  | 0.24 | 0.31 | 0.37 | 0.43 | 0.47 | 0.50 | 0.53 |
| 0.80 |  | 0.19 | 0.26 | 0.32 | 0.37 | 0.42 | 0.45 | 0.48 |
| 0.90 |  |  | 0.22 | 0.28 | 0.33 | 0.37 | 0.41 | 0.44 |
| 1.00 |  |  | 0.19 | 0.24 | 0.29 | 0.33 | 0.37 | 0.41 |
| 2.00 |  |  |  |  |  |  | 0.17 | 0.19 |
| 3.00 |  |  |  |  |  |  |  | 0.17 |
| 4.00 |  |  |  |  |  |  |  |  |

### 50 Steel

| $e_n/h$ | 0.01 | 0.02 | 0.03 | 0.04 | 0.05 | 0.06 | 0.07 | 0.08 |
|---|---|---|---|---|---|---|---|---|
| 0.10 | 0.88 | 0.96 | 1.05 | 1.14 | 1.22 | 1.31 | 1.39 | 1.48 |
| 0.20 | 0.69 | 0.77 | 0.84 | 0.92 | 0.99 | 1.07 | 1.14 | 1.21 |
| 0.30 | 0.53 | 0.61 | 0.69 | 0.76 | 0.82 | 0.89 | 0.95 | 1.01 |
| 0.40 | 0.41 | 0.50 | 0.57 | 0.63 | 0.69 | 0.75 | 0.81 | 0.86 |
| 0.50 | 0.30 | 0.42 | 0.49 | 0.55 | 0.60 | 0.65 | 0.70 | 0.75 |
| 0.60 | 0.22 | 0.34 | 0.42 | 0.48 | 0.53 | 0.57 | 0.62 | 0.67 |
| 0.70 | 0.18 | 0.27 | 0.36 | 0.43 | 0.47 | 0.51 | 0.56 | 0.60 |

**Steel Uniformly Distributed**

**Allowable $R_n$ at ultimate load** — Steel ratio $A_s/bh$

### Grade 40 Steel

| $e_n/h$ | 0.01 | 0.02 | 0.03 | 0.04 | 0.05 | 0.06 | 0.07 | 0.08 |
|---|---|---|---|---|---|---|---|---|
| 0.10 | 0.85 | 0.92 | 0.98 | 1.05 | 1.11 | 1.18 | 1.25 | 1.31 |
| 0.20 | 0.65 | 0.71 | 0.77 | 0.82 | 0.88 | 0.93 | 0.99 | 1.04 |
| 0.30 | 0.49 | 0.55 | 0.61 | 0.66 | 0.70 | 0.75 | 0.80 | 0.84 |
| 0.40 | 0.35 | 0.43 | 0.49 | 0.54 | 0.58 | 0.62 | 0.66 | 0.70 |
| 0.50 | 0.25 | 0.33 | 0.39 | 0.44 | 0.49 | 0.53 | 0.57 | 0.60 |
| 0.60 | 0.18 | 0.26 | 0.32 | 0.37 | 0.41 | 0.45 | 0.49 | 0.53 |
| 0.70 |  | 0.21 | 0.26 | 0.31 | 0.35 | 0.39 | 0.42 | 0.46 |
| 0.80 |  | 0.17 | 0.22 | 0.27 | 0.30 | 0.34 | 0.37 | 0.40 |
| 0.90 |  |  | 0.19 | 0.23 | 0.27 | 0.30 | 0.33 | 0.36 |
| 1.00 |  |  | 0.17 | 0.21 | 0.24 | 0.27 | 0.30 | 0.32 |

### 50 Steel

| $e_n/h$ | 0.01 | 0.02 | 0.03 | 0.04 | 0.05 | 0.06 | 0.07 | 0.08 |
|---|---|---|---|---|---|---|---|---|
| 0.10 | 0.87 | 0.95 | 1.03 | 1.12 | 1.20 | 1.28 | 1.36 | 1.44 |
| 0.20 | 0.67 | 0.74 | 0.81 | 0.88 | 0.94 | 1.01 | 1.08 | 1.14 |
| 0.30 | 0.50 | 0.57 | 0.64 | 0.70 | 0.75 | 0.81 | 0.87 | 0.92 |
| 0.40 | 0.37 | 0.46 | 0.52 | 0.57 | 0.62 | 0.67 | 0.72 | 0.77 |
| 0.50 | 0.27 | 0.36 | 0.43 | 0.48 | 0.53 | 0.57 | 0.61 | 0.66 |
| 0.60 | 0.20 | 0.29 | 0.35 | 0.41 | 0.45 | 0.49 | 0.53 | 0.57 |
| 0.70 |  | 0.24 | 0.30 | 0.35 | 0.39 | 0.43 | 0.47 | 0.51 |

| | | | | | | | | | | | | | | | | | |
|---|---|---|---|---|---|---|---|---|---|---|---|---|---|---|---|---|---|
| Grade | 0.80 | | 0.23 | 0.30 | 0.37 | 0.42 | 0.46 | 0.50 | 0.54 | | 0.20 | 0.25 | 0.30 | 0.34 | 0.38 | 0.42 | 0.45 |
| | 0.90 | | 0.20 | 0.26 | 0.32 | 0.38 | 0.42 | 0.46 | 0.49 | | 0.17 | 0.22 | 0.26 | 0.30 | 0.34 | 0.37 | 0.41 |
| | 1.00 | | 0.18 | 0.23 | 0.29 | 0.34 | 0.39 | 0.42 | 0.45 | | | 0.19 | 0.23 | 0.27 | 0.30 | 0.34 | 0.37 |
| | 2.00 | | | | | | 0.18 | 0.20 | 0.23 | | | | | | | | 0.18 |
| | 3.00 | | | | | | | | | | | | | | | | |
| | 4.00 | | | | | | | | | | | | | | | | |
| | $e_n/h$ | Allowable $R_n$ at ultimate load | | | | | | | | | | | | | | | |
| Grade 60 Steel | 0.10 | 0.89 | 1.00 | 1.11 | 1.21 | 1.31 | 1.42 | 1.52 | 1.62 | 0.88 | 0.98 | 1.08 | 1.18 | 1.27 | 1.37 | 1.47 | 1.57 |
| | 0.20 | 0.70 | 0.80 | 0.89 | 0.98 | 1.07 | 1.16 | 1.25 | 1.34 | 0.68 | 0.76 | 0.84 | 0.92 | 1.00 | 1.08 | 1.16 | 1.24 |
| | 0.30 | 0.55 | 0.64 | 0.73 | 0.81 | 0.89 | 0.96 | 1.04 | 1.12 | 0.51 | 0.59 | 0.66 | 0.73 | 0.80 | 0.87 | 0.93 | 1.00 |
| | 0.40 | 0.43 | 0.53 | 0.61 | 0.68 | 0.75 | 0.82 | 0.89 | 0.95 | 0.39 | 0.47 | 0.54 | 0.60 | 0.66 | 0.71 | 0.77 | 0.83 |
| | 0.50 | 0.33 | 0.44 | 0.52 | 0.58 | 0.65 | 0.71 | 0.77 | 0.83 | 0.29 | 0.38 | 0.45 | 0.50 | 0.55 | 0.60 | 0.65 | 0.70 |
| | 0.60 | 0.26 | 0.37 | 0.45 | 0.51 | 0.57 | 0.63 | 0.68 | 0.74 | 0.22 | 0.31 | 0.38 | 0.43 | 0.48 | 0.52 | 0.57 | 0.61 |
| | 0.70 | 0.21 | 0.31 | 0.40 | 0.46 | 0.51 | 0.56 | 0.61 | 0.66 | 0.18 | 0.26 | 0.32 | 0.38 | 0.42 | 0.46 | 0.50 | 0.54 |
| | 0.80 | 0.18 | 0.27 | 0.34 | 0.41 | 0.46 | 0.51 | 0.55 | 0.60 | | 0.22 | 0.28 | 0.33 | 0.37 | 0.41 | 0.45 | 0.49 |
| | 0.90 | | 0.23 | 0.30 | 0.37 | 0.42 | 0.46 | 0.50 | 0.55 | | 0.19 | 0.24 | 0.29 | 0.33 | 0.37 | 0.41 | 0.44 |
| | 1.00 | | 0.21 | 0.27 | 0.33 | 0.38 | 0.42 | 0.46 | 0.50 | | 0.17 | 0.22 | 0.26 | 0.30 | 0.34 | 0.37 | 0.40 |
| | 2.00 | | | | | 0.19 | 0.21 | 0.24 | 0.27 | | | | | | | 0.18 | 0.20 |
| | 3.00 | | | | | | | | 0.18 | | | | | | | | |
| | 4.00 | | | | | | | | | | | | | | | | |
| | $e_e/h$ | Elastic stress ratio $R_e$ for any elastic stress | | | | | | | | | | | | | | | |
| For All Steel | 0.10 | 0.73 | 0.83 | 0.93 | 1.03 | 1.14 | 1.24 | 1.34 | 1.43 | 0.71 | 0.80 | 0.89 | 0.97 | 1.06 | 1.15 | 1.24 | 1.32 |
| | 0.20 | 0.54 | 0.62 | 0.70 | 0.78 | 0.86 | 0.94 | 1.02 | 1.10 | 0.52 | 0.58 | 0.65 | 0.72 | 0.78 | 0.84 | 0.91 | 0.97 |
| | 0.30 | 0.39 | 0.47 | 0.54 | 0.61 | 0.68 | 0.75 | 0.82 | 0.88 | 0.36 | 0.42 | 0.47 | 0.53 | 0.58 | 0.63 | 0.68 | 0.73 |
| | 0.40 | 0.28 | 0.35 | 0.42 | 0.48 | 0.54 | 0.60 | 0.66 | 0.71 | 0.25 | 0.31 | 0.36 | 0.40 | 0.45 | 0.49 | 0.53 | 0.57 |
| | 0.50 | 0.22 | 0.28 | 0.34 | 0.39 | 0.45 | 0.50 | 0.55 | 0.60 | 0.19 | 0.24 | 0.28 | 0.32 | 0.36 | 0.40 | 0.43 | 0.47 |
| | 0.60 | 0.17 | 0.23 | 0.28 | 0.33 | 0.38 | 0.42 | 0.47 | 0.51 | 0.15 | 0.20 | 0.23 | 0.27 | 0.30 | 0.33 | 0.36 | 0.40 |
| | 0.70 | 0.15 | 0.20 | 0.25 | 0.29 | 0.33 | 0.37 | 0.41 | 0.45 | 0.13 | 0.17 | 0.20 | 0.23 | 0.26 | 0.29 | 0.32 | 0.34 |
| | 0.80 | 0.13 | 0.17 | 0.22 | 0.25 | 0.29 | 0.33 | 0.36 | 0.40 | 0.11 | 0.14 | 0.17 | 0.20 | 0.23 | 0.25 | 0.28 | 0.30 |
| | 0.90 | 0.11 | 0.16 | 0.19 | 0.23 | 0.26 | 0.30 | 0.33 | 0.36 | 0.09 | 0.13 | 0.15 | 0.18 | 0.20 | 0.23 | 0.25 | 0.27 |
| | 1.00 | 0.10 | 0.14 | 0.17 | 0.21 | 0.24 | 0.27 | 0.30 | 0.33 | 0.08 | 0.11 | 0.14 | 0.16 | 0.18 | 0.20 | 0.22 | 0.25 |
| | 2.00 | 0.05 | 0.07 | 0.09 | 0.11 | 0.12 | 0.14 | 0.16 | 0.17 | 0.04 | 0.06 | 0.07 | 0.08 | 0.09 | 0.10 | 0.12 | 0.13 |
| | 3.00 | 0.03 | 0.05 | 0.06 | 0.07 | 0.08 | 0.10 | 0.11 | 0.12 | 0.03 | 0.04 | 0.05 | 0.05 | 0.06 | 0.07 | 0.08 | 0.08 |
| | 4.00 | 0.02 | 0.04 | 0.05 | 0.05 | 0.06 | 0.07 | 0.08 | 0.09 | 0.02 | 0.03 | 0.03 | 0.04 | 0.05 | 0.05 | 0.06 | 0.06 |

There is no entry in the table for values of $R_n$ less than 0.168. Such members may be designed either as a beam carrying a small axial load or as a column carrying a dominant flexural load. The design procedures for a beam carrying a small axial load are simpler and are much preferred (see Chapter 9).

**Table A–12**     Development Length of Hooked Deformed Bars

*Development Length of Back of Hooks in Inches*

| Concrete Strength | Steel Grade | Bar Sizes | | | | | | | | |
|---|---|---|---|---|---|---|---|---|---|---|
| | | #3 | #4 | #5 | #6 | #7 | #8 | #9 | #10 | #11 |
| 3000 psi | 40 | 6 | 7 | 9 | 11 | 13 | 15 | 16 | 19 | 21 |
| | 50 | 7 | 9 | 11 | 14 | 16 | 18 | 21 | 23 | 26 |
| | 60 | 8 | 11 | 14 | 16 | 19 | 22 | 25 | 28 | 31 |
| 4000 psi | 40 | 6 | 6 | 8 | 9 | 11 | 13 | 14 | 16 | 18 |
| | 50 | 6 | 8 | 10 | 12 | 14 | 16 | 18 | 20 | 22 |
| | 60 | 7 | 9 | 12 | 14 | 17 | 19 | 21 | 24 | 27 |
| 5000 psi | 40 | 6 | 6 | 7 | 8 | 10 | 11 | 13 | 14 | 16 |
| | 50 | 6 | 7 | 9 | 11 | 12 | 14 | 16 | 18 | 20 |
| | 60 | 6 | 8 | 11 | 13 | 15 | 17 | 19 | 22 | 24 |

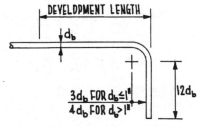

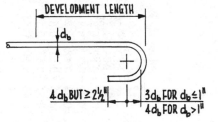

**Table A–13**     Development Length of Straight Deformed Reinforcing Bars

*Required Imbedment Length in Inches to Develop Full Strength*

| Ultimate Concrete Strength | Bar Size No. | Grade of Steel | Development Length | | | Tension Splice Length | | | | Compression Splice Length |
|---|---|---|---|---|---|---|---|---|---|---|
| | | | Tension | | Compression | Class A | | Class B | | |
| | | | Top | Other | All Bars | Top | Other | Top | Other | All Bars |
| | 3 | 40 | 12 | 12 | 8 | 12 | 12 | 14 | 12 | 12 |
| | | 50 | 13 | 12 | 8 | 13 | 12 | 17 | 13 | 12 |
| | | 60 | 16 | 12 | 8 | 16 | 12 | 21 | 16 | 12 |
| | 4 | 40 | 14 | 12 | 8 | 14 | 12 | 19 | 14 | 12 |
| | | 50 | 18 | 14 | 9 | 18 | 14 | 23 | 18 | 13 |
| | | 60 | 21 | 16 | 11 | 21 | 16 | 28 | 21 | 15 |
| | 5 | 40 | 18 | 14 | 9 | 18 | 14 | 23 | 18 | 13 |
| | | 50 | 22 | 17 | 11 | 22 | 17 | 29 | 22 | 16 |
| | | 60 | 27 | 21 | 14 | 27 | 21 | 35 | 27 | 19 |
| | 6 | 40 | 21 | 16 | 11 | 21 | 16 | 28 | 21 | 15 |
| | | 50 | 27 | 21 | 14 | 27 | 21 | 35 | 27 | 19 |
| | | 60 | 32 | 25 | 16 | 32 | 25 | 42 | 32 | 23 |
| 3000 psi | 7 | 40 | 25 | 19 | 13 | 25 | 19 | 32 | 25 | 18 |
| | | 50 | 31 | 24 | 16 | 31 | 24 | 40 | 31 | 22 |
| | | 60 | 37 | 29 | 19 | 37 | 29 | 49 | 37 | 26 |
| | 8 | 40 | 32 | 23 | 15 | 30 | 23 | 39 | 30 | 20 |
| | | 50 | 37 | 29 | 18 | 37 | 29 | 48 | 37 | 25 |
| | | 60 | 45 | 34 | 22 | 45 | 34 | 58 | 45 | 30 |
| | 9 | 40 | 38 | 29 | 16 | 38 | 29 | 49 | 38 | 23 |
| | | 50 | 47 | 36 | 21 | 47 | 36 | 62 | 47 | 28 |
| | | 60 | 57 | 44 | 25 | 57 | 44 | 74 | 57 | 34 |
| | 10 | 40 | 48 | 37 | 19 | 48 | 37 | 63 | 48 | 25 |
| | | 50 | 60 | 46 | 23 | 60 | 46 | 78 | 60 | 32 |
| | | 60 | 72 | 56 | 28 | 72 | 56 | 94 | 72 | 38 |
| | 11 | 40 | 59 | 46 | 21 | 59 | 46 | 77 | 59 | 28 |
| | | 50 | 74 | 57 | 26 | 74 | 57 | 96 | 74 | 35 |
| | | 60 | 89 | 68 | 31 | 89 | 68 | 116 | 89 | 42 |

**Table A–13**   *Continued*

*Required Imbedment Length in Inches to Develop Full Strength*

| Ultimate Concrete Strength | Bar Size No. | Grade of Steel | Development Length | | | Tension Splice Length | | | | Compression Splice Length All Bars |
|---|---|---|---|---|---|---|---|---|---|---|
| | | | Tension | | Compression All Bars | Class A | | Class B | | |
| | | | Top | Other | | Top | Other | Top | Other | |
| | 3 | 40 | 12 | 12 | 8 | 12 | 12 | 12 | 12 | 12 |
| | | 50 | 12 | 12 | 8 | 12 | 12 | 15 | 12 | 12 |
| | | 60 | 14 | 12 | 8 | 14 | 12 | 18 | 14 | 12 |
| | 4 | 40 | 12 | 12 | 8 | 12 | 12 | 16 | 12 | 12 |
| | | 50 | 15 | 12 | 8 | 15 | 12 | 20 | 15 | 13 |
| | | 60 | 18 | 14 | 9 | 18 | 14 | 24 | 18 | 15 |
| | 5 | 40 | 15 | 12 | 8 | 15 | 12 | 20 | 15 | 13 |
| | | 50 | 19 | 15 | 10 | 19 | 15 | 25 | 19 | 16 |
| | | 60 | 23 | 18 | 12 | 23 | 18 | 30 | 23 | 19 |
| | 6 | 40 | 18 | 14 | 9 | 18 | 14 | 24 | 18 | 15 |
| | | 50 | 23 | 18 | 12 | 23 | 18 | 30 | 23 | 19 |
| | | 60 | 28 | 21 | 14 | 28 | 21 | 36 | 28 | 23 |
| 4000 psi | 7 | 40 | 22 | 17 | 11 | 22 | 17 | 28 | 22 | 18 |
| | | 50 | 27 | 21 | 14 | 27 | 21 | 35 | 27 | 22 |
| | | 60 | 32 | 25 | 17 | 32 | 25 | 42 | 32 | 26 |
| | 8 | 40 | 26 | 20 | 13 | 26 | 20 | 34 | 26 | 20 |
| | | 50 | 32 | 25 | 16 | 32 | 25 | 42 | 32 | 25 |
| | | 60 | 39 | 30 | 19 | 39 | 30 | 50 | 39 | 30 |
| | 9 | 40 | 33 | 25 | 14 | 33 | 25 | 43 | 33 | 23 |
| | | 50 | 41 | 32 | 18 | 41 | 32 | 53 | 41 | 28 |
| | | 60 | 49 | 38 | 21 | 49 | 38 | 64 | 49 | 34 |
| | 10 | 40 | 42 | 32 | 16 | 42 | 32 | 54 | 42 | 25 |
| | | 50 | 52 | 40 | 20 | 52 | 40 | 68 | 52 | 32 |
| | | 60 | 62 | 48 | 24 | 62 | 48 | 81 | 62 | 38 |
| | 11 | 40 | 51 | 40 | 18 | 51 | 40 | 67 | 51 | 28 |
| | | 50 | 64 | 49 | 22 | 64 | 49 | 83 | 64 | 35 |
| | | 60 | 77 | 59 | 27 | 77 | 59 | 100 | 77 | 42 |

**Table A–13**    *Continued*

*Required Imbedment Length in Inches to Develop Full Strength*

| Ultimate Concrete Strength | Bar Size No. | Grade of Steel | Development Length | | | Tension Splice Length | | | | Compression Splice Length |
|---|---|---|---|---|---|---|---|---|---|---|
| | | | Tension | | Compression All Bars | Class A | | Class B | | All Bars |
| | | | Top | Other | | Top | Other | Top | Other | |
| | 3 | 40 | 12 | 12 | 8 | 12 | 12 | 12 | 12 | 12 |
| | | 50 | 12 | 12 | 8 | 12 | 12 | 13 | 12 | 12 |
| | | 60 | 12 | 12 | 8 | 12 | 12 | 16 | 12 | 12 |
| | 4 | 40 | 12 | 12 | 8 | 12 | 12 | 14 | 12 | 12 |
| | | 50 | 14 | 12 | 8 | 14 | 12 | 18 | 14 | 13 |
| | | 60 | 17 | 13 | 9 | 17 | 13 | 22 | 17 | 15 |
| | 5 | 40 | 14 | 12 | 8 | 14 | 12 | 18 | 14 | 13 |
| | | 50 | 17 | 13 | 9 | 17 | 13 | 22 | 17 | 16 |
| | | 60 | 21 | 16 | 11 | 21 | 16 | 27 | 21 | 19 |
| | 6 | 40 | 17 | 13 | 9 | 17 | 13 | 22 | 17 | 15 |
| | | 50 | 21 | 16 | 11 | 21 | 16 | 27 | 21 | 19 |
| | | 60 | 25 | 19 | 14 | 25 | 19 | 32 | 25 | 23 |
| 5000 psi | 7 | 40 | 19 | 15 | 11 | 19 | 15 | 25 | 19 | 18 |
| | | 50 | 24 | 19 | 13 | 24 | 19 | 31 | 24 | 22 |
| | | 60 | 29 | 22 | 16 | 29 | 22 | 38 | 29 | 26 |
| | 8 | 40 | 23 | 18 | 12 | 23 | 18 | 30 | 23 | 20 |
| | | 50 | 29 | 22 | 15 | 29 | 22 | 38 | 29 | 25 |
| | | 60 | 35 | 27 | 18 | 35 | 27 | 45 | 35 | 30 |
| | 9 | 40 | 29 | 23 | 14 | 29 | 23 | 38 | 29 | 23 |
| | | 50 | 37 | 28 | 17 | 37 | 28 | 48 | 37 | 28 |
| | | 60 | 44 | 34 | 20 | 44 | 34 | 57 | 44 | 34 |
| | 10 | 40 | 37 | 29 | 15 | 37 | 29 | 48 | 37 | 25 |
| | | 50 | 47 | 36 | 19 | 47 | 36 | 61 | 47 | 32 |
| | | 60 | 56 | 43 | 23 | 56 | 43 | 73 | 56 | 38 |
| | 11 | 40 | 46 | 35 | 17 | 46 | 35 | 60 | 46 | 28 |
| | | 50 | 57 | 44 | 21 | 57 | 44 | 75 | 57 | 35 |
| | | 60 | 69 | 53 | 25 | 69 | 53 | 90 | 69 | 42 |

# B

# DETAILED DERIVATION OF COLUMN CONSTANTS

The ACI column formula is introduced in Chapter 13:

$$P_o = 0.85 f'_c (A_g - A_{st}) + f_y A_{st} \tag{13-2}$$

where  $P_o$ is the concentric load at full compressive yield of both steel and concrete

$A_g$ is the gross area of concrete, $bh$

$A_{st}$ is the total area of all longitudinal reinforcement

$0.85 f'_c$ is the idealized yield stress of concrete

$f_y$ is the yield stress of the reinforcement

It is again observed that $P_o$ is a concentric load. It is further observed that the cross section cannot sustain any moment when it is in full compressive yield as given by the ACI formula.

Code limits the maximum allowable magnitude of the design load $P_n$ to no more than 80% of the load $P_o$ at full compressive yield; that is,

$$P_n \leq 0.80 P_o \tag{B-1}$$

Where the value of $P_n$ is less than $0.80 P_o$, the section will have extra capacity; this extra capacity may be used to sustain a moment $M_n$. Determination of the magnitude of such a moment $M_n$ is the topic of the following derivations.

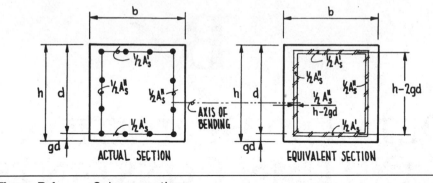

**Figure B-1**   Column section.

## Derivation of Column Constants

A typical column section is shown in Figure B-1. The total area of steel is designated as $A_{st}$ and its steel ratio as $\rho$. That part of the steel distributed to the flanges is $A'_s$ and its steel ratio is $\rho'$. That part of the steel distributed to the web is $A''_s$ and its steel ratio is $\rho''$.

In the equivalent section of Figure B-1, the web steel $A''_s$ is replaced by an equivalent imaginary strip of steel having a finite width, as shown. All other symbols are the same as those used and defined in Chapter 13.

An elevation of a column section is shown in Figure B-2. The line OA represents the case of fully plastic deformation with a strain of 0.003. The yield strain of steel is shown as $0.003s$ and of concrete as $0.003c$. By definition, the yield strain of concrete is $0.85f'_c/E_c$. The two values of strain are equated and the result solved for $c$, yielding

$$c = 333\frac{0.85f'_c}{E_c} \qquad \text{(B-2a)}$$

Similarly for steel,

$$s = 333\frac{f_y}{E_s} \qquad \text{(B-2b)}$$

The section is now allowed to rotate up to line OB. At any angle of rotation within the sector 1, all steel and all concrete are in full plastic yield. Stresses are therefore unchanged regardless of what angle of rotation is imposed; the section simply cannot develop any resistance to moment.

Anywhere within sector 1, the concentric axial force $P_{n1}$ required to

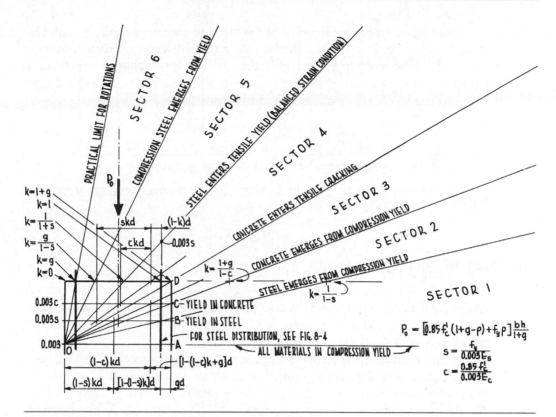

**Figure B-2**    Column rotations.

maintain the state of strain is seen to be the same as the concentric load given by the ACI column formula:

$$P_{n1} = P_o = 0.85 f'_c (A_g - A_{st}) + f_y A_{st} \qquad \text{(B-3a)}$$

Equivalently, in terms of steel ratios,

$$P_{n1} = \left[ 0.85 f'_c (1 + g - \rho' - \rho'') + f_y (\rho' + \rho'') \right] \frac{bh}{1+g} \qquad \text{(B-3b)}$$

where $A_{st} = A'_s + A''_s$, $\rho = A_{st}/bd$, $\rho' = A'_s/bd$, $\rho'' = A''_s/bd$ and $d = h/(1 + g)$.

The second subscript for $P_{n1}$ denotes the sector number, a practice that will be followed hereafter to denote the sector. The superscripts $(')$ and $('')$ denote flange steel and web steel, respectively.

Refer again to Figure B-2. For the section to rotate beyond the line OB into sector 2 requires that the steel on the right side emerge from compression yield. The force in that steel is thus reduced; the change produces a moment about the centerline of the section. In terms of the variable $k$, the stress in the steel in sector 2 is found by ratios to be

$$f_{s2} = 0.003E_s \frac{1-k}{k} \qquad (B-4)$$

where the sign is negative when stress is being reduced.

The change in force $\Delta P'_{s2}$ in sector 2 is the change in stress times the steel area, where the steel area is $\frac{1}{2}\rho' bd$:

$$\Delta P'_{s2} = \frac{1}{2}\left(f_y + f_{s2}\right)\rho' b \frac{h}{1+g} \qquad (B-5)$$

The moment arm of the force $\Delta P'_{s2}$ to centerline of column is given by

$$a'_{s2} = \frac{h}{2} - gd = \frac{1}{2}\left(\frac{1-g}{1+g}\right)h \qquad (B-6)$$

Also in sector 2, there is a change in force on the steel in the web, $A''_s$. The maximum change in stress on the voided triangular stress block is seen to be the same as that derived for $f_{s2}$:

$$f_{s2} = 0.003E_s \frac{1-k}{k} \qquad (B-7)$$

The total force is the area of the voided triangular stress block:

$$\Delta P''_{s2} = \frac{f_y + f_{s2}}{2}\rho'' \frac{1-(1-s)k}{1-g} b \frac{h}{1+g} \qquad (B-8)$$

The moment arm for this force to the centerline of the column is

$$a''_{s2} = \left[\frac{1}{2} - \frac{1-(1-s)k+3g}{3(1+g)}\right]h \qquad (B-9)$$

In sector 2, the final force on the section is then

$$P_{n2} = P_o - \Delta P'_{s2} - \Delta P''_{s2} \qquad (B-10)$$

and the final moment is:

$$M_{n2} = \Delta P'_{s2}\, a'_{s2} + \Delta P''_s\, a''_{s2} \qquad (B-11)$$

The eccentricity ratio $e_{n2}/h$ is given by

$$\frac{e_{n2}}{h} = \frac{M_{n2}}{P_{n2}} \qquad \text{(B-12)}$$

When the equations for $P_n$ and $M_n$ in sector 2 are written out completely, they become quite cumbersome. Even so, they can be solved readily by a small computer and the results tabulated for easy reference. Before introducing the final design tables, however, further development of these highly discontinuous equations is warranted.

As the rotation is increased beyond the line OC into sector 3, a new discontinuity is introduced. A part of the concrete emerges from yield and enters the elastic range. It is noted that Eq. (B-4) through (B-9), previously derived for sector 2, remain valid in sector 3, changing only their subscript from 2 to 3.

As before, the stress in the concrete in sector 3 is found by ratios:

$$f_{c3} = 0.003E_c \frac{1 + g - k}{k} \qquad \text{(B-13)}$$

It is noted that the sign of Eq. (B-12) will be negative when $k > 1 + g$ and the stress is being reduced. The voided area of the stress block which has emerged from yield is found to be

$$A_{c3} = \left[1 - (1 - c)k + g\right]\frac{h}{1 + g} \qquad \text{(B-14)}$$

The change in force on the concrete is then the volume of the voided stress block:

$$\Delta P_{c3} = \tfrac{1}{2}(0.003cE_c + f_{c3})A_c \qquad \text{(B-15)}$$

The moment arm of this force is

$$a_{c3} = \frac{h}{2} - \frac{1 - (1 - c)k + g}{3(1 + g)}h \qquad \text{(B-16)}$$

The final force on the section then is

$$P_{n3} = P_o - \Delta P'_{s3} - \Delta P''_{s3} - \Delta P_{c3} \qquad \text{(B-17)}$$

The final moment is 8

$$M_{n3} = \Delta P'_{s3}\, a'_{s3} + \Delta P''_{s3}\, a''_{s3} + \Delta P_{c3}\, a_{c3} \qquad \text{(B-18)}$$

The eccentricity ratio $e_{n3}/h$ is given by

$$\frac{e_{n3}}{h} = \frac{M_{n3}}{P_{n3}} \qquad \text{(B-19)}$$

**Table B-1**    Summary of Column Equations

| | | Sector | 2 | 3 | 4 | 5 | 6 |
|---|---|---|---|---|---|---|---|
| Flange Steel | Fixed Side | $\Delta P'_s$ | | | | | $0.003E_s \dfrac{g-k(1-s)}{2k(1+g)}\rho' bh$ |
| | | Arm | | | | | $\dfrac{1-g}{2(1+g)}h$ |
| | Elevated Side | $\Delta P'_s$ | | $\left[f_y + 0.003E_s \dfrac{1-k}{k}\right]\dfrac{\rho'}{2(1+g)}bh$ | | | $f_y \dfrac{\rho'}{(1+g)}bh$ |
| | | Arm | | $\dfrac{1-g}{2(1+g)}h$ | | | $\dfrac{1-g}{2(1+g)}h$ |
| Web Steel | Triangle Shape | $\Delta P''_s$ | | $\left[f_y + 0.003E_s \dfrac{1-k}{k}\right]\dfrac{1-(1-s)k}{2(1-g^2)}\rho'' bh$ | | | $f_y \dfrac{2sk}{1-g^2}\rho'' bh$ |
| | | Arm | | $\dfrac{1-3g+2k(1-s)}{6(1+g)}h$ | | | $\dfrac{2k(3+s)-3(1+g)}{6(1+g)}h$ |
| | Rectangle Shape | $\Delta P''_s$ | | | | | $2f_y \dfrac{1-(1+s)k}{1-g^2}\rho'' bh$ |
| | | Arm | | | | | $\dfrac{(1+s)k-g}{2(1+g)}h$ |
| Concrete | Triangle Shape | $\Delta P_c$ | | $0.003E_c \dfrac{(1-(1-c)k+g)^2}{2k(1+g)}bh$ | | $0.003E_c \dfrac{c^2 k}{2(1+g)}bh$ | |
| | | Arm | | $\dfrac{1+g+2k(1-c)}{6(1+g)}h$ | | $\dfrac{2k(3-c)-3(1+g)}{6(1+g)}h$ | |
| | Rectangle Shape | $\Delta P_c$ | | | | $0.003E_c \dfrac{1+g-k}{1+g}bh$ | |
| | | Arm | | | | $\dfrac{k}{2(1+g)}h$ | |

The foregoing procedure is repeated for each sector, using the equivalent section shown in Figure B-1 for all derivations. A new sector begins at each discontinuity; a discontinuity occurs whenever materials enter or emerge from yield and when concrete cracks in tension. A summary of the resulting equations is presented in reduced form in Table B-1.

When the concrete cracks in tension (sector 4 and beyond), the voided part of the stress block is no longer a triangle but becomes a rectangle plus a triangle. These two shapes are listed separately for the concrete in Table B-1. The same shapes occur in the web steel when the steel enters tensile yield.

The line separating sector 4 from sector 5 in Figure B-2 represents the "balanced strain" condition for columns. At that degree of rotation, the concrete strain is seen to be 0.003 in./in. and the tensile steel is exactly at yield stress. In sectors 1 through 4 up to this line, rotations have progressed relatively slowly and predictably, with rotations at least reasonably proportional to moment. From this line onward into sectors 5 and 6, rotations will suddenly become much larger, since the tensile steel can undergo an increase in strain with no increase in stress. It is therefore a feature of the strength design method that after the balanced strain condition is exceeded, a small increase in moment will produce an inordinately large increase in rotation.

## Parametric Design Tables

In all the equations developed in the foregoing derivations, the load $P_n$ and the moment $M_n$ can be expressed in parametric form,

$$R_n = \frac{P_n/bh}{0.085f_c'} = f\left(k, \rho', \rho'', g, E_c, E_s, f_c', f_y\right) \tag{B-20}$$

$$R_n = \frac{M_n/bh}{0.085f_c'} = f\left(k, \rho', \rho'', g, E_c, E_s, f_c', f_y\right) \tag{B-21}$$

From these parameters, the eccentricity ratio $e_n/h$ can be found as

$$\frac{e_n}{h} = \frac{R_m}{R_n} \tag{B-22}$$

In all cases, the parameters $R_n$ and $R_m$ can be found readily as functions of the section constants $k$, $\rho'$, $\rho''$, and $g$, along with the material constants $E_c$, $E_s$, $f_c'$ and $f_y$. The parametric design tables A-9, A-10, and A-11 given in Appendix A were developed in this way. The parameter $R_n$ is called a "stress ratio" in those tables, but it is in fact no more than a useful parameter suggested by the ACI design tables based on the 1971 Code.

In the development of the tables, the expressions for $P_n$ and $M_n$ just derived for each sector were evaluated numerically,

$$\text{Sector 1:} \quad P_{n1} = \left[0.85f_c'(1+g-\rho'-\rho'') + f_y(\rho'+\rho'')\right]\frac{bh}{1+g} \qquad \text{(B-2b)}$$

$$M_{n1} = 0$$

$$\text{Sector 2:} \quad P_{n2} = P_o - \Delta P_{s2}' - \Delta P_{s2}'' \qquad\qquad \text{(B-10)}$$

$$M_{n2} = \Delta P_{s2}'a_{s2}' + \Delta P_s''a_{s2}'' \qquad\qquad \text{(B-11)}$$

$$\text{Sector 3:} \quad P_{n3} = P_o - \Delta P_{s3}' - \Delta P_{s3}'' - \Delta P_{c3} \qquad\qquad \text{(B-17)}$$

$$M_{n3} = \Delta P_{s3}'a_{s3}' + \Delta P_{s3}''a_{s3}'' + \Delta P_{c3}a_{c3} \qquad\qquad \text{(B-18)}$$

In all sectors, the eccentricity ratio is found in the same way,

$$\frac{e_n}{h} = \frac{M_{ni}}{P_{ni}} \qquad\qquad \text{(B-12), (B-19)}$$

The limiting value of the tables occurs when $P_n$ reaches its limiting value, $P_n = 0.10f_c'bh/\phi$, at which point $R_n$ reaches a fixed value, $R_n = 0.168$. Thereafter, a new variable $\phi$ must be included, where $\phi$ goes from 0.7 for columns to 0.9 for beams. The design tables in Appendix A are stopped at $R_n = 0.168$. For small values of $P_n$, that is, for $P_n \leq 0.10f_c'bh/\phi$, the design procedures for a beam carrying a small axial load are simpler than the procedures for a column carrying a dominant flexural load and are used throughout this text.

Computations for the design tables A-9, A-10, and A-11 were performed on a small personal computer, using 6 subroutines corresponding to the 6 sectors. The values of the materials constants ($E_c$, $E_s$, $f_c'$, and $f_y$) were entered first, then the section constants ($\rho'$, $\rho''$, and $g$) were entered next. The value of $k$ was then entered and varied until the desired value of $e_n/h$ was achieved. For this value of $k$, the value of $R_n$ was entered in the table. New values of section constants $\rho'$ and $\rho''$ were then entered and the loop repeated until the table was completed, the yield stress in steel, $f_y$, being varied as the table progressed.

Results of the tabulation were checked against the ACI design tables and were found to correspond quite closely to the ACI design standards. The validity of the approach and of the equations of Table B-1 are therefore accepted.

# GLOSSARY
# OF NOTATIONS

$A$ = basic generic symbol for area

$A_b$ = cross-sectional area of a reinforcing bar, square inches

$A_s$ = area of tension reinforcement, square inches

$A_g$ = gross area of column section, square inches

$A'_s$ = area of compression reinforcement in beams, or, area of longitudinal reinforcement at the flexural faces of a column, square inches

$A''_s$ = area of longitudinal reinforcement at the web faces of a column, square inches

$A_{st}$ = total area of longitudinal column reinforcement, square inches

$A_{tr}$ = area of transverse reinforcement for anchorage requirements

$A_1$ = loaded area

$A_2$ = maximum area of the portion of the supporting surface that is geometrically similar to and concentric with the loaded area

$A_v$ = area of shear reinforcement within a distance $s$, square inches

$b$ = width of compression face of member, inches

$b'$ = effective width of a tee beam flange, inches

$b_1$ = required width of a rectangular beam to sustain only the flexural portion of loads when the beam is subject both to flexure and a small axial load

$b_2$ = required width of a rectangular beam to sustain only the axial load when the beam is subject both to flexure and a small axial load

$b_3$ = reduction or increase in width of rectangular beam to compensate for eccentricity of the axial load when the beam is subject both to flexure and a small axial load

$b_m$ = least flexural width, $b_1$–$b_3$, inches

$b_o$ = length of perimeter subject to shear stress punch-out in footings, inches

$b_w$ = width of the stem of a tee beam, inches

$C$ = resultant of axial compressive forces acting on a section

$c$ = perpendicular distance from the neutral axis of bending to the outermost compression fiber, inches

$D$ = dead load effects

$d$ = effective depth of a flexural section, measured from the outermost compression fiber to the center of tensile reinforcement, inches

$d_b$ = diameter of a reinforcing bar, inches

$E$ = earthquake load effects

$E_c$ = elastic modulus of elasticity of concrete, psi

$E_s$ = elastic modulus of elasticity of steel, psi

$e$ = ratio of moment to axial force M/P, or, eccentricity of axial load

$e_e$ = value of $e$ in the elastic range of loads

$e_n$ = value of $e$ at ultimate levels of load

$f$ = basic generic symbol for direct stress

$f_c$ = elastic stress in concrete, psi

$f'_c$ = specified ultimate compressive strength of concrete, psi

$f_s$ = elastic stress in steel, psi

$f_{sc}$ = elastic stress in steel that is located in a compression zone, psi

$f_y$ = yield stress in steel, psi

$g$ = factor defining concrete cover over longitudinal reinforcement

$h$ = overall depth of a member, measured normal to the axis of bending, in.

$I$ = basic generic symbol for moment of inertia in bending, inches$^4$

$I_{cr}$ = moment of inertia of a cracked concrete section in bending, inches$^4$

$I_e$ = effective moment of inertia of a section after corrections, inches$^4$

$I_g$ = moment of inertia of the gross uncracked section of concrete neglecting reinforcement, inches$^4$

$j$ = factor defining the distance from the center of compressive forces to the center of tensile forces in a flexural section

$k$ = factor defining the distance from the outermost compression face to the neutral axis of bending, measured normal to the neutral axis of bending

$k_b$ = value of $k$ at the balanced strain condition

$k_n$ = value of $k$ at nominal ultimate levels of load

$k_{sv}$ = value of $k$ at service levels of stress

$L$ = length of span, measured between centerlines of bearing

$L$ = live load effects

$L_n$ = length of clear span, measured face-to-face of supports

$L_u$ = unsupported length of a compression member

$\ell_a$ = embedment length beyond center of support or point of inflection, inches

$\ell_d$ = development length of a reinforcing bar, inches

$\ell_{db}$ = basic development length of a reinforcing bar, before modifiers or multipliers are applied, inches

$\ell_{dh}$ = development length of standard hook in tension, inches

$\ell_{hb}$ = basic development length of a standard hook in tension, before modifiers or multipliers are applied, inches

$M$ = basic generic symbol for moment acting on a section

$M_a$ = moment acting on a section at the time deflections are being calculated

$M_{DL}$ = that portion of the moment caused by dead loads

$M_E$ = that portion of the moment caused by earthquake

$M_e$ = moment at elastic levels of stress

$M_{LL}$ = that portion of the moment caused by live loads

$M_1$ = value of smaller end moment on a compression member, positive if member is in single curvature, negative if in double curvature

$M_2$ = value of larger end moment on a compression member, always positive

$M_n$ = nominal ultimate moment acting on a section, or, nominal ultimate resisting moment that the section must be able to develop

$M_{sv}$ = moment acting on a section at service levels of load

$M_u$ = factored moment acting on a section

$M_w$ = that portion of the moment caused by wind

$N$ = number of bars within a layer that are being spliced

$P$ = basic generic symbol for axial load acting on a section

$P_c$ = critical buckling load in the Euler formula

$P_{DL}$ = that portion of the axial load caused by dead loads

$P_E$ = that portion of the axial load caused by earthquake

$P_e$ = axial load at elastic levels of stress

$P_{LL}$ = that portion of the axial load caused by live loads

$P_n$ = nominal ultimate axial load acting on a section

$P_o$ = ultimate concentric axial load on a compression section at full plasticity

$P_u$ = factored axial load acting on a section

$P_w$ = that portion of the axial load caused by wind

$p$ = basic generic symbol for pressure

$p_a$ = allowable increase in soil pressure due to footing loads

$p_{DL}$ = that portion of the soil pressure caused by dead loads

$p_{LL}$ = that portion of the soil pressure caused by live loads

$R_e$ = column stress ratio at elastic levels of stress

$R_n$ = column stress ratio at ultimate load

$S$ = basic generic symbol for section modulus, I/c, inches$^3$

$S_c$ = section modulus taken to the outermost compression fiber, inches$^3$

$s$ = stirrup spacing longitudinally, inches

$T$ = resultant of axial tension loads acting on a section

$T_{DL}$ = that portion of the axial tension caused by dead loads

$T_{LL}$ = that portion of the axial tension caused by live loads

$T_n$ = nominal ultimate axial tension load acting on a section

$t$ = overall thickness of a concrete slab, inches

$U_n$ = load components at nominal ultimate load

$U_{sv}$ = load components at service levels of load

$u$ = bond stress, either at elastic levels of stress or at ultimate levels of load

$V$ = basic generic symbol for shear force acting on a section

$V_c$ = shear force resisted by a concrete section unreinforced in shear

$V_{DL}$ = that portion of the shear force caused by dead loads

$V_E$ = that portion of the shear force caused by earthquake

$V_{LL}$ = that portion of the shear force caused by live loads

$V_s$ = that portion of the shear force that is carried by stirrups

$V_{sv}$ = shear force acting on a section at service levels of load

$V_n$ = nominal ultimate shear force acting on a section, or, nominal ultimate resistance to shearing force that the section must be able to develop

$V_w$ = that portion of the shear force caused by wind

$v$ = basic generic symbol for shear stress, psi

$v_c$ = that portion of the shear stress carried only by concrete, psi

$v_{sv}$ = average shear stress on a section, to include effects of reinforcement, under service levels of stress, psi

$v_n$ = average shear stress on a section, to include effects of reinforcement, under nominal ultimate levels of load, psi

$W$ = wind load effects

$w$ = uniformly distributed load, force per unit length or per unit area

$w_{DL}$ = that portion of the total uniform load caused by dead loads

$w_{LL}$ = that portion of the total uniform load caused by live loads

$y$ = factor defining the distance from the neutral axis to the center of compression loads on the concrete, measured normal to the neutral axis

$y_n$ = factor $y$ at nominal ultimate levels of load

$y_{sv}$ = factor $y$ at service levels of stress

$Z$ = basic generic symbol for plastic section modulus, inches$^3$

$Z_c$ = plastic section modulus at the outermost compression fiber, inches$^3$

$\beta$   =   ratio of the long side of a footing to the short side

$\beta_1$   =   factor defining the height of the ACI equivalent stress block

$\gamma$   =   factor for correcting deflections in concrete for effects of time and for effects of compressive reinforcement

$\Delta$   =   basic generic symbol for deflection

$\Delta_{DL}$   =   deflections under dead load only

$\Delta_{LL}$   =   deflections under live load only

$\Delta_{sv}$   =   deflections under service levels of load

$\varepsilon$   =   basic generic symbol for strain, inches per inch

$\varepsilon_c$   =   strain in concrete, inches per inch

$\varepsilon_s$   =   strain in steel, inches per inch

$\phi$   =   strength reduction factor

$\rho$   =   ratio of tensile steel area to effective cross-sectional area $bd$ in a flexural section

$\rho_b$   =   value of $\rho$ at balanced strain conditions

$\rho'$   =   ratio of compressive steel area to effective cross sectional area $bd$ in a flexural section, or, ratio of longitudinal steel area at flexural faces of a column to the effective area $bd$ of the column

$\rho''$   =   ratio of longitudinal steel area at the web faces of a column to the effective area $bd$ of the column

$\Sigma$   =   basic generic symbol for summation

$\xi$   =   empirical time factor for computing long-term deflections in concrete

# INDEX